IET MATERIALS, CIRCUITS AND DEVICES SERIES 39

Functionality-Enhanced Devices

Other volumes in this series:

Volume 2 **Analogue IC Design: The current-mode approach** C. Toumazou, F.J. Lidgey and D.G. Haigh (Editors)

Volume 3 **Analogue–Digital ASICs: Circuit techniques, design tools and applications** R.S. Soin, F. Maloberti and J. France (Editors)

Volume 4 **Algorithmic and Knowledge-based CAD for VLSI** G.E. Taylor and G. Russell (Editors)

Volume 5 **Switched Currents: An analogue technique for digital technology** C. Toumazou, J.B.C. Hughes and N.C. Battersby (Editors)

Volume 6 **High-frequency Circuit Engineering** F. Nibler *et al.*

Volume 8 **Low-power High-frequency Microelectronics: A unified approach** G. Machado (Editor)

Volume 9 **VLSI Testing: Digital and mixed analogue/digital techniques** S.L. Hurst

Volume 10 **Distributed Feedback Semiconductor Lasers** J.E. Carroll, J.E.A. Whiteaway and R.G.S. Plumb

Volume 11 **Selected Topics in Advanced Solid State and Fibre Optic Sensors** S.M. Vaezi-Nejad (Editor)

Volume 12 **Strained Silicon Heterostructures: Materials and devices** C.K. Maiti, N.B. Chakrabarti and S.K. Ray

Volume 13 **RFIC and MMIC Design and Technology** I.D. Robertson and S. Lucyzyn (Editors)

Volume 14 **Design of High Frequency Integrated Analogue Filters** Y. Sun (Editor)

Volume 15 **Foundations of Digital Signal Processing: Theory, algorithms and hardware design** P. Gaydecki

Volume 16 **Wireless Communications Circuits and Systems** Y. Sun (Editor)

Volume 17 **The Switching Function: Analysis of power electronic circuits** C. Marouchos

Volume 18 **System on Chip: Next generation electronics** B. Al-Hashimi (Editor)

Volume 19 **Test and Diagnosis of Analogue, Mixed-signal and RF Integrated Circuits: The system on chip approach** Y. Sun (Editor)

Volume 20 **Low Power and Low Voltage Circuit Design with the FGMOS Transistor** E. Rodriguez-Villegas

Volume 21 **Technology Computer Aided Design for Si, SiGe and GaAs Integrated Circuits** C.K. Maiti and G.A. Armstrong

Volume 22 **Nanotechnologies** M. Wautelet *et al.*

Volume 23 **Understandable Electric Circuits** M. Wang

Volume 24 **Fundamentals of Electromagnetic Levitation: Engineering sustainability through efficiency** A.J. Sangster

Volume 25 **Optical MEMS for Chemical Analysis and Biomedicine** H. Jiang (Editor)

Volume 26 **High Speed Data Converters** Ahmed M.A. Ali

Volume 27 **Nano-Scaled Semiconductor & Devices** E.A. Gutiérrez-D (Editor)

Volume 29 **Nano-CMOS and Post-CMOS Electronics: Devices and modelling** Saraju P. Mohanty and Ashok Srivastava

Volume 30 **Nano-CMOS and Post-CMOS Electronics: Circuits and design** Saraju P. Mohanty and Ashok Srivastava

Volume 32 **Oscillator Circuits: Frontiers in design, analysis and applications** Y. Nishio (Editor)

Volume 33 **High Frequency MOSFET Gate Drivers** Z. Zhang and Y. Liu

Functionality-Enhanced Devices

An alternative to Moore's law

Edited by
Pierre-Emmanuel Gaillardon

The Institution of Engineering and Technology

Published by The Institution of Engineering and Technology, London, United Kingdom

The Institution of Engineering and Technology is registered as a Charity in England & Wales (no. 211014) and Scotland (no. SC038698).

First published 2018

The Institution of Engineering and Technology
Michael Faraday House
Six Hills Way, Stevenage
Herts, SG1 2AY, United Kingdom

www.theiet.org

British Library Cataloguing in Publication Data
A catalogue record for this product is available from the British Library

ISBN 978-1-78561-558-0 (hardback)
ISBN 978-1-78561-559-7 (PDF)

Typeset in India by MPS Limited

Contents

Chapter 1
Introduction to functionality-enhanced devices

Pierre-Emmanuel Gaillardon[1]

For more than four decades, the semiconductor industry answered the demand for an increasingly higher level of integration and performance by following Moore's law [1], which predicts that the number of transistors and thus the complexity of circuits that can be integrated economically doubles every 18–24 months. Moore's law led us today to manufactured devices with dimensions of few tens of nanometers [2–5]. However, while the reduction of device dimensions increases the computing density, i.e., the maximal possible number of computations per unit area and time, the research community commonly admits that Moore's law is at its twilight and that innovations are required toward a more sustainable route [6–8].

1.1 General background

The arrival of post-Complementary Metal-Oxide-Semiconductor (CMOS) nanotechnologies has brought new logic devices and new computational paradigms. Device level innovations, including novel geometries and materials, introduce new logic devices [6,9], such as carbon nanotubes [10], 2D materials (e.g., graphene [11] and MoS_2 [12]), spintronics [13] or devices with improved functionalities with regard to traditional transistors [14]. New computational paradigms are exemplified by quantum computing [15], adiabatic computation [16] and neurocomputing [17]. In this section, we introduce the background relevant to *Nanosystems* exploiting devices with enhanced functionalities.

1.1.1 Advanced transistor scaling

Since the introduction of *metal-oxide-semiconductor field-effect transistors* (MOSFETs), the semiconductor industry continues to advance the capabilities of computing systems by constantly scaling down the device dimensions and introducing many device-level innovations. During the last decade, novel device geometries, such as Intel's tri-gate transistor, a.k.a. FinFETs [2,4], or *fully depleted silicon-on-insulator* (FDSOI) transistor [3] technologies, have been successfully employed to improve

[1]Department of Electrical and Computer, The University of Utah, USA

the current density, reduce the leakage floor and reduce the short-channel effects at advanced technology nodes. Concurrently, complex materials, such as high-κ gate stacks, silicon-germanium channels or copper-based interconnects, are now ubiquitous and allow for device scalability down to the contemporary 14-nm technology node [4] and near-future 7- and 5-nm nodes. It is widely admitted that the near-future generations of electronic devices will continue to rely on this trend, especially by exploiting advanced gate-all-around geometries and 1D channels, a.k.a., nanowires [5,18], and materials with improved electronic properties, e.g., III–V semiconductor compounds [19].

1.1.2 Emerging devices

In recent years, nanoelectronics introduced a variety of devices that are considered disruptive to regular CMOS and are usually tagged as *beyond-CMOS*. Though all beyond-CMOS devices are worth detailed discussion, this section will introduce further only a small subset of this broad portfolio that are relevant to the proposal.

1.1.2.1 Unconventional channel materials

Although significant scientific progress has recently been made toward identifying alternative routes to complement or even replace CMOS electronics [9,20,21], there is still the largely unexplored route of increasing the basic switching primitive of the elementary transistors, i.e., enhancing their functionality rather than focusing only on reducing their size and/or improving their performances. This CAREER proposal is framed by these considerations and seeks to develop a viable path toward functionality-enhanced devices as a transformative way to design computing systems. Introduced as a material science revolution, *carbon nanotubes* (CNTs) [22] and 2D crystals, e.g., graphene [23], MoS_2 [12], *h*-BN [24] or black phosphorus [25], have drawn considerable attention in electronics due to their superior electrical, thermal and mechanical properties. CNTs have been considered promising channel materials for building *carbon nanotube field effect transistors* (CNFETs) [26,27] that can serve as extensions to silicon MOSFETs. Recent work has demonstrated exceptional CNFETs scalability with sub-10 nm channel lengths [10], complementary *n*-type and *p*-type CNFETs [28] and improved contact resistances [29]. After the discovery of graphene [11], 2D-layered semiconductors received major attention as their conduction and scaling properties are highly desirable within a FET device. In particular, the 2D materials from the *transition metal di-chalcogenides* (TMDs) family [30], such as molybdenum disulfide (MoS_2) and tungsten diselenide (WSe_2), have a band gap and have been shown to exhibit excellent electrical properties [12,31–40]. As a result, TMDs are currently drawing considerable attention as viable candidates for ultra-deeply scaled (5 nm and beyond) electronics [40–46].

1.1.2.2 Sub-60 mV/decade swing FETs

Steep-slope transistors are a subset of beyond-CMOS devices that rely on innovative switching mechanisms to offer a sub-threshold slope below the CMOS intrinsic thermal limit of 60 mV/decade, resulting in increased on-current and reduced leakage current at low-supply voltage, and opening up the possibility of drastically reducing the power supply and the power consumption [9]. Several technologies have been proposed. For instance, *tunnel field effect transistors* (TFETs) switch by modulating tunneling through a barrier. While there is a large body of work about TFET devices, e.g., gnrTFET [47], GaNTFET [48] or Thin-TFET [49], most of the demonstrated devices do only achieve steep-slope in narrow ranges of current, strongly impeding the average reported slope. *Impact-ionization MOS* achieves sub-5 mV/decade conduction by relying on an avalanche breakdown effect, though they require high V_{DS} and suffer from reliability issues [50]. Another approach called *feedback FET* was demonstrated capable of ~2 mV/decade and high I_{on}/I_{off} ratio but does suffer from the need to be "programmed" at large bias and an asymmetric behavior (slope and hysteresis) between turn-*on* and turn-*off* [51]. Finally, *negative-capacitance FETs*, whose gate dielectric of a conventional MOSFET is replaced by a ferroelectric capacitor [52,53] or phase-transition material [54], promise to achieve steeper transitions without restricting the current drive but do present an hysteresis between turn-*on* and turn-*off*.

1.1.2.3 Functionality-enhanced devices

In parallel to the focus on device-level improvements and scaling, an alternative approach promises to increase the functionalities of the basic transistors by means of additional gate controls. In particular, several worldwide research efforts have sought transistors able to exploit a multiple-independent-gate structure to control elementary device polarity at run time. These transistors utilize a large variety of channel materials and geometries, such as carbon nanotubes [55], graphene [56], *silicon nanowires* (SiNWs) [57–60], silicon fins [61] and 2D TMDs [62–65]. Increasing the basic function primitive of a single device from negation to *Exclusive OR* (XOR) or *MAJority* (MAJ) operations can lead to multiple advantages from a design perspective [66].

1.1.3 Toward nanosystems

The exploration of novel solid-state devices for computing has led to the emergence of *nanosystems* [20,67]. Nanosystems are integrated systems exploiting nanodevice technologies. Emerging technologies can bring unprecedented possibilities from a design perspective to create ultra-low-power computing systems that will be capable of sustaining the future needs for information processing. Nanosystems have recently drawn increasingly larger attention. A catalyzing example is the *Nano-Engineered Computing Systems Technology* (N3XT) approach [20] that capitalizes on several recent nanotechnology breakthroughs. To enable these electronic systems based on unconventional technologies, advances in processing and fabrication techniques alone are not sufficient. Indeed, nanosystem research also requires studying system organization, architectures and design tools that allow us to realize nanotechnology-enabled systems capable of outperforming current integrated systems in both energy-efficiency and performance. As a result, design methodologies at large and *electronic design automation* (EDA) in particular are asked to meet emerging technologies to make these demonstrations real. A practical example of the importance of specialized EDA is the imperfection immune design methodology [68,69] developed for the N3XT platform that made possible the recent experimental demonstration of complex nanosystems [70,71].

1.2 Book organization

Although significant scientific progress has recently been made toward identifying alternative routes to complement or even replace CMOS electronics [9,20,21], the route of increasing the basic switching primitive of the elementary transistors, i.e., enhancing their functionality rather than focusing only on reducing their size and/or improving their performances, has still been largely unexplored. This book is framed by these considerations and seeks to propose a comprehensive overview of functionality-enhanced devices as a transformative way to design computing systems. The book is organized in two parts.

Part I focuses on materials and device research related to functionality-enhanced devices. Chapter 2 from Weber *et al.* presents state-of-the-art Germanium-based functionality-enhanced transistors reporting very clean *n*-type and *p*-type symmetries. Chapter 3 from Gopalan *et al.* introduces the two-dimensional materials as a strong opportunity for the future of electronic systems, while Chapter 4 from Resta *et al.* and Chapter 5 from Nakaharai *et al.* both present results of functionality-enhanced devices made out of 2D materials. While all the previous chapters report polarity control, Chapter 6 from Romero-González *et al.* describes a promising mode of operation achieved by functionality-enhanced devices allowing to reach super steep slope behaviors, and Chapter 7 from Hasan *et al.* discusses about the use of this effect to achieve super-sensitive THz detection.

Part II moves higher up in the design hierarchy and focuses on applications and design techniques of functionality-enhanced devices. Chapter 8 from Hu *et al.* introduces a compact model able to capture the unique behavior of functionality-enhanced devices. Chapter 9 from Zografos *et al.* proposes novel physical design techniques to mitigate the effect of the additional gate terminals. Chapter 10 from Romero-González *et al.* presents a comprehensive benchmarking study to evaluate the benefits of such technologies. Chapter 11 from Amaru *et al.* looks into the EDA challenges brought by functionality-enhanced devices and proposes novel techniques to fully unlock their capabilities. Chapter 12 from Tang *et al.* does look into applications in the reconfigurable logic domain, while Chapter 13 from Bi *et al.* discusses the opportunities of emerging technologies for hardware security.

Acknowledgments

The editor would like to thank all the contributors of the book and particularly his former mentors Prof. Ian O'Connor from Ecole Centrale of Lyon and Prof. Giovanni De Micheli from EPFL for their support. This effort has been partly supported by the National Science Foundation under Grant No. 1644592 and under CAREER Award No. 1751064.

References

[1] G. E. Moore, "Cramming more components onto integrated circuits," Electronics, 38(8):114–117, 1965.

[2] C. H. Jan, U. Bhattacharya, R. Brain, *et al.*, "A 22 nm SoC platform technology featuring 3-D tri-gate and high-k/metal gate, optimized for ultra low power, high performance and high density SoC applications," IEDM Tech. Dig., pp. 3.1.1–3.1.4, 2012.

[3] N. Planes, O. Weber, V. Barral, *et al.*, "28 nm FDSOI technology platform for high-speed low-voltage digital applications," VLSI Technology Symposium, pp. 133–134, 2012.

[4] S. Natarajan, M. Agostinelli, S. Akbar, *et al.*, "A 14 nm logic technology featuring 2nd-generation FinFET, air-gapped interconnects, self-aligned double patterning and a 0.0588 μm^2 SRAM cell size," IEDM Tech. Dig., pp. 71–73, 2014.

[5] C. Dupré, A. Hubert, S. Becu, *et al.*, "15 nm-diameter 3D stacked nanowires with independent gates operation: ΦFET," IEDM Tech. Dig., 2008.

[6] K. Bernstein, R. K. Cavin, W. Porod, A. Seabaugh and J. Welser, "Device and architecture outlook for beyond CMOS switches," Proceedings of the IEEE, 9(12):2169–2184, 2010.

[7] M. M. Waldrop, "The chips are down for Moore's law," Nature News, 530(7589):144–147, 2016.

[8] T. N. Theis and H. -S. P. Wong, "The end of Moore's law: a new beginning for information technology," Computing in Science & Engineering, 19(2):41–50, 2017.

[9] D. E. Nikonov and I. A. Young, "Benchmarking of beyond-CMOS exploratory devices for logic integrated circuits," IEEE Journal on Exploratory Solid-State Computational Devices and Circuits, 1:3–11, 2015.

[10] A. D. Franklin, M. Luisier, S. -J. Han, *et al.*, "Sub-10 nm carbon nanotube transistor," Nano Letters, 12(2):758–762, 2012.

[11] K. S. Novoselov, A. K. Geim, S. V. Morozov, *et al.*, "Electric field effect in atomically thin carbon films," Science, 306(5696):666–669, 2004.

[12] B. Radisavljevic, A. Radenovic, J. Brivio, V. Giacometti and A. Kis, "Single-layer MoS_2 transistors," Nature Nanotechnology, 6(3):147–150, 2011.

[13] S. A. Wolf, D. D. Awschalom, R. A. Buhrman, *et al.*, "Spintronics: a spin-based electronics vision for the future," Science, 294(5546):1488–1495, 2001.

[14] P.-E. Gaillardon, L. Amarú, J. Zhang, and G. De Micheli, "Advanced system on a chip design based on controllable-polarity FETs," DATE Tech. Dig., 2014.

[15] D. Deutsch, "Quantum theory, the Church-Turing principle and the universal quantum computer," Proceedings of the Royal Society of London – A. Mathematical and Physical Sciences, 400(1818):97–117, 1985.

[16] E. Farhi, J. Goldstone, S. Gutmann and M. Sipser, "Quantum computation by adiabatic evolution," arXiv eprint quant-ph/0001106, 2000. https://arxiv.org/abs/quant-ph/0001106

[17] F. M. Ham and I. Kostanic, "Principles of neurocomputing for science and engineering," New York, McGraw-Hill Higher Education, 2000.

[18] S. Bangsaruntip, G. M. Cohen, A. Majumdar, *et al.*, "High performance and highly uniform gate-all-around silicon nanowire MOSFETs with wire size dependent scaling," IEDM Tech. Dig., 2009.

[19] J. A. del Alamo, "Nanometre-scale electronics with III–V compound semiconductors," Nature, 479(7373):317–323, 2011.

[20] M. Sabry Aly, M. Gao, G. Hills, *et al.*, "Energy-efficient abundant-data computing: the N3XT 1,000X," IEEE Computer, Special Issue on Rebooting Computing Computer, 48(12):24–33, 2015.

[21] R. Perricone, X. S. Hu, J. Nahas and M. Niemier, "Can beyond-CMOS devices illuminate dark silicon?," DATE Tech. Dig., 2016.

[22] R. Saito, G. Dresselhaus and M. Dresselhaus, "Physical properties of carbon nanotubes," Imperial College Press, 1998.

[23] F. Schwierz, "Graphene transistors," Nature Nanotechnology, 5(7):487–496, 2010.

[24] A. Lipp, K. A. Schwetz and K. Hunold, "Hexagonal boron nitride: fabrication, properties and applications," Journal of the European Ceramic Society, 5(1): 3–9, 1989.

[25] L. Li, Y. Yu, G. J. Ye, *et al.*, "Black phosphorus field-effect transistors," Nature Nanotechnology, 9(5):372–377, 2014.

[26] S. J. Tans, A. R. M. Verschueren and C. Dekker, "Room-temperature transistor based on a single carbon nanotube," Nature, 393(6680):49–52, 1998.

[27] A. Bachtold, P. Hadley, T. Nakanishi and C. Dekker, "Logic circuits with carbon nanotube transistors," Science, 294(5545):1317–1320, 2001.

[28] L. Suriyasena, X. Xu, G. Pitner, Z. Bao and H.-S. Philip Wong, "VLSI-compatible carbon nanotube doping technique with low work-function metal oxides," Nano Letters, 14(4):1884–1890, 2014.

[29] Q. Cao, S.-J. Han, J. Tersoff, *et al.*, "End-bonded contacts for carbon nanotube transistors with low, size-independent resistance," Science, 350(6256):68–72, 2015.

[30] J. A. Wilson and A. D. Yoffe, "The transition metal dichalcogenides discussion and interpretation of the observed optical, electrical and structural properties," Advances in Physics, 18(73):193–335, 1969.

[31] S. Das, H. Y. Chen, A. V. Penumatcha and J. Appenzeller, "High performance multilayer MoS_2 transistors with scandium contacts," Nano Letters, 13(1):100–105, 2012.

[32] Q. H. Wang, K. Kalantar-Zadeh, A. Kis, J. N. Coleman and M. S. Strano, "Electronics and optoelectronics of two-dimensional transition metal dichalcogenides," Nature Nanotechnology, 7(11):699–712, 2012.

[33] H. Fang, S. Chuang, T. C. Chang, K. Takei, T. Takahashi and A. Javey, "High-performance single layered WSe_2 p-FETs with chemically doped contacts," Nano Letters, 12(7):3788–3792, 2012

[34] M. Tosun, S. Chuang, H. Fang, *et al.*, "High-gain inverters based on WSe_2 complementary field-effect transistors," ACS Nano, 8(5):4948–4953, 2014.

[35] J.-K. Huang, J. Pu, C. L. Hsu, *et al.*, "Large-area synthesis of highly crystalline WSe_2 monolayers and device applications," ACS Nano, 8(1):923–930, 2013.

[36] W. Liu, J. Kang, D. Sarkar, Y. Khatami, D. Jena and K. Banerjee, "Role of metal contacts in designing high-performance monolayer n-type WSe_2 field effect transistors," Nano Letters, 13(5):1983–1990, 2013.
[37] S. Das and J. Appenzeller, "WSe_2 field effect transistors with enhanced ambipolar characteristics." Applied Physics Letters, 103(10):103501, 2013.
[38] S. Das, M. Dubey and A. Roelofs, "High gain, low noise, fully complementary logic inverter based on bi-layer WSe_2 field effect transistors," Applied Physics Letters, 105(8):083511, 2014.
[39] L. Yu, A. Zubair, E. J. Santos, *et al.*, "High-performance WSe_2 complementary metal oxide semiconductor technology and integrated circuits," Nano Letters, 15(8):4928–4934, 2015.
[40] H.-J. Chuang, B. Chamlagain, M. Koehler, *et al.*, "Low-resistance 2D/2D ohmic contacts: a universal approach to high-performance WSe_2, MoS_2, and $MoSe_2$ transistors," Nano Letters, 16:1896–1902, 2016.
[41] G. Fiori, F. Bonaccorso, G. Iannaccone, *et al.*, "Electronics based on two-dimensional materials," Nature Nanotechnology, 9(10):768–779, 2014.
[42] D. Jariwala, V. K. Sangwan, L. J. Lauhon, T. J. Marks and M. C. Hersam, "Emerging device applications for semiconducting two-dimensional transition metal dichalcogenides," ACS Nano, 8(2):1102–1120, 2014.
[43] H. Wang, L. Yu, Y. H. Lee, *et al.*, "Integrated circuits based on bilayer MoS_2 transistors," Nano Letters, 12(9):4674–4680, 2012.
[44] A. C. Ferrari, F. Bonaccorso, V. Fal'Ko, *et al.*, "Science and technology roadmap for graphene, related two-dimensional crystals, and hybrid systems," Nanoscale, 7(11):4598–4810, 2015.
[45] Y. Yoon, K. Ganapathi and S. Salahuddin, "How good can monolayer MoS2 transistors be?," Nano Letters, 11(9):3768–3773, 2011.
[46] C. S. Lee, B. Cline, S. Sinha, G. Yeric and H. S. P. Wong, "32-bit processor core at 5-nm technology: analysis of transistor and interconnect impact on VLSI system performance," IEDM Tech. Dig., pp. 28.3.1–28.3.4, 2016.
[47] M. Luisier and G. Klimeck, "Performance analysis of statistical samples of graphene nanoribbon tunneling transistors with line edge roughness," Applied Physics Letters, 94(22):223505, 2009.
[48] W. Li, S. Sharmin, H. Ilatikhameneh, *et al.*, "Polarization-engineered III-nitride heterojunction tunnel field-effect transistors," IEEE Journal on Exploratory Solid-State Computational Devices and Circuits, 1:28–34, 2015.
[49] M. Li, D. Esseni, J. J. Nahas, D. Jena and H. G. Xing, "Two-dimensional heterojunction interlayer tunneling field effect transistors (Thin-TFETs)," IEEE Journal of the Electron Devices Society, 3(3):200–207, 2015.
[50] E. H. Toh, G. H. Wang, M. Zhu, *et al.*, "Impact ionization nanowire transistor with multiple-gates, silicon-Germanium impact ionization region, and sub-5 mV/decade subtheshold swing," IEDM Tech. Dig., 2007.
[51] A. Padilla, C. W. Yeung, C. Shin, C. Hu and T.-J. K. Liu, "Feedback FET: a novel transistor exhibiting steep switching behavior at low bias voltages," IEDM Tech. Dig., 2008.

[52] S. Salahuddin and S. Datta, "Use of negative capacitance to provide voltage amplification for low power nanoscale devices," Nano Letters, 8(2):405–410, 2008.

[53] A. Rusu, G. A. Salvatore, D. Jiménez and A. M. Ionescu, "Metal-ferroelectric-meta-oxide-semiconductor field effect transistor with sub-60 mV/decade subthreshold swing and internal voltage amplification," IEDM Tech. Dig., 2010.

[54] B. Grisafe, N. Shukla, M. Jerry and S. Datta, "A steep slope phase-FET based on 2D MoS2 and the electronic phase transition in VO_2," International Symposium on VLSI Technology, Systems and Application (VLSI-TSA), 2017.

[55] Y.-M, Lin, J. Appenzeller, J. Knoch and P. Avouris, "High-performance carbon nanotube field-effect transistor with tunable polarities," IEEE Transactions on Nanotechnology, 4(5):481–489, 2005.

[56] H. Yang, J. Heo, S. Park, *et al.*, "Graphene barristor, a triode device with a gate-controlled Schottky barrier," Science, 336(6085):1140–1143, 2012.

[57] M. De Marchi, D. Sacchetto, S. Frache, *et al.*, "Polarity control in double-gate, gate-all-around vertically stacked silicon nanowire FETs," IEDM Tech. Dig., pp. 8.4.1–8.4.4, 2012.

[58] J. Zhang, M. De Marchi, D. Sacchetto, P.-E. Gaillardon, Y. Leblebici and G. De Micheli, "Polarity-controllable silicon nanowire transistors with dual threshold voltages," IEEE Transactions on Electron Devices, 61(11):3654–3660, 2014.

[59] A. Heinzig, S. Slesazeck, F. Kreupl, T. Mikolajick and W. M. Weber, "Reconfigurable silicon nanowire transistors," Nano Letters, 12(1):119–124, 2012.

[60] S. Pregl, A. Heinzig, L. Baraban, G. Cuniberti, T. Mikolajick and W. M. Weber, "Printable parallel arrays of Si nanowire Schottky-barrier-FETs with tunable polarity for complementary logic," IEEE Transactions on Nanotechnology, 15(3):549–556, 2016.

[61] J. Zhang, M. De Marchi, P.-E. Gaillardon and G. De Micheli, "A Schottky-barrier silicon FinFET with 6.0 mV/dec subthreshold slope over 5 decades of current," IEDM Tech. Dig., pp. 339–342, 2014.

[62] S. Nakaharai, M. Yamamoto, K. Ueno, Y. F. Lin, S. L. Li and K. Tsukagoshi, "Electrostatically reversible polarity of ambipolar α-$MoTe_2$ transistors," ACS Nano, 9(6):5976–5983, 2015.

[63] S. Tanachutiwat, J. U. Lee, W. Wang and C. Y. Sung, "Reconfigurable multi-function logic based on graphene p–n junctions," Design Automation Conference (DAC), 2010.

[64] Y.-F. Lin, Y. Xu, S. T. Wang, *et al.*, "Ambipolar $MoTe_2$ transistors and their applications in logic circuits," Advanced Materials, 26(20):3263–3269, 2014.

[65] G. V. Resta, S. Sutar, Y. Balaji, *et al.*, "Polarity control in WSe_2 double-gate transistors," Scientific Reports, 6:29448, 2016.

[66] J. Romero-González and P.-E. Gaillardon, "BCB evaluation of high-performance and low-leakage three-independent-gate field-effect transistors,"

IEEE Journal on Exploratory Solid-State Computational Devices and Circuits, 4:35–43, 2018.

[67] P.-E. Gaillardon, L. Amarù, S. Bobba, M. De Marchi, D. Sacchetto and G. De Micheli, "NanoWire systems: technology and design," Philosophical Transactions of the Royal Society of London A, 372(2012):1–15, 2014.

[68] S. Bobba, J. Zhang, A. Pullini, D. Atienza and G. De Micheli, "Design of compact imperfection-immune CNFET layouts for standard-cell-based logic synthesis," DATE Tech. Dig., 2009.

[69] J. Zhang, A. Lin, N. Patil, *et al.*, "Robust digital VLSI using carbon nanotubes," IEEE Transactions on Computer-Aided Design of Integrated Circuits and Systems, 31(4):453–471, 2012.

[70] M. M. Shulaker, G. Hills, N. Patil, *et al.*, "Carbon nanotube computer," Nature, 501(7468):526–530, 2013.

[71] M. M. Shulaker, G. Hills, R. S. Park, R. T. Howe, K. Saraswat, H.-S.P. Wong and S. Mitra, "Three-dimensional integration of nanotechnologies for computing and data storage on a single chip," Nature, 547:74–78, 2017.

Part I

Materials and device research related to functionality-enhanced devices

Chapter 2
Germanium-based polarity-controllable transistors

Walter M. Weber[1,2], *Jens Trommer*[1,2], *André Heinzig*[2] *and Thomas Mikolajick*[1,2]

2.1 Introduction

Performance enhancement and reduction of power consumption are important figures of merit of electronic devices and circuits. This also applies to the emerging family of polarity-controllable or reconfigurable transistors that merge p- and n-type operations of field-effect transistors into a single programmable transistor with different low-dimensional semiconductor channel materials [1–9]. Different to conventional metal-oxide-semiconductor field-effect transistors (MOSFETs) polarity-controllable devices search for the ability to reduce latencies and combat power consumption at the circuit level, rather than at the device level [10–12]. To this end, they exploit their multifunctionality since novel and efficient circuit topologies can be applied targeting the reduction of transistor count and shortening critical paths [13]. Unnecessary parallel operations that would otherwise have a performance and power cost are thereby eliminated. Such benefits are impossible to be realized in complementary metal-oxide-semiconductor (CMOS) technology.

However, as of today, the CMOS components, MOSFETs, do exhibit higher performance and lower dynamic power consumption at the device level as compared to the polarity-controllable transistors. To exploit the full benefit of reconfigurable or polarity-controllable circuits, it is necessary to both utilize efficient circuit designs and to search for the performance enhancement and power reduction at the device level. Although the source to drain leakage at the off-state is lower than in CMOS counterparts, the supply voltages and therefore the dynamic power consumption are usually higher than in CMOS, since the threshold voltage for the individual conduction types cannot be adjusted independently from each other.

The fact that reconfigurable FETs (RFETs) offer lower performance than CMOS FETs lies in the nature of polarity-controllable or reconfigurable transistors. To allow

[1]NaMLab gGmbH, Germany
[2]Center for Advancing Electronics Dresden (CfAED), TU Dresden, Germany

for both transistor polarities, i.e., electron and hole-dominated characteristics, an undoped or lowly doped semiconductor with Schottky source/drain junctions is employed. Different to MOSFETs, the injection of charge carriers in the on-state is thus given by quantum mechanical tunneling through the Schottky junction. There, the current injection decays exponentially with linearly increasing barrier heights.

Therefore, it makes sense to search for solutions with enhanced tunneling probabilities, i.e., to study the use of semiconductors with small bandgaps and low-effective masses for electrons and holes. In a similar manner, the threshold and supply voltages and power consumption for gates coupling to the Schottky junctions can be reduced by lowering the bandgap of the semiconductor channel.

In this respect, germanium and silicon–germanium compounds are interesting semiconductors for polarity-controllable or reconfigurable transistors, because of their low-bulk bandgap as low as 0.66 eV and comparatively low-effective masses for both electrons and holes and due to the high compatibility with modern CMOS-integrated technologies as implemented today for miniaturized p-type FETs. In comparison to conventional MOSFETs made from low bandgap materials, where the off-state is strongly degraded by a high static current and associated power loss, in RFETs, the off-currents can be drastically reduced by the blocking potential induced by the program gate.

In Section 2.2, we will show device metric predictions as determined by device simulations, and Sections 2.3 and 2.4 will present experimental demonstrator results in terms of fabrication and electrical characterization, respectively. Measurements and simulations show that in comparison to Si RFETs, the supply voltage can be reduced by a factor of 2 and dynamic power consumption can be ~4 times lower compared to silicon-based RFETs. In addition, on-currents can be boosted by up to a factor of 10 without degradation of capacitances, bringing a benefit in the intrinsic delay. In Section 2.5, performance and power consumption metrics were extracted for different device geometries and benchmarked with modern conventional devices. Thereto, we will show that scaled Ge RFETs are competitive compared to modern low standby- and low operating power technologies. The performance boosting at the device level combined with the circuit capabilities of RFETs holds the promise of enabling new circuit applications.

2.2 Theory and device simulations

Figure 2.1 shows a schematic of the dual independent gated germanium nanowire RFET together with a cross section transmission electron micrograph (TEM) of an experimental demonstrator, reported in [9]. First, a series of finite-element drift-diffusion simulations was set up employing two- and three-dimensional computer-aided design (TCAD Sentaurus Device) software. To first illustrate the impact of the bandgap and barrier height of different semiconductors on the drive currents and on threshold voltages, simulations on RFETs with varying bandgaps were analyzed. In the first approach, it was assumed that the source and drain electrode work functions

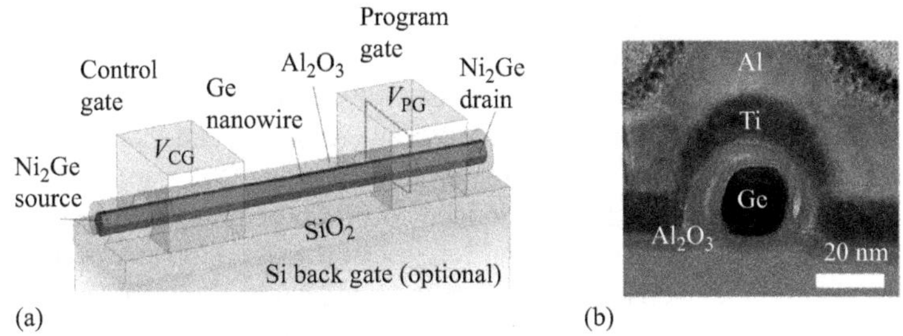

Figure 2.1 Germanium nanowire reconfigurable FETs (RFET). (a) Schematic of transistor with independent junction gating. (b) Cross section transmission electron micrograph of device demonstrator of gated semiconductor region, e.g., indicated by the dashed box in (a). A conformal Al_2O_3/Ti/Al stack forms an omega gate

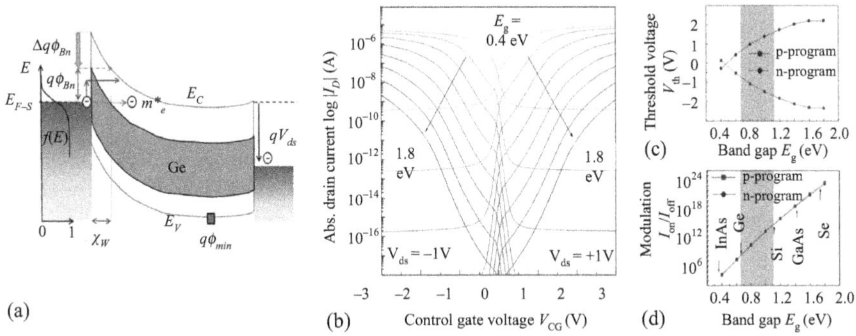

Figure 2.2 (a) Schematic band diagram of dually gated RFET in the on-state, comparing Si and Ge channels. For comparability, the complete band diagrams from Ge and Si are shifted to align at the same Fermi level of the source/drain electrodes. Change to Ge ideally leads to smaller barrier heights $q\phi_{Bn}$ and widths x_W as well as to ideally smaller tunneling masses for charge carriers m_e^*. (b)–(d) Generalized and hypothetic trend for n- and p-program of dual independent gated RFETs with semiconductor materials of different bandgaps. TCAD simulation results with source/drain metal Fermi levels aligned to $E_g/2$. (b) Subthreshold transfer characteristics. Extracted threshold voltages (c) and total drain current modulation (d) in dependence of the bandgap. Gray region denotes the one covered by Si_xGe_{1-x} compounds

can be chosen to align to the bandgap middle of the semiconductor. Figure 2.2(b) shows the calculated subthreshold transfer characteristics for hypothetic energy gaps ranging between 0.4 and 1.8 eV. The smaller the bandgap is, the more the characteristics for n- and p-shift toward the gate voltage axis origin and strongly enhance on-currents, especially at low gate voltages. Additionally, it is evident that for the smallest bandgaps, the off-state degrades. However, that occurs at a much smaller degree than the on-current enhancement due to the blocking nature of the polarity-gated junction. Extracted values of threshold voltage and of total modulation (I_{on}/I_{off}) are plotted in Figures 2.2(c) and (d) for different bandgaps. It can be seen that the $Si_xGe_{(1-x)}$ and Ge nanowire channels indeed cover a broad range, since the bandgap can be tuned smoothly between 0.66 and 1.12 eV by the stoichiometry of the material.

In the next step, we consider pure Ge RFETs and apply a surround gate geometry and scaled dimensions which are in principle patternable in a 45-nm technology. The nanowire thickness d_{NW} is set to 6 nm as it can be achieved by line trimming. The length of the germanium region to 48 nm, the length of the individual gates to 24 nm and the intra-gate spacing to 16 nm, e.g., realizable by double patterning techniques like spacer etching. As a gate dielectric, a 5.6-nm-thick HfO_2 layer that corresponds to an effective oxide thickness (EOT) of 1 nm in a planar capacitor geometry was considered. The gate electrode material is titanium. As to the choice of the source and drain metallic contacts, Mn_3Ge_5 contacts were considered, since it has been reported that these exhibit a hole Schottky barrier height of approximately 0.25 eV, thereto being close to the midgap energy of Ge [14]. It must be noted that no quantum confinement effects were applied as 6 nm is considered to be larger than the Fermi wave length of germanium [15].

Figure 2.3 shows the corresponding subthreshold transfer characteristics for p- and n-type operation, considering an operation voltage of 1.2 V, see full lines. For comparison, the transfer curves of a silicon-based nanowire RFET model with identical geometric dimensions and $NiSi_2$ electrodes, as considered in [3,16], are plotted as dash-dotted curves in the same graph. For the Ge RFET, both n- and p-program types on-currents of more than 2.516 mA/μm as normalized to the nanowire diameter were determined. When employed in a specified technology node, the final current values per chip width will depend on the number of integrated nanowires in parallel as well as on the spacing between nanowires. Compared to Si RFETs, Ge RFETs indeed exhibit a current enhancement by more than a factor of 10×. An inverse subthreshold slope of 80 and 85 mV/dec is obtained for n- and p-program, respectively. The threshold voltages are $V_{TH,n}/V_{TH,p} = 0.4$ V/–0.3 V. The inherent feature of RFETs with independent junction gating is able to keep off-currents (source-drain leakage) low due to the high energy blocking barrier formed at drain. The off-state drain current of 0.2 nA/μm results in low values compared to state-of-the-art germanium-based conventional MOSFETs. The barrier height within the off-state could be determined to be 0.33 eV. In addition, on/off current ratio of $1 \cdot 10^6$ for both operation types can be obtained.

2.3 Fabrication of polarity-controllable germanium nanowire transistors

Device demonstrators of dually gated Ge nanowire RFETs were reported by Trommer *et al.* [9]. As a prototyping vehicle bottom – upgrown germanium nanowires were

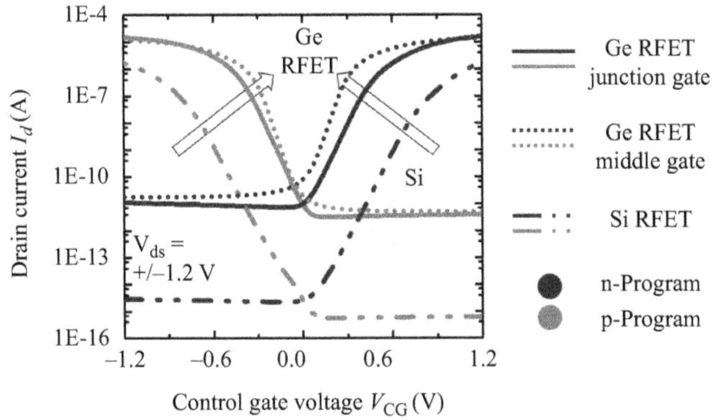

Figure 2.3 Comparison of germanium and silicon nanowire surround gate RFETs with 48 nm channel length, 6 nm diameter and 1 nm EOT. The TCAD simulation results of the subthreshold characteristics show a substantial on-current enhancement and reduction of threshold voltage. $V_{ds} = \pm 1.2\,V$ *and* $V_{PG} = \pm 1.2\,V$*. For the Ge nanowire RFET, both the characteristics for the program gate acting in the middle of the channel (enhanced subthreshold swing-dotted lines) and gating at the junction (full lines) are shown, whereas for the Si RFET (dash, dot, dot), only the junction gating is shown*

employed. The employed nanowires were catalytically grown from monogermane (GeH_4) by an Au particle-assisted vapor–liquid–solid growth mechanism. Due to the different type of electronic properties of germanium, depending on the crystal orientation, two types of common orientations were studied, ⟨1 1 0⟩ as well as ⟨1 1 1⟩ axis-oriented nanowires. For further processing and to provide a stable and well-defined semiconductor/dielectric interface, the as-grown germanium nanowires were immediately subject to a thermal desorption step of native GeO_x in vacuum, directly followed with the deposition of a protective Al_2O_3 layer. The alumina was grown by atomic layer deposition from trimethylaluminium and ozone (O_3) precursors. Due to an ozone-first treatment, an in situ and controlled GeO_x passivation layer grows at the germanium–alumina interface [17]. The presence of this oxide with a thickness below 1 nm and with a mixed composition of GeO_2 and GeO was verified by X-ray photoelectron spectroscopy measurements [18]. Planar germanium on silicon samples were co-processed as reference samples in order to extract the layer and interface properties.

Source and drain contacts were provided by intruded metallic Ni_2Ge contacts. Thereto, nickel was sputtered after local removal of the GeO_x/Al_2O_3 oxide with NHF_4 buffered HF. Annealing was performed in a rapid-thermal-annealing furnace. The nickel germanide was formed within the passivated germanium nanowire core giving an abrupt metal to germanium junction [19] similar to the sharp junction between silicon and $NiSi_2$ in silicon nanowires [20]. These heterostructure transistors were already

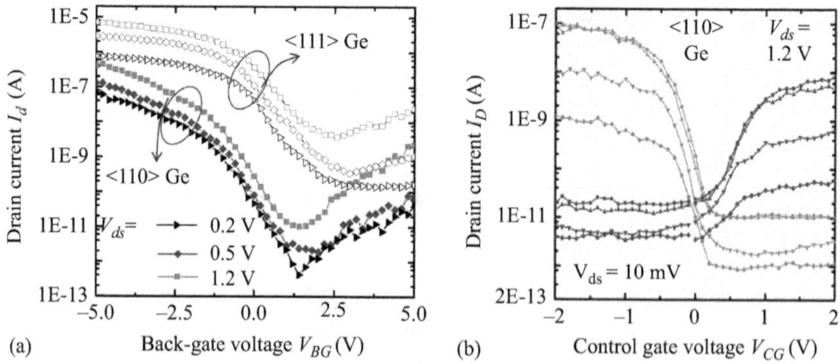

Figure 2.4 *(a) Comparison of subthreshold transfer characteristic of Ge nanowire Schottky FETs with nickel germanide contacts with either ⟨1 1 1⟩ or ⟨1 1 0⟩ nanowire orientation. Substantially higher junction transmissibility for ⟨1 1 1⟩ germanium wires. (b) Programmable characteristics of reconfigurable Ge nanowire FET with ⟨1 1 0⟩ orientation*

characterized by a common back gate in a conventional Schottky FET configuration, prior to the finalization of the device with top gates. The final reconfigurable device is finalized by additional deposition of Al_2O_3 and the patterning of two independent titanium/aluminum gates per device, each overlapping one Schottky junction. The final structure of an RFET device with ⟨1 1 0⟩ Ge channel can be seen in the cross-sectional TEM of Figure 2.1(b). The device revealed to have an ~18-nm-thick germanium channel, and the high-resolution TEM image shown in Figure 2.1(b) confirms the ⟨1 1 0⟩ crystal orientation along the wire axis. The Al_2O_3 gate dielectric thickness below the top gates is ~7 nm thick.

2.4 Electrical characteristics of polarity-controllable germanium nanowire transistors

First, the common, back-gated structures are characterized electrically prior to the top gate fabrication in order to understand the influence of nanowire crystal orientation. Later, the characteristics of the reconfigurable device demonstrators with dual top gates are presented.

Common back gate measurements. In order to prevent hysteresis from quasi-stationary measurements related to border traps or defects within the oxide [21], a pulsed measurement scheme proposed by Mattmann *et al.* was applied [22]. Thereto, the individual measurement pulses are followed by a de-trapping pulse of the same magnitude but opposite polarity. The transfer characteristics are shown in Figure 2.4 for representative devices with an assumed ⟨1 1 0⟩ and ⟨1 1 1⟩ germanium channel orientation with different source to drain voltages ranging between 0.2 and 1.2 V. In this configuration, a typical ambipolar behavior characteristic for Schottky junction

FETs is observed. For the ⟨1 1 0⟩ oriented germanium nanowire, at a drain-source voltage level $|V_{ds}|$ of 1.2 V on-currents of $4.6 \cdot 10^{-7}$ and $2.2 \cdot 10^{-9}$ A are achieved for the p- and n-type branch at $V_{BG} = 5$ V and $V_{BG} = -5$ V, respectively. This leads to on/off ratios of $2.2 \cdot 10^2$ and $4.1 \cdot 10^4$ for the n- and p-branch, respectively. For the ⟨1 1 1⟩ oriented nanowires significantly, higher total on-currents are observed, corresponding to $6.9 \cdot 10^{-6}$ A (~385 μA/μm) and $2.5 \cdot 10^{-8}$ A (~1.38 μA/μm) for the p- and n-type branch, respectively. For both device types, the effective Schottky barrier height was extracted employing electrical measurements with thermal activation. Since the potential of the semiconductor cannot be directly controlled like in a conventional Schottky diode with a degenerately doped semiconductor, the extrapolation methods as described previously in [23,24] were applied. An effective Schottky barrier height for electrons $q\phi_{Be} \sim 0.4$ eV and for holes $q\phi_{Bp} \sim 0.2$ eV for electrons could be extracted for the ⟨1 1 0⟩ germanium nanowire. The barrier values are in good agreement with literature data for Ni_2Ge reported in [25]. The formation of this nickel germanide phase could also be confirmed by quantitative energy-dispersive X-ray analysis in a TE microscopy setup. The tunneling probability T_{WKB} as approximated by the Wentzel–Kramers–Brillouin (WKB) approximation for a triangular barrier is given by

$$T_{WKB} \approx \exp\left[\frac{-4x_w\sqrt{2m_{n,p}^*}(q\phi_{Bn,p})^{3/2}}{3q\hbar V}\right] \tag{2.1}$$

for electrons and holes, respectively, where $\hbar$ is the reduced Planck constant. Accordingly, the tunneling current density through the Schottky junction strongly depends on the effective tunneling masses for electrons and holes (m_n^*, m_p^*), the barrier heights ($q\phi_{Bp}$, $q\phi_{Bn}$) and the barrier thickness x_W given by the voltage drop (V) or field (V/x_W) across the junction. As the barrier thickness is comparable for both germanium nanowire orientations, the strong differences in current are attributed to significantly lower effective masses along the ⟨1 1 1⟩ direction [26] and to lower barrier heights as different interfaces and even silicide phases can form for different germanium nanowire crystal orientations [15].

Dual and independently gated germanium nanowire RFET. The ambipolar conduction of back-gated Schottky transistors generally leads to high off-state leakage currents. In order to reduce the leakage current, two independent omega-shaped top-gates were added to the nanowire structure aligned on top of the two Schottky junctions, as shown in Figure 2.1(a). The gate electrode coupling to the source contact is named the control gate as it steers the transistor on or off. The drain sided gate electrode is labeled the polarity or program gate as it determines the charge carrier type that flows through the channel and thus controls the polarity of the device. In this respect, the potential at the polarity gate induces an energy barrier at the drain-sided Schottky junction to effectively block the injection of charge carriers with the unwanted polarity into the channel. The applied source drain bias ensures that this blocking is only needed at the drain junction. In this demonstrator, the back-gate is used together with the drain-sided top-gate for programing thereby eliminating possible charge trapping at the interface and/or border traps. However, in an ideal device, this potential would not be needed. The transfer subthreshold characteristics can be seen in Figure 2.4(b) for control gate voltages V_{CG} ranging from −2 to +2 V showing successful polarity control for different V_{ds}. Applying a program gate voltage of $V_{PG} = +3$ V n-type and for $V_{PG} = -3$ V p-type characteristics can be programed. A total on/off-ratio of $1{\cdot}10^4$ and $6{\cdot}10^2$ is achieved for p-type and n-type mode, respectively. For the measured ⟨1 1 0⟩ oriented germanium nanowire RFET maximum on-currents of $1{\cdot}10^{-7}$ A ($1.8{\cdot}10^{-6}$ A/μm normalized to the nanowire's Ge-core diameter) and $8.5{\cdot}10^{-9}$ A ($1.5{\cdot}10^{-7}$ A/μm) were achieved for p-type and n-type program at $|V_{ds}| = 2$ V, respectively. This corresponds to maximum current densities of 38 and 3.5 kA/cm^{-2}, respectively. A peak transconductance of 1.8 μS/μm was calculated for the p-type program, and 160 nS/μm for the n-type program. The respective off-currents for p-type and n-type operation of the measured device were $1{\cdot}10^{-11}$ A (~$0.9{\cdot}10^{-10}$ A/μm) and $1.3{\cdot}10^{-11}$ A (~$0.73{\cdot}10^{-10}$ A/μm). The off-state value currents can be solely attributed to thermionic-emitted charge carriers, while other common reverse junction leakage mechanisms, which would normally lead to an ambipolar conduction, are sufficiently suppressed. Although the achieved off-currents are significantly larger than those typically achieved in silicon-based MOSFETs, the measured off-currents are about three orders of magnitude lower compared to other germanium-based devices, such as fully depleted germanium-on-insulator [27], germanium finFETs [28], and germanium/silicon–core/shell quantum-well heterostructures [29], which typically show off-currents of ~$1{\cdot}10^{-7}$ A/μm at $V_{ds} = 0.5$ V. This remarkable result is achieved by the energy barrier that is introduced into the channel via the dual top gate approach. The threshold voltages are $V_{th,p} = -0.2$ V and $V_{th,n} = +0.4$ V, respectively for p- and n-program characteristics. These values are considerably lower than the previously reported values for reconfigurable silicon nanowire transistors or other polarity-controllable devices with independent Schottky junction gating. The germanium-based RFET is able to operate at approximately the half of the supply voltage, as compared to silicon-based RFETs with similar dimensions. This would lead to a reduction of the dynamic power consumption to one-fourth of the silicon values, since the gate capacitance values do not change strongly.

Finally, a minimal subthreshold swing of 150 and 215 mV/dec could be extracted for p- and n-program, respectively. Ideally, 60 mV/dec should be able to be reached at 300 K, if the gate oxide capacitance C_{ox} dominates over C_{semi} and C_{IT}, where C_{semi} is the semiconductor body capacitance and C_{IT} the capacitance of interface traps and if the semiconductor body is sufficiently small, i.e., if a sufficiently small screening or natural length can be reached.

2.5 Benchmark and perspectives

The presented experimental data of the ⟨1 1 0⟩ germanium nanowire RFET show the feasibility of substantially reducing the supply voltage and power consumption. Additionally, it was shown through back gate measurements that the ⟨1 1 1⟩ oriented germanium nanowire can boost on-currents for both carrier types. Therefore, the simulations shown in Section 2.2 are plausible. To improve conductance in ⟨1 1 0⟩ germanium nanowires and to induce symmetry of the IV characteristics, strain engineering can be applied as it has been successfully used to provide symmetry in bottom-up and more recently top-down based silicon RFETs [16,30,31]. Alternatively, different metal germanide materials can be used, e.g., Mn_5Ge_3 [32] to adjust the barrier height.

Next, we benchmark the silicon and germanium nanowire RFET technology at the device level in terms of intrinsic delay and dynamic power consumption. We compare the performance of a germanium RFET based on the geometry of the simulated 48 nm channel length structure from Section 2.2 [18]. Through mixed mode TCAD simulations, it is possible to calculate the intrinsic delay of RFET inverters. Thereto, the inverter propagation delay for various fan outs was calculated [33,34]. The considered intrinsic delay is taken from the slope of the delay versus fan out plot and is thus a fan out independent entity. For comparison, the same procedure is done for a 48-nm silicon-based RFET technology [34], keeping all dimensions identical (Figure 2.5). We clearly see that the intrinsic delay can be substantially reduced, especially for low supply voltages. Indeed, the Ge nanowire RFET technology reaches comparable values to modern low standby power Si MOSFET technologies. However, the performance is lower compared to the International Roadmap for Devices and Systems (IRDS) roadmap predictions for scaled MOSFET devices dedicated to high performance. Next we analyze the power consumption of the germanium RFETs. The use of Ge channels enables the reduction of dynamic power consumption by allowing a stronger reduction of Vds compared to the case with Si channels. As stated in Section 2.1, the true benefit of RFETs comes into play, when both disruptive circuit synthesis and design paradigms apply and they are combined with performance boosters like the approach with germanium nanowires elucidated here (Figure 2.5).

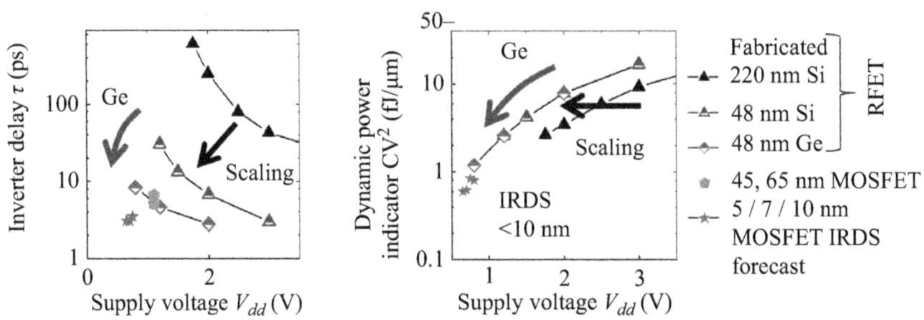

Figure 2.5 Device level benchmarking of performance and power consumption of silicon and germanium nanowire RFETs with conventional CMOS and IRDS roadmap predictions [35] for nodes below 10 nm. The black triangles correspond to the geometries of a fabricated silicon nanowire RFET reported in [16]. The 48-nm silicon RFET was calculated in [34], and the 48 nm germanium RFET with the dimensions discussed above was calculated in [18] considering a fan out independent delay. The high performance MOSFET delay data are taken from a single stage ring oscillator with fan out of 2 in the 5-nm [36] and 65-nm [37] nodes

Although additional number of gates limit downscaling of transistors, substantial functionality is added to the device and circuit level by operating the same device as both n- or p-type transistor upon demand. Simple logic gate examples are runtime reconfigurable NAND/NOR/Majority circuits where all three combinational functions can be executed by a single cell composed of only six transistors without the need for multiplexers [37]. Combining both cells, a 1-bit full adder can be built with a reduced amount of hardware components as compared to CMOS. Likewise, efficient XOR/XNOR reconfigurable cells can be built [34,38]. The number of transistors required to provide a specific function can be further reduced if more than two independent gates are cramped into a single device as proposed in [13,39] and as recently shown in top-down processed and symmetric silicon nanowire RFETs with omega gate [31].

2.6 Conclusions

In this chapter, it could be demonstrated that germanium nanowire-based reconfigurable transistors can eliminate the issue of high reverse junction leakage typically present in low-bandgap semiconductor devices. The limitations given by the smaller bandgap energy can be overcome by opposite band-bending at the drain contact vs. the bending at the source contact as induced by a program gate potential. Thereto, the static energy consumption is strongly reduced. Further lowering of the supply voltages has been proven, resulting in the reduction of dynamic power consumption, as compared to silicon-based RFETs. Finally, in terms of technological process complexity, the device concept does not require any doping, wells or complex isolations. Indeed, germanium and silicon germanium are already integrated into modern CMOS flows for the enhancement of p-FETs. Thus, the integration barriers toward exploitation in semiconductor industry are comparatively low.

Acknowledgments

This work was funded by the German Research Foundation (Deutsche Forschungsgemeinschaft) within the project "ReproNano III" (WE4853/1-3) and the Center for Advancing Electronics Dresden "CfAED."

References

[1] M. De Marchi, D. Sacchetto, S. Frache, *et al.*, "Polarity control in double-gate, gate-all-around vertically stacked silicon nanowire FETs", in *Electron Devices Meeting (IEDM), 2012 IEEE International*, 2012, p. 8.4.1–8.4.4.

[2] W. M. Weber, L. Geelhaar, L. Lamagna, *et al.*, "Tuning the polarity of Si-nanowire transistors without the use of doping", in *Nanotechnol. 2008 NANO 08 8th IEEE Conf. on*, p. 580–581, 2008.

[3] A. Heinzig, S. Slesazeck, F. Kreupl, T. Mikolajick, and W. M. Weber, "Reconfigurable silicon nanowire transistors", *Nano Lett.*, Vol. 12, No. 1, p. 119–124, Jan. 2012.

[4] G. V. Resta, S. Sutar, Y. Balaji, *et al.*, "Polarity control in WSe2 double-gate transistors", *Sci. Rep.*, Vol. 6, No. 29448, p. 1–6, Sep. 2016.

[5] J. Zhang, M. De Marchi, D. Sacchetto, P.-E. Gaillardon, Y. Leblebici, and G. De Micheli, "Polarity-controllable silicon nanowire transistors with dual threshold voltages", *IEEE Trans. Electron Devices*, Vol. 61, No. 11, p. 3654–3660, Nov. 2014.

[6] F. Wessely, T. Krauss, and U. Schwalke, "Dopant-independent and voltage-selectable silicon-nanowire-CMOS technology for reconfigurable logic applications", in *Solid-State Device Research Conference (ESSDERC), 2010 Proceedings of the European*, 2010, p. 365–367.

[7] S. Nakaharai, T. Iijima, S. Ogawa, *et al.*, "Electrostatically-reversible polarity of dual-gated graphene transistors with He ion irradiated channel: Toward reconfigurable CMOS applications", in *Electron Devices Meeting (IEDM), 2012 IEEE International*, 2012, p. 4.2.1–4.2.4.

[8] T. Krauss, F. Wessely, and U. Schwalke, "Electrostatically doped planar field-effect transistor for high temperature applications", *ECS J. Solid State Sci. Technol.*, Vol. 4, No. 5, p. Q46–Q50, 2015.

[9] J. Trommer, A. Heinzig, U. Mühle, *et al.*, "Enabling energy efficiency and polarity control in germanium nanowire transistors by individually gated nanojunctions", *ACS Nano*, Vol. 11, No. 2, p. 1704–1711, Feb. 2017.

[10] P.-E. Gaillardon, L. G. Amaru, S. Bobba, M. De Marchi, D. Sacchetto, and G. De Micheli, "Nanowire systems: technology and design", *Philos. Trans. R. Soc. Math. Phys. Eng. Sci.*, Vol. 372, No. 2012, p. 20130102–20130102, Feb. 2014.

[11] W. M. Weber, A. Heinzig, J. Trommer, D. Martin, M. Grube, and T. Mikolajick, "Reconfigurable nanowire electronics—a review", *Solid-State Electron.*, Vol. 102, p. 12–24, Dec. 2014.

[12] T. Mikolajick, A. Heinzig, J. Trommer, T. Baldauf, and W. M. Weber, "The RFET—a reconfigurable nanowire transistor and its application to novel electronic circuits and systems", *Semicond. Sci. Technol.*, Vol. 32, No. 4, p. 043001, Apr. 2017.

[13] J. Trommer, A. Heinzig, T. Baldauf, *et al.*, "Reconfigurable nanowire transistors with multiple independent gates for efficient and programmable combinational circuits", In *Proc. Des. Autom. Test Eur.*, 2016.

[14] P. Goley and M. Hudait, "Germanium based field-effect transistors: challenges and opportunities", *Materials*, Vol. 7, No. 3, p. 2301–2339, Mar. 2014.

[15] W. M. Weber and T. Mikolajick, "Silicon and germanium nanowire electronics: physics of conventional and unconventional transistors", *Rep. Prog. Phys.*, Vol. 80, No. 6, p. 066502, Jun. 2017.

[16] A. Heinzig, T. Mikolajick, J. Trommer, D. Grimm, and W. M. Weber, "Dually active silicon nanowire transistors and circuits with equal electron and hole transport", *Nano Lett.*, Vol. 13, No. 9, p. 4176–4181, Sep. 2013.

[17] D. Kuzum, T. Krishnamohan, A. J. Pethe, *et al.*, "Ge-interface engineering with ozone oxidation for low interface-state density", *IEEE Electron Device Lett.*, Vol. 29, No. 4, p. 328–330, Apr. 2008.

[18] J. Trommer, A. Heinzig, S. Slesazeck *et al.*, "Reconfigurable germanium transistors with low source-drain leakage for secure and energy-efficient doping-free complementary circuits", *2017 75th Annu. Device Res. Conf. DRC*, 2017.

[19] J. Tang, C.-Y. Wang, M.-H. Hung, *et al.*, "Ferromagnetic germanide in Ge nanowire transistors for spintronics application", *ACS Nano*, Vol. 6, No. 6, p. 5710–5717, Jun. 2012.

[20] W. M. Weber, L. Geelhaar, A. P. Graham, *et al.*, "Silicon-nanowire transistors with intruded nickel-silicide contacts", *Nano Lett.*, Vol. 6, No. 12, p. 2660–2666, 2006.

[21] D. M. Fleetwood, P. S. Winokur, R. A. Reber Jr., *et al.*, "Effects of oxide traps, interface traps, and "border traps" on metal-oxide-semiconductor devices", *J. Appl. Phys.*, Vol. 73, No. 10, p. 5058–5074, May 1993.

[22] M. Mattmann, D. Bechstein, C. Roman, K. Chikkadi, and C. Hierold, "Reduction of gate hysteresis above ambient temperature via ambipolar pulsed gate sweeps in carbon nanotube field effect transistors for sensor applications", *Appl. Phys. Lett.*, Vol. 97, No. 15, p. 153103, 2010.

[23] J. Beister, A. Wachowiak, A. Heinzig, J. Trommer, T. Mikolajick, and W. M. Weber, "Temperature dependent switching behaviour of nickel silicided undoped silicon nanowire devices", *Phys. Status Solidi C*, Vol. 11, No. 11–12, p. 1611–1617, Nov. 2014. DOI: 10.1002/pssc.201400055

[24] D.-Y. Jeon, S. Pregl, S.-J. Park, *et al.*, "Scaling and graphical transport-map analysis of ambipolar Schottky-barrier thin-film transistors based on a parallel array of Si nanowires", *Nano Lett.*, Vol. 15, No. 7, p. 4578–4584, Jul. 2015.

[25] J. Tang, C.-Y. Wang, F. Xiu, *et al.*, "Single-crystalline $Ni_2Ge/Ge/Ni_2Ge$ nanowire heterostructure transistors", *Nanotechnology*, Vol. 21, No. 50, p. 505704, Dec. 2010.

[26] Y. M. Niquet, A. Lherbier, N. H. Quang, M. V. Fernández-Serra, X. Blase, and C. Delerue, "Electronic structure of semiconductor nanowires", *Phys. Rev. B*, Vol. 73, No. 16, p. 165319, Apr. 2006.

[27] X. Yu, J. Kang, M. Takenaka, and S. Takagi, "Experimental study on carrier transport properties in extremely-thin body Ge-on-insulator (GOI) p-MOSFETs with GOI thickness down to 2 nm", in *2015 IEEE International Electron Devices Meeting (IEDM)*, 2015, p. 2.2.1–2.2.4.

[28] Y. C. Yeo, X. Gong, M. J. H. van Dal, G. Vellianitis, and M. Passlack, "Germanium-based transistors for future high performance and low power logic applications", in *2015 IEEE International Electron Devices Meeting (IEDM)*, 2015, p. 2.4.1–2.4.4.

[29] J. Xiang, W. Lu, Y. Hu, Y. Wu, H. Yan, and C. M. Lieber, "Ge/Si nanowire heterostructures as high-performance field-effect transistors", *Nature*, Vol. 441, No. 7092, p. 489–493, May 2006.

[30] T. Baldauf, A. Heinzig, J. Trommer, T. Mikolajick, and W. M. Weber, "Stress-dependent performance optimization of reconfigurable silicon nanowire transistors", *IEEE Electron Device Lett.*, Vol. 36, No. 10, p. 991–993, Oct. 2015.

[31] M. Simon, J. Trommer, Liang, *et al.*, "A wired-AND transistor: polarity controllable FET with multiple inputs", *Accept. IEEE Dev Res Conf 2018*, 2018.

[32] J. Tang, C.-Y. Wang, L.-T. Chang, *et al.*, "Electrical spin injection and detection in Mn_5Ge_3/Ge/Mn_5Ge_3 nanowire transistors", *Nano Lett.*, Vol. 13, No. 9, 4036–4043, Sep. 2013.

[33] J. Trommer, A. Heinzig, S. Slesazeck, T. Mikolajick, and W. M. Weber, "Elementary aspects for circuit implementation of reconfigurable nanowire transistors", *IEEE Electron Device Lett.*, Vol. 35, No. 1, p. 141–143, Jan. 2014.

[34] J. Trommer, A. Heinzig, T. Baldauf, S. Slesazeck, T. Mikolajick, and W. M. Weber, "Functionality-enhanced logic gate design enabled by symmetrical reconfigurable silicon nanowire transistors", *IEEE Trans. Nanotechnol.*, Vol. 14, No. 4, p. 689–698, Jul. 2015.

[35] "International Roadmap for Devices and Systems (IRDS), Release 2017".

[36] K. Mistry, C. Allen, C. Auth, *et al.*, "A 45 nm logic technology with high-k+metal gate transistors, strained silicon, 9 Cu interconnect layers, 193 nm dry patterning, and 100% Pb-free packaging", 2007, p. 247–250.

[37] S. Tyagi, C. Auth, P. Bai, *et al.*, "An advanced low power, high performance, strained channel 65 nm technology", 2005, p. 245–247.

[38] M. De Marchi, J. Zhang, S. Frache, *et al.*, "Configurable logic gates using polarity-controlled silicon nanowire gate-all-around FETs", *IEEE Electron Device Lett.*, Vol. 35, No. 8, p. 880–882, Aug. 2014.

[39] J. Zhang, X. Tang, P.-E. Gaillardon, and G. De Micheli, "Configurable circuits featuring dual-threshold-voltage design with three-independent-gate silicon nanowire FETs", *IEEE Trans. Circuits Syst. Regul. Pap.*, Vol. 61, No. 10, p. 2851–2861, Oct. 2014.

Chapter 3
Two-dimensional materials for functionality-enhanced devices

Prashanth Gopalan[1] and Berardi Sensale-Rodriguez[1]

3.1 2D materials for functionality-enhanced devices

3.1.1 Introduction

The field of two-dimensional materials, albeit studied in past, was propelled to prominence by the discovery of atomically thin sheet of graphene from mechanical exfoliation of graphite in 2004 [1]. Graphene's linear E–k dispersion (Figure 3.1), ultra-high electron mobility in suspended monolayers ($\sim$200,000 cm^2/V s) [2], thermal conductivity [3] and its ability to support high current capacity [4] made it a very exciting material for both the science and engineering communities. A number of applications such as interconnects, stretchable electronics and plasmonics were quickly envisioned and demonstrated. Additionally, the reduced dimensionality resulted in greatly enhanced sensitivity to adsorbates and adatoms making it suitable for gas- and bio-sensing [5–8]. This ability of graphene to overarch between different fields made it hugely popular and model a "two-dimensional" material. Despite the plethora of promising and desirable properties, graphene's lack of bandgap precludes its widespread application. Although bandgap in graphene has been demonstrated through patterning [9,10], introducing dopant atoms/domains [11–13], uniaxial strain [14], chemical functionalization [15,16] and high electric fields [17–19], these methods increase processing difficulty, inhibit easy integration with standard fabrication procedures and materials, compromise the pristine nature of graphene and result in at most a few hundreds of meV of bandgap.

It is important to note that carbon-based nanostructures (carbon nanotubes, nanoribbons, nanofibers and fullerenes) have been widely studied over the past several decades. Building upon this foundation, researchers recognized the plethora of possibilities, both in structure and electronic properties that could stem from two-dimensional compound semiconductors. For example, in transition metal compounds, HfS_2 has a large bandgap [21], molybdenum ditelluride (1T-$MoTe_2$) has

[1]Department of Electrical and Computer Engineering, University of Utah, USA

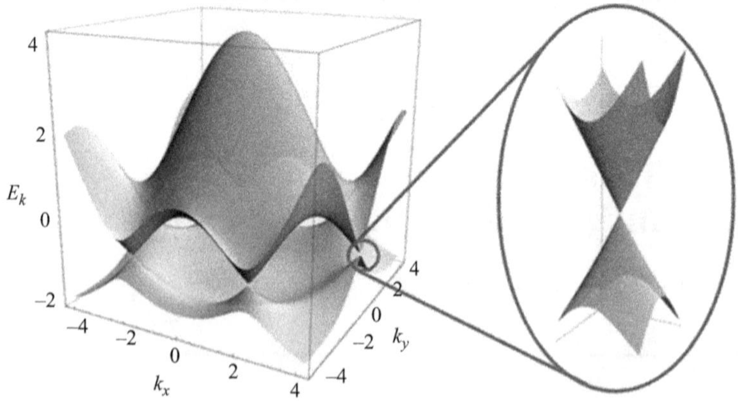

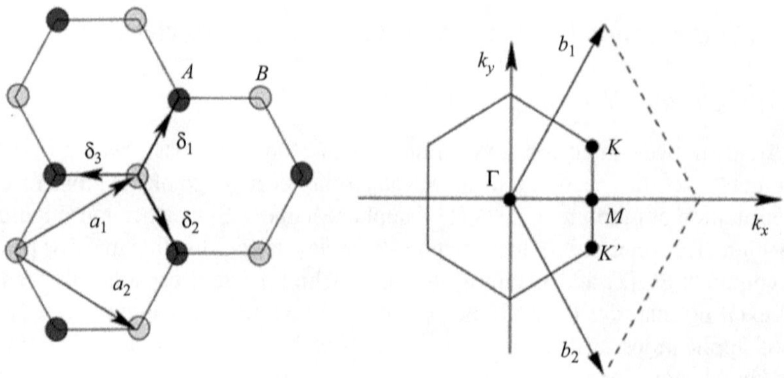

Figure 3.1 (a) Energy dispersion of graphene showing the linear E–k relation (zoomed region), (b) lattice structure of graphene comprising two interpenetrating triangular lattice with a_1 and a_2 representing the unit vectors. (Right) The corresponding Brillouin zone with Dirac cones at K and K′. © 2018 by the American Physical Society. Reprinted, with permission from, Ref [20]

semimetallic nature [22] while MoS_2, WS_2 and WSe_2 [23–25] exhibit semiconducting properties. Spurred by the discovery of graphene and also in part to discover materials with finite bandgap, the field of layered and atomically thin materials has rapidly expanded in the last 5 years to include a wide variety of compounds such as transition metal dichalcogenides (TMDs), hexagonal boron nitride (h-BN) [26] and topological materials (bismuth selenide (Bi_2Se_3), bismuth telluride (Bi_2Te_3), etc.) [27,28] and non-carbon graphene analogue such as phosphorene [29]. Similar to graphite, in their bulk form, these materials possess strong in-plane covalent bonding and van-der-Waals (vdW) interactions between adjacent layers. These semiconducting 2D materials have drawn a considerable amount of attention toward applications in beyond CMOS technology. Being layered compounds, as opposed to bulk semiconductors, they can be made atomically flat with no dangling bonds which would suppress trap generation and carrier scattering. They exhibit exceptional electrostatic control allowing for flatronics with robust performance. Aggressive scaling down of transistor dimensions has brought on the additional issue of variability in channel doping concentrations. In this regard, contrary to the conventional practice of suppressing ambipolar carrier transport in CMOS devices (via doping), functionality enhanced devices that provide electrostatic control over the polarity of charge carriers in the channel (i.e., device can be configured as n- or p-type with an electrostatic field) without having to dope the semiconductor would alleviate dopant concentration variations in short channels. Such devices with their bipolar nature intrinsically support XOR function and potentially allow for device building with reduced number of transistors. This enables us to push the extent of computational capabilities with compact devices that go beyond the limits of scaling. This plays perfectly well into the capabilities and properties of 2D TMDs as their high ON–OFF ratio, immunity to short channel effects and abrupt switching make them ideal candidates for high performance logic gates.

The following sections will provide a brief overview of some of the prominent features of non-carbon 2D materials that are currently being investigated and predicted to play significant role in the development of ultrathin electronic and optoelectronic devices in the coming years that could push the boundaries of current CMOS technology.

3.1.2 Non-carbon 2D materials with bandgaps

3.1.2.1 Hexagonal boron nitride

2D h-BN is an isomorph of graphene with layered structure containing alternating atoms of B and N atoms arranged in a honeycomb lattice and held together by strong "σ" bonds formed through sp^2 orbitals (Figure 3.2). Similar to graphene, weak vdW forces hold different h-BN planes together allowing for micromechanical exfoliation to obtain monolayers. And, unlike graphene, h-BN has a large bandgap of ~5.5–6 eV [30–32] which makes it a potential candidate for use as barrier and active layer in tunneling devices [33] and possible candidate for deep ultraviolet emission (215 nm) sources [30,31]. 2D h-BN, on the other hand, has been envisioned to be employed as an atomically flat dielectric platform without charge traps or dangling bonds for graphene based electronic devices. In graphene devices, an uneven dielectric platform can cause wrinkles in the transferred graphene sheet and presence of impurities can act as charge traps or scattering centers, drastically reducing carrier mobility. Possessing a lattice mismatch of about 1.5%–1.7% ($a = 0.250$ nm and $c = 0.666$ nm) with graphene, stability at high temperatures (up to 1,000°C in air), high breakdown field strength of 0.7 V/nm [34] and optical phonon energy of twice of that of SiO_2 [34] makes it an excellent substrate for graphene-based electronics. The current research in 2D h-BN aims toward *(i) developing efficient large-scale growth and transfer techniques, (ii) successful integration with other 2D materials (heterojunctions) and (iii) effectively modulate carrier concentrations and energy bands.*

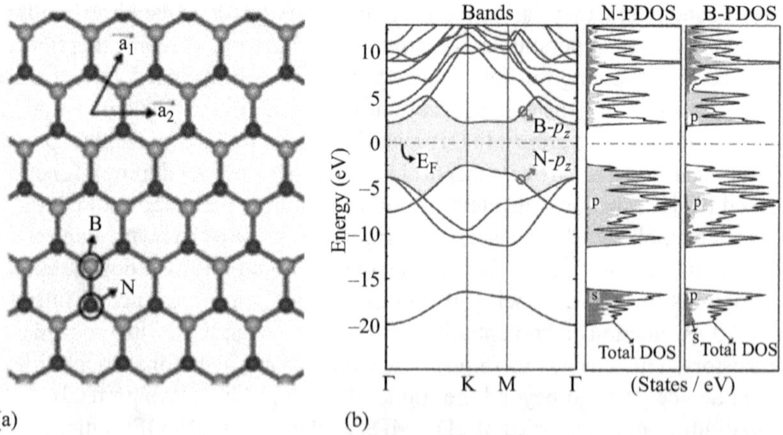

(a) (b)

Figure 3.2 (a) Top view of h-BN's honeycomb crystal lattice (a_1 and a_2 are lattice unit vectors) and (b) calculated band structure of 2D h-BN with TDOS and PDOS (right) on boron and nitrogen atoms. © 2018 by the American Physical Society. Reprinted, with permission from, Ref [35]

3.1.2.2 Phosphorene

Phosphorene is an atomic layer of black phosphorus (stable semiconducting allotrope of phosphorus). Similar to other layered materials the 2D sheets are held together by weak vdW forces and in-plane bonds are formed by sp^3 hybridization resulting in a puckered honeycomb structure. Phosphorene displays both armchair (AM) and zigzag (ZZ) structures. The bond angle and P–P bond length along the former is 98.1° and 2.164 Å while those along ZZ direction are 103.3° and 2.207 Å [36]. The bandgap of phosphorene has been estimated to be around 1.5 eV [37] and changes with the number of layers due to change in interlayer hybridization while still maintaining the band topology unlike TMDs, i.e., both monolayer and few-layer phosphorene are direct bandgap semiconductors with minima at Γ-point [37]. This structural arrangement (Figure 3.3) is the source of anisotropy in electronic band structure, carrier transport and thermoelectric and mechanical properties. For example, increased hole effective mass (~8 times) and larger group velocity were observed along ZZ direction as compared to AM [37]. Phosphorene has been the subject of intensive research ranging from understanding fundamental electronic transport properties [38,39]; engineering these by edge oxidation, chemical functionalization and adatom doping [40–42]; and discovering novel physical phenomena such as superconductivity, quantum Hall effect, etc. [43]. Despite the plethora of promising properties that phosphorene has to offer, large-scale application has been hindered by its poor ambient stability. Uncoupled electron pairs in each P atom (due to sp^3 hybridization) and its enormous surface-to-volume ratio makes it highly susceptible to attack by oxygen, water and light or their combination [44,45]. Robust passivation techniques either with dielectric layers

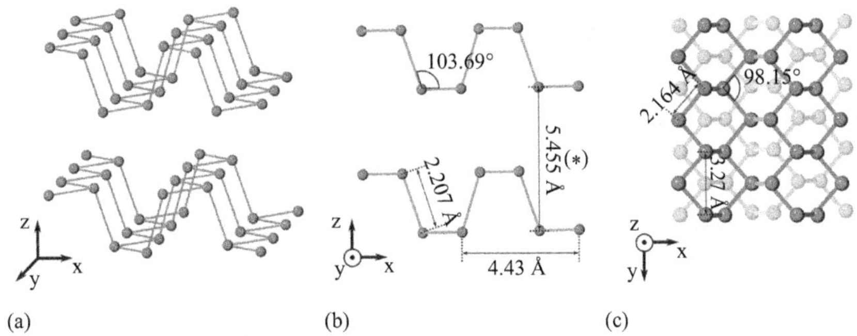

Figure 3.3 *Schematic of crystalline structure of BP obtained through relaxation of the structure in DFT calculations. (a) 3D representation of lattice, (b) side-view showing sheets of phosphorus atoms arranged in a puckered honeycomb structure containing bond angles, lengths and inter-planar separation. Adjacent layers are stacked on top of each other in ABA order and held together by weak vdW forces akin to other layered materials and (c) top-view of BP lattice. © 2018. Reprinted, with permission from, Ref [36]*

[46], atomically precise oxygen plasma treatment of upper layers along with dielectric passivation [47], solvents [48,49] or vdW [50] structures become imperative for actual device applications.

3.1.2.3 Transition metal dichalcogenides

TMDs are layered materials that contain transition metal atom (M = Mo, W, Hf) sandwiched between two atomic planes of chalcogen (X = S, Se, Te) forming a unit cell denoted by MX_2. The unit cell can exist as trigonal prismatic (H phase), octahedral (T phase) and a distorted form (T′ phase) as shown in Figure 3.4(a). In the H-phase, the transition metal atom is bonded to six chalcogen atoms forming two tetrahedrons in the +*z* and -*z* direction (Figure 3.4(a)—left). In T-phase, the tetrahedrons on either side of the metal atom are rotated by 180° with respect to each other (Figure 3.4(a)—middle). Finally, in the T′ phase, the chalcogen atoms are further displaced in the *z*-direction leading to the distortion as compared with the previous two phases (Figure 3.4(a)—right). The X–M–X sandwich is considered a monolayer, and the unit cells are held by vdW forces in the *z*-direction which allows for mechanical exfoliation from bulk crystals.

The driving force behind (re)discovering TMDs has been the need for finite and usable bandgap. As expected, they exhibit a range of bandgaps spanning most of the visible and near-IR region (as shown in Figure 3.4(b)). Additionally, the nature (direct/indirect) of bandgap can also be tuned within the same system of MX_2. Another degree of control stems from the fact that the 2H and 1T phases of these materials exhibit semiconducting and semimetallic properties, respectively. The transition

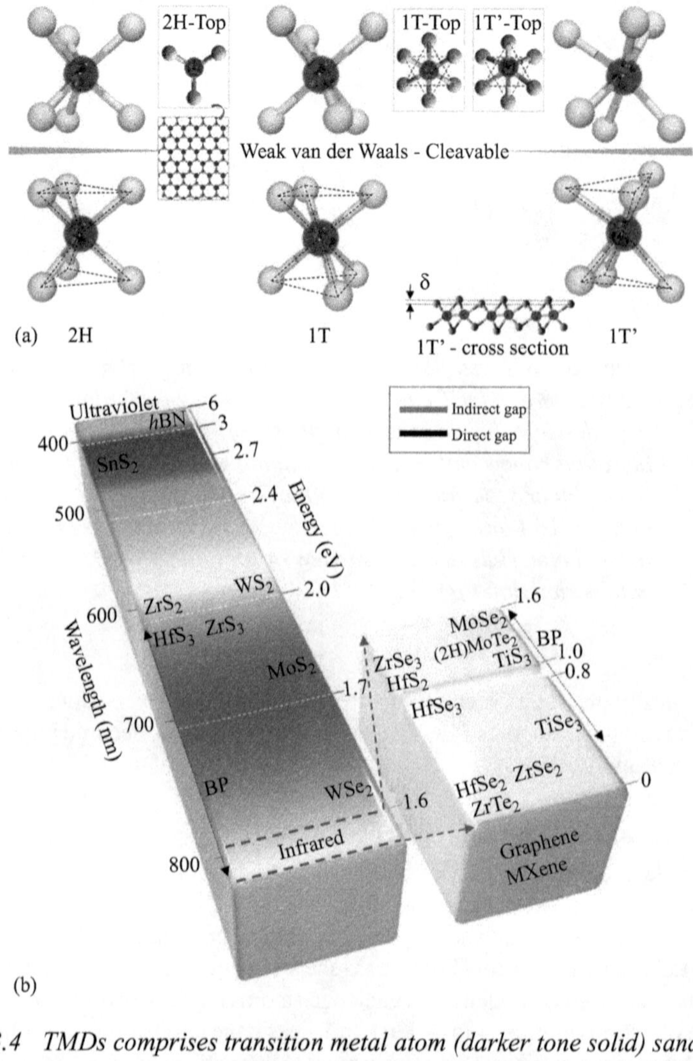

Figure 3.4 TMDs comprises transition metal atom (darker tone solid) sandwiched between two atomic-layers of chalcogen atom (lighter shade solid). (a) Different structural phases of TMDs (2H, 1T and 1T') illustrating the differences in arrangement of chalcogen atoms with adjacent layers held together by vdW forces and (b) variation in the bandgaps of different 2D materials ranging from zero bandgap in graphene to wide bandgap in h-BN. Direct and indirect bandgap materials are represented by darker and lighter tones, respectively. © 2017 Elsevier. Reprinted, with permission from, Ref [51]

between these two phases can be triggered through laser irradiation [52] or through plasmonic hot electrons [53].

3.2 Large area growth of 2D materials

3.2.1 Introduction

Studies of fundamental physics, electronic and optical properties of 2D materials have relied primarily on exfoliated samples due to their high quality and pristine nature. However, exfoliation imposes an inherent limitation on flake size that prohibits large area device design. As a result, in the recent years, considerable effort has been devoted to developing both top-down and bottom-up techniques for large area and scalable growth of high-quality 2D materials. Some of these are CVD (or CVT), MOCVD and molecular beam epitaxy (MBE). Each of these techniques have their own advantages and limitations, which would be briefly discussed in the following sections along with their basic principles of operation and significant results reported in literature utilizing these methods.

3.2.1.1 Chemical vapor deposition

Transition metal dichalcogenides

Atomically thin films of TMDs through CVD have been widely demonstrated. Limited only by the size of the reactor tube, this process could potentially achieve large area substrate coverage. In the most common version of CVD, the vapor phase reaction between metal and chalcogen precursors at high temperatures (700–900°C) in a furnace results in thin film deposition on required substrates. The precursors commonly employed for growth of TMDs are metal oxides or chlorides (e.g., $WO_3/MoCl_5$) and elemental chalcogen (e.g., S, Se or Te). By employing a carrier gas (Ar or N_2), the elemental chalcogen located upstream (at a temperature close to its melting point) is transported to the furnace center containing the (volatile) metal precursor and substrates (Figure 3.5). Optimizing the growth temperature, flux of reactants and the surface energy of the substrate are crucial for high-quality film growth. Although TMDs with high crystallinity have been reported with this process, the coverage is not uniform as the growth mechanism is self-limiting and produces randomly oriented domains as opposed to continuous films. Triangular flakes of MoS_2 grown with oxide powders exhibited an average mobility of 4.3 cm^2/V s with ON/OFF ratios exceeding $\sim 10^6$ [54]. $MoCl_5$ precursor-based growth produced large area coverage albeit very low field effect mobility ~0.003–0.03 cm^2/V s could be attained [55]. Similarly, Chang *et al.* [56] reported electron and hole mobility of ~23 and ~17 cm^2/V s in CVD grown $MoSe_2$ flakes employing oxide precursors.

CVT is a two-step variant of CVD that involves a thin transition metal film (M = Mo, W) deposited by sputtering or evaporation and subsequent conversion to MX_2 by supplying the required chalcogen (X = S, Se, Te) either in a closed system (sealed ampoule) or in a carrier gas-flow type configuration. Although this method can produce continuous films, it also poses a limit on the lowest achievable film

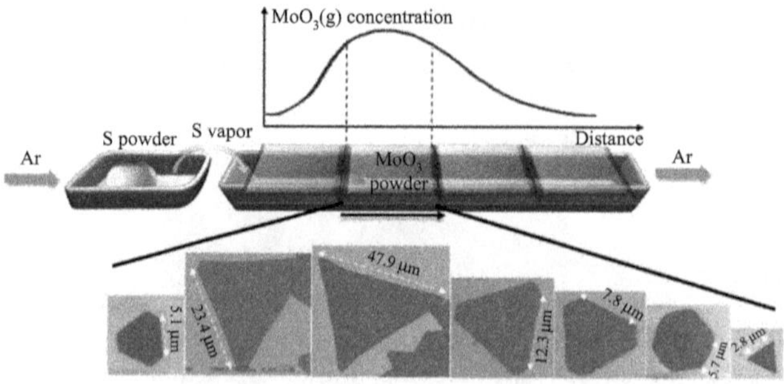

Figure 3.5 Schematic of CVD growth process with Ar carrier gas transporting sulfur located upstream to the furnace center containing the MoO_x powder and the substrates maintained at growth temperature. The bottom AFM images show the shape evolution of MoS_2 domains with different Mo precursor concentration (shown above) along the length of the furnace. Control of reactant flux and growth temperature is crucial for achieving desired domain shape and size with high crystallinity. © 2018 American Chemical Society. Reprinted, with permission from, Ref [57]

thickness (ultimately dependent upon the initial metal film thickness). Laskar *et al.* [58] reported ampoule growth of multilayer MoS_2 thin film on sapphire by using the powdered MoS_2 as the source of sulfur. MoS_2 source releases the sulfur slowly at temperatures exceeding 900°C. Minimizing the difference between chalcogen source decomposition point and growth temperature allowed for low nucleation density and significant lateral grain growth at elevated growth temperature. Rhyee *et al.* [59] reported $MoSe_2$ thin films transistors with a mobility of ~121 cm^2/V s and grain sizes exceeding several hundreds of microns with a specially modified chemical vapor transport process.

h-BN and phosphorene

CVD is an extremely cost-effective and versatile approach for growing h-BN monolayers. Unlike TMD's growth on semiconducting or insulating substrates, h-BN requires metallic substrates that act as catalysts for growth. Different metal substrates such as Cu [60], Ni [61], Pt [62], Fe [63] and Co [64] have been employed successfully. Also, the use of a variety of starting precursors such as diborane and ammonia [65], borazine [66], ammonia borane [67] and trimethylamine borane [68] has been reported. In the most commonly adopted growth process, ammonia borane (BH_3NH_3) powder is employed. Heating the precursor to its decomposition point (70–140°C) releases hydrogen, borazine and aminoborane. The by-products are transported to the furnace center (maintained at ~1,050°C) by carrier gas, adsorb onto the metal surface (e.g., Cu) wherein aminoborane is dehydrogenated and forms BN dimers whose agglomeration creates h-BN domains. The nucleation of BN domains on the metal substrate is greatly influenced by grain boundaries, surface irregularities, dislocation threads and impurities. Pretreatment of metal surface such as electrochemical polishing significantly influences the domain size and subsequent film quality. Chang *et al.* demonstrated an oxide-assisted process that produced large-area continuous monolayer h-BN films with increased domain sizes up to 20 μm without the use of expensive surface pretreatment techniques [69]. The oxidized surface reduced nucleation density by three orders of magnitude (from 10^6 to 10^3 per mm^2) and allowed for significant lateral grain growth. Also, Lu *et al.* [70] drastically reduced their nucleation density to just 60 per mm^2 by introducing 10%–20% Ni in their starting Cu film and attained extremely large domain area (up to 7,500 μm^2).

Fundamental studies and device demonstration of phosphorene have primarily relied on flakes from mechanical or liquid phase exfoliation (i.e., top-down methods). Reports of bottom-up methods (such as CVD) to achieve mono-/few-layers have been very minimal mostly owing the lack of substrates, extreme growth conditions and poor atmospheric stability. One such work from Smith *et al.* [71] reported the synthesis of four-layer 2D BP from red phosphorus directly on a Si substrate with average grain area exceeding 3 μm^2 and thicker samples with grain areas exceeding 100 μm^2 using two-step CVD-like process. In their method, thin film of red phosphorus was initially deposited on Si substrates through CVT (from bulk red phosphorus or BP) at 600°C under vacuum. These were later heated to 950°C with Sn/SnI_4 mineralizing agents in an Ar atmosphere at 400 psi to obtain BP thin films. The two-step conversion process utilizing the mineralizing agents was found to be necessary as direct vapor transport of BP onto Si proved unsuccessful.

3.2.1.2 Metal–organic chemical vapor deposition

MOCVD, simply put, is a variant of CVD that uses metal–organic precursors for growth. The organic (gas phase) precursor is flown over the substrate wherein decomposition results in the deposition of target atom (Figure 3.6) followed by the volatilization of the organic molecules in the precursor to avoid its incorporation in the film. This makes precursor purity highly critical to minimize contamination. For example, 99.999% precursor purity as compared to 99% leads to a drastically reduced carbon contamination of the film. This method affords very precise control over the partial pressure of reactants (metal and chalcogen) and thin film thickness and helps achieve wafer scale (4 in.) uniformity.

Transition metal dichalcogenides

The MOCVD has been recently adopted to grow TMDs with wafer scale uniformity. Kang *et al.* [72] reported growth of MoS_2 and WS_2 mono- and few-layered films on fused silica substrates employing $Mo(CO)_6$, $W(CO)_6$, $(C_2H_5)_2S$ and H_2 gas phase precursors. This allowed for batch fabrication of more 8,100 FET devices (Figure 3.7) with a room temperature field effect electron mobility of ~30 and ~114 cm^2/V s at 90K in MoS_2-based devices. Large area mono- and few-layer WSe_2 growth using $W(CO)_6$ and $(CH_3)_2Se$ was reported by Eichfeld *et al.* [73]. It was found that the temperature, pressure, Se:W precursors ratio and the growth substrate had a huge impact on WSe_2 morphology. Among the substrates employed (epitaxial graphene, CVD graphene, BN and sapphire), largest domain sizes of 5–8 μm were obtained

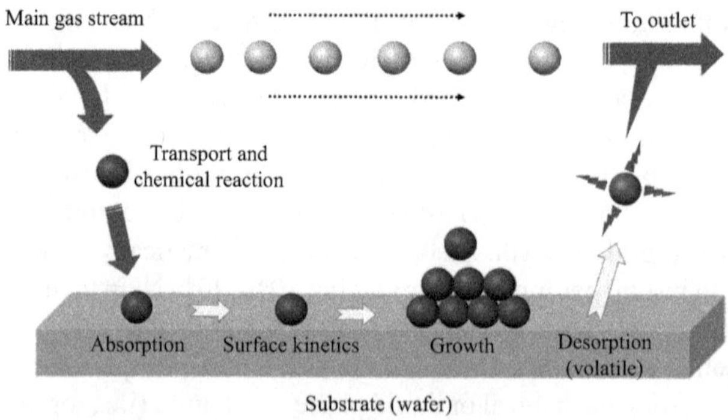

Figure 3.6 General schematic of reactions in a typical MOCVD process. The metal–organic precursors thermally decompose, are adsorbed onto the surface and react to form thin films, while volatile organic part of the precursors desorbs from the surface. High precursor purity is imperative for high-quality TMD film with less organic contaminants. © 2018 Elsevier. Reprinted, with permission from, Ref [51]

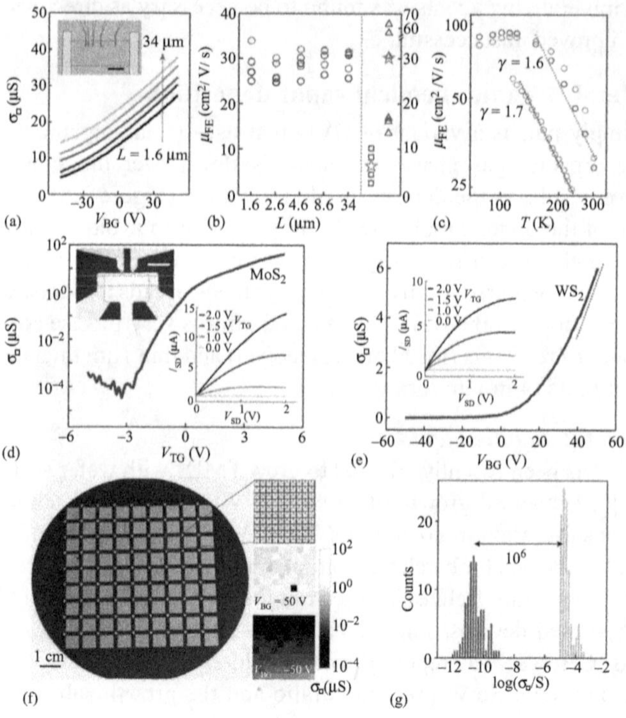

Figure 3.7 Continued

on sapphire while amorphous BN yielded the smallest nucleation density of WSe_2 with a strong preference for vertical growth rather than lateral growth. This is indicative of significant precursor–substrate interaction. They also reported the growth of WSe_2/graphene heterostructures that exhibited a tunnel barrier. Additionally, Lin *et al.* [74] reported resonant tunneling of charge carriers in MoS_2/WSe_2/graphene heterostructures (grown with MOCVD) leading to room temperature NDR (Figure 3.8). It is quite evident that MOCVD is quite versatile for growing heterostructures with a high degree of control on stoichiometry over a large area.

Hexagonal boron-nitride

Growth of h-BN using MOCVD techniques have been reported although they have mainly focused on relatively thick films (i.e., not atomically thin). Kobayashi and Akasaka [75] reported the growth of 300 nm epitaxial h-BN films on (0 0 0 1) sapphire substrates using MOVPE utilizing $(C_2H_5)_3B$ and NH_3 precursors at ~1,080°C. Also, Li *et al.* [76] reported the growth of 1 μm h-BN epilayer on sapphire at 1,300°C using the same precursors. Reports of achieving mono- to few-layers h-BN using MOCVD

Figure 3.7 (a) Gate voltage (back-gate indicated by V_{BG}) dependent sheet conductance of MoS_2 monolayer devices with different channel lengths (vertically displaced for clarity). Inset contains optical image of the device (scale bar is 10 μm). All devices show near identical n-type behavior, (b) comparison of field-effect mobility of FETs at five different locations in the batch (indicated by the ellipses), with triangles and squares representing the mobility measurements of exfoliated samples (from [77]) and CVD grown (from [78])—their median value indicated by purple stars. Device mobility measurements at different areas are nearly identical, indicating large-area homogeneity of MoS_2 thin film and nearly the same as exfoliated samples, (c) shows temperature dependent mobility of FETs of channel lengths 1.6 μm (solid-fill circles) and 8.6 μm (hollow circle markers) and exfoliated samples (square shaped markers) with the MOCVD devices having higher mobility than exfoliated ones at lower temperatures, (d) sheet conductance (MoS_2 devices) as a function top-gate (V_{TG}) voltage showing high ON/OFF ratio (~10^6). Upper inset shows device geometry while lower inset shows $I_{SD} - V_{SD}$ curves with a good current saturation and ohmic behavior, (e) similar measurements on WS_2 devices showing sheet conductance vs V_{BG} dependence and current saturation and ohmic behavior in $I_{SD} - V_{SD}$ curves (inset), (f) batch fabricated 8,100 MoS_2 FETs on a 4 in. Si wafer. Zoomed area shows devices in a single square also with the color maps of sheet conductance at $V_{BG} = 50$ V and $V_{BG} = -50$ V, respectively, (g) histogram of ON-state and OFF-state sheet conductance (top-gated behavior) showing a median ON–OFF ratio of 10^6. © 2018 Springer Nature. Reprinted, with permission from, Ref [72]

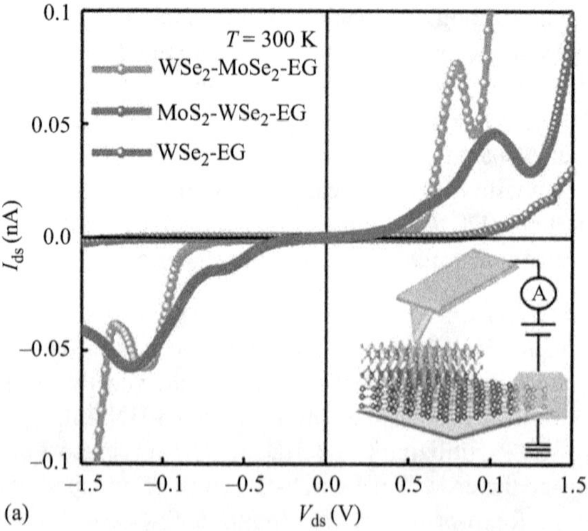

(a)

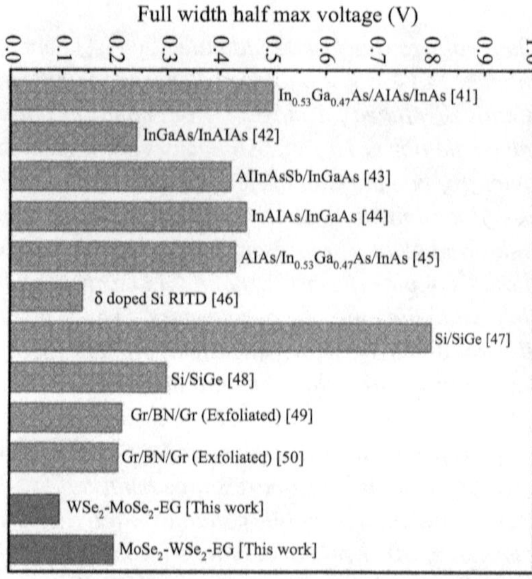

(b)

Figure 3.8 (a) $I_{DS} - V_{DS}$ of different heterostructures showing the NDR behavior at room temperature. Inset shows the device geometry used in measurements and (b) comparison of FWHM voltage at NDR region of different heterostructures reported in literature. © 2018 Creative Commons. Reprinted, with permission from, Ref [74] have been minimal. Paduano *et al.* [79] reported the growth of highly crystalline atomically smooth few layers (~1.6 nm) of h-BN on sapphire in a low pressure MOCVD setup at ~1,050°C using the previously mentioned gaseous precursors. Additionally, few-layer growth of h-BN on sapphire with wafer scale uniformity was also reported by Kim *et al.* [80]. Grown at 1,050°C, with $(C_2H_5)_3B$ and NH_3 precursors, the h-BN films were found to have an average thickness of about 3.5 nm (from XRR measurements) and were also found to have nitrogen vacancies. The authors found that increasing the reactor pressure during growth serves to reduce nitrogen vacancies and also improve the film's crystallinity. Additionally, the authors also demonstrate a PMMA-assisted transfer technique that would allow its easy integration with other semiconducting materials (i.e., vertical heterostructures for building 2D tunneling field effect transistors (TFETs)).

3.2.1.3 Molecular beam epitaxy

MBE is an epitaxial growth method with an extremely high degree of control over the thickness and composition, primarily employed for the growth of single crystal compound semiconductor thin films because of their high technological value to the semiconductor industry. In this process, a beam of constituent atoms or molecules is made to react and deposit on crystalline substrate (ultra-clean and atomically smooth) in an ultra-high vacuum (10^{-8}–10^{-12} Torr) chamber. The crystalline thin films are formed in registry with the substrate, i.e., epitaxial film. These films are of the highest quality and composition and come closest to the ideal models employed in theoretical calculations. Figure 3.9 shows the general setup of an MBE—source ovens are used to evaporate constituent atoms from ultra-pure sources (>99.99999%), and mechanical shutters control their entry into the growth chamber. The beam of constituent atoms condenses after a single pass through the chamber (use of chilled walls) and keeps the background pressure very low, thereby ensuring the highest quality [81]. This system is generally fitted with RHEED gun for in situ measurement of diffraction pattern as compared to the traditional post-growth characterization as with any other growth method. This is allowed for continuous surface analysis to directly observe the effect of growth conditions on the film crystallinity.

Although multiple reports of MBE grown TMDs are present in literature with the growth of $MoSe_2$ being the most reported, reports of charge-carrier mobility measurements on MBE films are nonexistent. Zhang *et al.* reported ARPES measurements of bandgap evolution in $MoSe_2$ layers grown on 6H-SiC substrates using MBE [82]. Vishwanath *et al.* [83] studied the growth mechanisms and quality of $MoSe_2$ on graphene, HOPG and CaF_2 substrates. More interestingly, Barton *et al.* [84] demonstrated the growth of TMD/h-BN heterostructures with an atomically abrupt interface with no detectable misfit-dislocation or strain despite large lattice mismatch. Further studies and optimization of TMD growth is required especially to account for differences in the mobility of adatoms. Despite its prohibitively high initial cost, MBE can provide an excellent route for developing high-quality vdW heterostructures and help take a step toward practical device integration of 2D-layered materials.

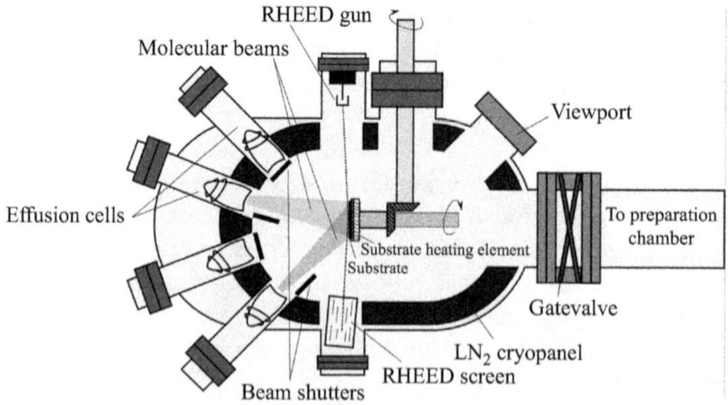

Figure 3.9 Typical schematic of MBE system shows a chamber maintained at UHV with cooled walls to condense the molecular beam after a single pass and keep a low background pressure. The sublimation of ultra-pure sources is performed in effusion cells. Substrate is mounted on a heated stage, and beam shutters expose to or isolate from the molecular beams from sources. Additionally, RHEED system allow for in situ characterization of the thin films. Image obtained from Wikimedia Commons

Doping of 2D materials

Doping (controlling the type of charge carriers and their concentration) is a crucial tool in semiconductor device design. Substitutional doping has been predominantly employed in bulk semiconductors, i.e., introducing dopant atoms (e.g., B, As, P, Mg, Si, etc.) that replace the host atoms (e.g., Si, Ga, N, etc.) in the crystal lattice. Contrary to traditional semiconductors, several doping mechanisms in mono-/few-layer TMDs such as substitutional and defect doping (direct alteration the crystal lattice), chemical functionalization with electron donor/acceptor layers and physisorption/chemisorption of gaseous molecules have been utilized. Controllable substitutional doping would be the most preferred and stable method with the dopant atoms incorporated and covalently bonded to the lattice. Surface adsorption of molecules, although functional, can be unstable due to desorption at different operating conditions (ambient, temperature, etc.). Chemical functionalization presents a midway route to achieve stable doping in TMDs. Laskar *et al.* [85] reported degenerate p-type doping of large area few layer MoS_2 grown by CVT by incorporating Nb into the starting Mo film. These films exhibited a 3.1×10^{20} cm^{-3} with a Hall mobility of 8.5 cm^2/V s at room temperature. Niobium belonging to Period V in the periodic table, similar in size and possessing one less electron than molybdenum, had been suggested to be a suitable acceptor for MoS_2. Fang *et al.* [86] reported n-type vapor phase surface charge doping of few-layers MoS_2 and WSe_2 using potassium atoms. Potassium atoms degenerately doped the TMDs and showed a sheet charge density (extracted) of ~1×10^{13} and ~2.5×10^{12} cm^{-2} for MoS_2 and WSe_2, respectively. Similarly, Lin *et al.* [87] reported chemical functionalization of MoS_2 with Cs_2CO_3 to achieve n-type behavior in MoS_2. With the extent of doping modulated with the thickness of Cs_2CO_3 layer, the authors measured ~9 times increase in carrier concentration as compared to pristine MoS_2. Nipane *et al.* [88] demonstrated p-type doping of MoS_2 by shallow phosphorus implantation by a CMOS compatible PIII process with a higher degree of control over dopant concentration and area. This plasma assisted process was found to result in a combination of substitutional (P atoms) and adsorbent (P adatoms and PH_3) doping mechanism. Fang *et al.* [89] showed degenerate p-type doping of WSe_2 monolayers through chemisorption of NO_2 molecules with an estimated doping concentration ~3.1×10^{19} cm^{-3} with a peak hole mobility of ~250 cm^2/V s. Because of their oxidizing property, NO_2 acts as electron pumps and was utilized for blanket doping of WSe_2 or selective doping at source-drain contacts to improve current injection by reducing contact resistance. As mentioned earlier, the NO_2 doping is reversible, and desorption of gas molecules happens when the devices are exposed to ambient air.

A significant number of reports on electrical and magnetic doping of phosphorene exist in literature that explore the possibility with first-principle DFT simulations [90–92]. Xiang *et al.* [94] demonstrated both n- and p-type doping employing Cs_2CO_3 and MoO_3 via surface charge transfer. Performed on few layers of BP, Cs_2CO_3 was found to increase electron concentration by ~3 times from its intrinsic value. Additionally, surface functionalization was also found to improve the mobility by ~25 times. The authors believe that presence of large number of charge trapping sites in undoped BP results in trapped charge scattering and increase in carrier concentration through doping results in filling of these sites and also screen the trapped charges, thereby increasing their mobility. Similarly, MoO_3 was found to impart p-type doping to BP layers with almost ~3 times increase in hole concentration while retaining its mobility value. Electron doping of BP with Cu adatoms (not limited just to Cu) deposited by sputtering can also be achieved [95]. Cu adatoms shifted the threshold voltage (of undoped BP) by an average of -18.1 V corresponding to an additional charge of 12.7×10^{11}/cm^{-2}. Yu *et al.* [96] utilized benzyl viologen as an electron dopant to dope a portion of p-type BP to form an in-plane p–n junction. Covalent functionalization with aryl diazonium has been shown to imbue p-type doping in BP layers [97]. In addition to affording controllable hole doping, aryl chemistry also successfully passivate the surface of BP, with the devices showing no surface degradation after air-exposure for several weeks, along with improving ON–OFF ratio and improving hole-carrier mobility.

3.3 Metal–semiconductor contacts and 2D heterostructures

Careful understanding of metal semiconductor contact such as the type (Ohmic or Schottky) and the resistance value is essential for device applications. Interestingly, researchers working on TMDs have long felt that the measured low-mobility of TMDs might not actually reflect the intrinsic property of the semiconducting film but rather become skewed by unfavorable contacts. Although several low-resistance ohmic contacts to TMD (primarily MoS_2 and WSe_2) have been reported in literature [97,99], a systematic study of metal–TMD contacts is absent. Understanding the basic parameters influencing contact formation is crucial. In this regard, Kang *et al.* [100] and Popov *et al.* [101] reported computational modeling of different metal contacts (Au, Pd, In, Sc, Zr and Ti). By starting off with standard constraints such as metal work function (WF), high or low enough, to obtain an n-type or p-type Schottky contact, minimal lattice mismatch to maximize orbital overlap (with TMDs) and presence of d-orbitals to facilitate better mixing with those in transition metals, their study revealed that a proper contact is not solely dependent on the WF but rather is an interplay of different parameters including (i) the geometry, i.e., side-contact (SC) or end-contact (EC), (ii) partial-DOS (i.e., DOS on specified atoms or orbitals), (iii) interface electron density and (iii) effective potential. The results of DFT calculations summarized below highlight the trends followed by metal–semiconductor contacts in various geometries

(1) In SCs, Pd and Ti provided the smallest physical separation among high-WF and low-WF metals, respectively. On the other hand, metals in EC had small separation and formed covalent bond with the TMD.
(2) Calculations showed n-type doping of MoS_2 with ohmic contact and SB using Ti and In metals, respectively. Similarly, p-type doping in WSe_2 can be achieved with Au or Pd contacts. Interestingly, partial DOS were found to be much higher than SCs which would aid in reducing contact resistance.
(3) High-interface electron densities (indicative of greater orbital overlap) were observed in EC configuration, even higher than those any SC.
(4) Finally, EC were found to provide a small narrow tunnel barrier as compared to SC (e.g., Au-MoS_2 SC (0.31 eV) as compared EC (0.14 eV)).

It has been found that Au, a commonly used electrode metal, is not efficient in electron injection as compared to others such as Ti. Au contacts shift the Fermi level (combined system) upward by ~0.1 eV (as compared to bare MoS_2) with the states near E_F are predominantly Mo4d_{z^2} with minimal contributions from 's' and 'p' orbitals of sulfur while the Ti contacts shift the combined Fermi level upwards by ~0.25 eV (resulting in increased DOS near E_F) with significant contributions from *S3p* and *Mo4*d_{xy} as well as resulting in better electron injection. Interestingly, other members of the 2D materials family can be employed to form low-resistance contacts. Leong *et al.* [102] utilized an etched graphene–Ni hybrid to form contact with MoS_2 whose resistance value was as small as 200 Ω μm (comparable to semimetallic phase contacts). Etched graphene contained hexagonal pits with exposed zigzag edges that formed a low resistance EC with the subsequent Ni metallization and serves to reduce the SB height (from ~0.804 eV in Ni-MoS_2 to ~0.299 eV in Ni-SLG-MoS_2). This results in increased tunneling probabilities and improved current injection, in addition to improving mobility (from 27 to 80 cm^2/V s). Experimentally, Sc contacts to MoS_2 were observed to reduce the SB height to just 30 meV and boosting the mobility up to 184 cm^2/V s with Ti close behind with a 50 meV SB and mobility of 125 cm^2/V s. Alternatively, local phase engineering of TMDs with intercalation of organolithium can be employed to reduce contact resistance. Kappera *et al.* [103] demonstrated phase engineering of MoS_2 from semiconducting 2H phase to metallic 1T phase near the Au contacts to decrease resistance from 0.7 to 10 kΩ μm to just 200–300 Ω μm. This was attributed to the atomically sharp interface between the two phases and the WF of the 1T phase and the CB energy w.r.t. vacuum level being very similar (~4.2 eV). Similarly, a lower contact resistance to $MoTe_2$ with locally patterning 1T phase near the Au electrodes was reported by Cho *et al.* [22].

Heterojunctions are interfaces formed between two dissimilar semiconductors, generally, of unequal bandgap. Such heterostructures are basic building blocks of semiconductor manufacturing and have found widespread use in designing devices such as semiconductor lasers, BJTs, HEMTs, etc. In a similar fashion, vdW heterostructures with 2D materials such as TMDs, h-BN, graphene or phosphorene can be designed to augment their functionality. Such systems are quite different from their 3D counterparts because *(i) atomically abrupt interfaces can be obtained by simple stacking of layers without necessarily requiring MBE growth and (ii) each layer acts as the bulk and the interface simultaneously.* Stacking layers potentially leads to significant charge redistribution and/or structural changes between neighboring crystals. This can be effectively controlled by adjusting the stacking orientation between individual building blocks. However, developing (vertical) heterostructures requires clean and sharp interfaces that cannot be easily achieved with mechanical transfer due to high possibility of contamination, very little control over stacking orientation and low-yield. In-plane heterostructures can be only created directly during growth. Ideally, a one-step growth would be the best way to proceed in building 2D heterostructures for which MBE or MOCVD would be preferred choices. Interestingly, CVD-based parallel stitching of TMDs, graphene and h-BN has been demonstrated [104]. Utilizing PTAS molecules for seeding in selective heterostructures of metal–semiconductor (graphene–MoS_2), semiconductor–semiconductor (WS_2–MoS_2) and insulator–semiconductor (h-BN–MoS_2) was fabricated. Growth of high-quality, in-plane heterostructures, using an easily scalable and large-area process such as CVD, represents a significant step in fabricating building blocks which could be used in logic devices and ICs.

3.4 2D materials for steep slope devices

3.4.1 Band to band carrier tunneling (BTBT) in 2D systems

This section provides a brief description and discussion about the band-to-band tunneling transport through a quantum mechanical barrier as a prelude to steep-slope devices. Analytical expressions describing tunneling transport have long since been established for analyzing tunnel junctions in semiconductor devices. These expressions could be successfully applied to 2D systems (with bandgap) properly taking into account the reduced DOS [105]. Our discussion is based on the general configuration of a TFET consisting of a standard p–i–n junction, with doping levels designed to align their Fermi level at respective band edge (i.e., $E_v^p = E_n^c$). Under zero bias, no

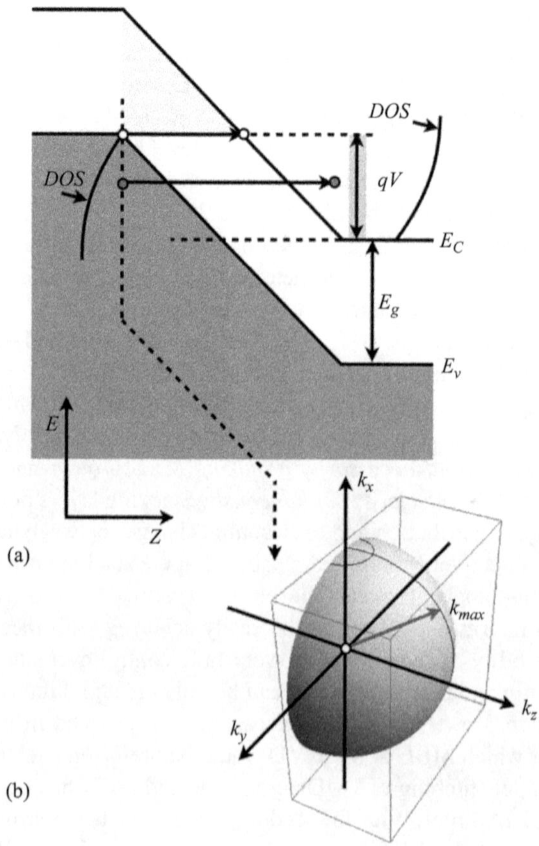

Figure 3.10 (a) Reverse bias p–i–n junction diode using Zener tunneling for current injection when the applied bias reduces the barrier width, (b) restricted volume of k-space hemisphere as described by the equation $2k_\perp^2 + k_{zp}^2 \leq k_{max}^2$. © 2018 IEEE. Reprinted, with permission from, Ref [103]

current flow exists, whereas upon application of reverse bias, a finite barrier is formed ($E_v^p - E_n^c = qV$) that allows electrons within this energy window to tunnel from VB to the CB (Figure 3.10). The resulting tunneling current can be calculated from the summation of contribution of each k-state electron and can be written as

$$I_T = q\frac{g_s g_v}{L}\sum_k v_g(k)\,(f_v - f_c) \times T$$

where g_s and g_v are spin and valley degeneracy respectively, L is the macroscopic length along the electric field, $v_g(k)$ is the group velocity of carriers in the band $E(k)$, f_c and f_v are the Fermi–Dirac occupation factors of VB and CB, respectively, and T is the tunneling probability. The summation extends over all the possible

$k-$states the carriers can tunnel. An analytical expression for the tunneling probability can be obtained through Wentzel–Kramers–Brillouin (WKB) approximation [106]. Assuming that electrons in the p-doped side possess a transverse kinetic energy $E_{\perp} = \hbar^2 k_{\perp}^2 / 2m_v^*$ the tunneling probability is then given by

$$T_{\mathrm{WKB}} = \exp\left[-\frac{4\sqrt{2m_R^*}\left(E_g + E_{\perp}\right)^{3/2}}{3q\hbar F} \right] \approx T_o \exp\left(-\frac{E_{\perp}}{\overline{E}} \right)$$

$$\text{where } T_o = \exp\left[-\frac{4\sqrt{2m_R^*}\left(E_g\right)^{3/2}}{3q\hbar F} \right],$$

$$\overline{E} = \frac{q\hbar F}{\sqrt{2m_R^* E_g}} \text{ and}$$

$$m_R^* = \frac{m_c^* m_v^*}{\left(m_c^* + m_v^*\right)}$$

In the above expressions, F is the electric field at the junction, E_g is the bandgap and m_R^* is the reduced effective mass. Each electronic k-state has a tunneling probability as evaluated above and their summation would yield the total current. In order obtain an estimate of the allowed carrier momentum that could tunnel through the barrier, we employ momentum and energy conservation principle with the following assumptions: *(i) tunneling process is elastic (absence of phonons), (ii) lateral momentum is conserved and (iii) bands are symmetric (i.e.* $m_c^* = m_v^* = 2m_R^*$). Based on these assumptions, we obtain the following expressions:

$$E_v^p - \frac{\hbar^2}{2m_v^*}\left(k_{xp}^2 + k_{yp}^2 + k_{zp}^2\right) = E_c^n - \frac{\hbar^2}{2m_c^*}\left(k_{xn}^2 + k_{yn}^2 + k_{zn}^2\right)$$

$$2k_{\perp}^2 + k_{zp}^2 = \frac{4qVm_R^*}{\hbar^2} - k_{zn}^2$$

$$k_{\perp}^2 = k_{xp}^2 + k_{yp}^2 \quad \text{and}$$

$$k_{\max}^2 = \frac{4qVm_R^*}{\hbar^2}$$

For the electron to emerge on the other side of the barrier k_{zn}^2 must be nonzero which results in the constraint $2k_{\perp}^2 + k_{zp}^2 \leq k_{\max}^2$ that describes the restricted volume of k-space hemisphere (for a 3D case) that satisfies the requirements for tunneling. In a 2D material system, the restricted k-space volume is given simply by$2k_y^2 + k_x^2 \leq k_{\max}^2$. The interband tunneling current density per unit width is then given by

$$J_T^{\mathrm{2D}} = \frac{qg_v g_s \sqrt{2m_R^* \overline{E}}}{2\pi^2 \hbar^2} \times T_o \times \left[(qV - \overline{E}) \times \sqrt{\pi} \times \operatorname{erf}\left[\sqrt{\frac{qV}{2\overline{E}}} \right] + \sqrt{qV.2\overline{E}} \exp\left[\frac{-qV}{2\overline{E}} \right] \right]$$

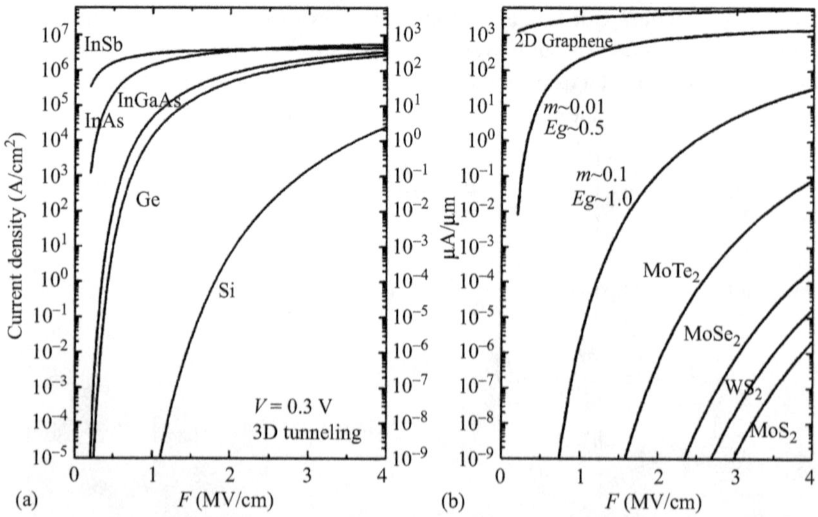

Figure 3.11 (a) Calculated tunneling current densities for traditional semiconductor p–n homojunctions, (b) tunneling current per unit width for 2D materials such as graphene and TMDs. It is observed that larger the bandgap of semiconductor, smaller is the I_{ON}. © 2018 IEEE. Reprinted, with permission from, Ref [105]

At very low reverse bias, $J_T^{2D} \sim V^{3/2}$, while for larger bias voltages, the tunneling current exhibits a more linear dependence. Figure 3.11(a) and (b) shows the comparison of the tunneling current densities of traditional bulk semiconductors and TMDs respectively. The above expressions accurately describe in-plane tunneling currents, and from Figure 3.11(b), it is evident that smaller bandgap semiconductors exhibit a higher ON-state current. This issue can nevertheless be easily addressed in TMDs by employing multilayer films. Interestingly, in recent years, a counterpart to in-plane tunneling, out-of-plane tunneling in vertically stacked heterostructures has become an area of intense research. Unlike in-plane mechanism, conventional analysis of this mechanism is not feasible, and instead quantitative calculations such as Bardeen-transfer Hamiltonian approach [107] used in STM can be utilized.

3.4.2 Tunnel field effect transistors

Constant scaling of transistor channel dimensions has provided tremendous increase in speed, afforded more functionality and higher transistor density in today's microprocessors. In conjunction, scaling power supply in MOSFET circuits has attracted significant interest since scaling supply voltages without degrading I_{ON}/I_{OFF} ratio can result in considerable power savings. A traditional MOSFET that relies on thermionic injection of carriers over an energy barrier has a fundamental limit on the steepness of slope during ON–OFF transition which is reflected in the subthreshold swing and is given by

$$S = \frac{dV_g}{d(\log_{10} I_D)} = (\ln 10) \times \left(1 + \frac{C_d}{C_{ox}}\right) \times (K_B T/q)$$
$$\sim 60 \text{ mV/decade at } T = 300\text{K}$$

The above expression gives us the change in gate voltage (V_g) required for a change in drain current (I_D) by an order. Here, KT/q is the thermal voltage, C_d and C_{ox} are depletion and oxide capacitances, respectively. To fully realize the potential of transistor channel length reduction, reduction in supply voltage (V_{DD}) and threshold voltage (V_T) would also be necessary in order to keep the overdrive voltage ($V_{DD} - V_T$) high. This would result in an increase in I_{OFF} due to irreducible subthreshold slope (Figure 3.12(a)) and result in degrading ON–OFF ratio while also increasing leakage power in traditional MOSFET devices. In principle, supply voltage scaling can be achieved, in principle, by reducing the subthreshold swing. To realize steep-slope design one has to base their current injection mechanism on a different physical phenomenon, i.e., band-to-band tunneling instead of thermionic injection. This quantum mechanical effect, first identified by Zener in 1934, can be abruptly switched on and off by controlling the band bending with an applied voltage. In this regard, TFETs can provide significant power savings by reducing supply voltages, lower leakage power and provide sharper switching by relying on quantum mechanical tunneling. This makes them excellent candidates for low-power logic devices or circuits. Figure 3.12(b) compares the performance of multi-gate device, high mobility III–V devices, and bulk-Si MOSFETs with two operation points "*A*" and "*B*" marked on the figure. At normal operating conditions, TFETs offer higher $I_{\text{ON}}/I_{\text{OFF}}$ and offer higher I_{ON} for a smaller gate voltage as compared to the other devices, whereas at higher performance requirements, MOSFET indicated by point "*B*" MOSFET becomes a better choice. The total energy consumed in a logic gate can be expressed as a sum of "*dynamic*" and "*leakage*" components. With "α" being logic activity factor, "L_d," the logic depth, "C" being the switched capacitance and time delay "τ_{delay}," the total energy consumed can be given by [108]

$$E_{total} = E_{dynamic} + E_{leakage} = \alpha L_d C V_{DD}^2 + L_d V_{DD} I_{\text{OFF}} \tau_{delay}$$

$$\approx \alpha L_d C V_{DD}^2 = L_d C V_{DD}^2 \frac{I_{\text{OFF}}}{I_{\text{ON}}} = L_d C V_{DD}^2 \left(\alpha + \frac{I_{\text{OFF}}}{I_{\text{ON}}}\right)$$

$$E_{total} \approx L_d C V_{DD}^2 (\alpha + 10^{-V_{DD}/S})$$

From the above equation, we can easily infer that a device design that has a smaller I_{OFF}/I_{ON} is more energy efficient per logic cycle. Now, the operation frequency "f" is given by $1/L_d \tau_{delay}$ and being empirically proportional to the supply voltage [109]. Therefore, the total power dissipated can be expressed as

$$P = \alpha L_d C V_{DD}^2 f = I_{\text{OFF}} V_{DD}^3$$

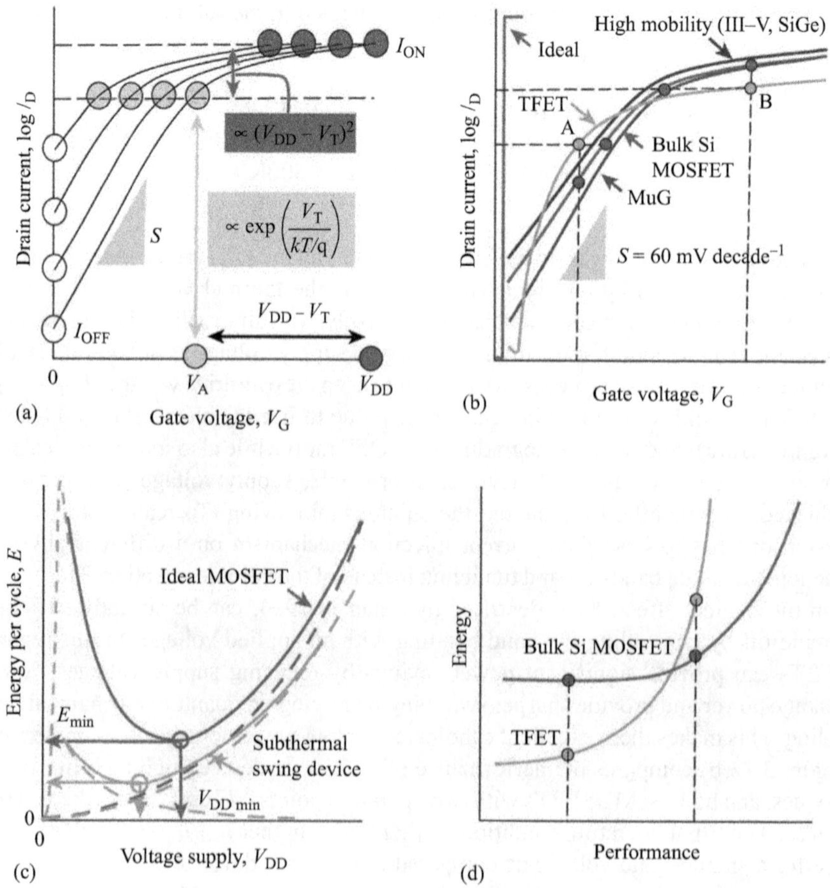

Figure 3.12 *(a) Transfer characteristics I_D vs V_G of conventional MOSFET showing an exponential increase in I_{OFF} due to incompressible subthreshold slope. Simultaneous scaling of V_{DD} and V_T would be required to maintain same I_{ON} by keeping overdrive constant, (b) comparison of multi-gate device, Si-MOSFET, high-mobility III–V and SiGe and TFET. At operating point "A," TFET offers better ON–OFF ratio and superior performance. At point "B" (required for higher performance), MOSFET becomes a better option, (c) comparison of minimum switching energy and V_{DD} for a MOSFET and TFET at same I_{ON}/I_{OFF} shows the considerable energy savings and voltage scaling feasible with a TFET. (d) Comparison of switching energy and performance for a TFET and MOSFET showing energy savings at low and moderate performance region of the device.* *© 2018 Springer Nature. Reprinted, with permission from, Ref [108]*

From this equation, one can observe that a five-fold voltage scaling without compromising on leakage current can lead to reducing power dissipation by almost 125 times.

As mentioned in the previous section, the most common TFET configuration is a p–i–n device wherein a bias voltage is used to alter the band bending allowing carriers to tunnel. Although inherently an ambipolar device, the doping profiles can be made asymmetric or heterostructures can be employed to suppress ambipolarity and make it a dominant n- or p-type. The arrangement of the bands is such that the VB edge of the channel is located below the CB edge of the source leading to suppression of band to band carrier tunneling (BTBT) and very low I_{OFF}. Applying a revers bias lifts the channel VB over the CB of the source and opens up a conduction channel. Since only the carriers with an energy window "$\Delta\Phi$" can have states to tunnel into, the higher energy tail of the Fermi distribution (of charge carriers) are filtered out. The device is "*cooled down*" effectively acting as a MOSFET operating at a lower temperature. As a consequence of BTBT mechanism, the subthreshold swing is not a constant but is a function of gate-source voltage. Although it can be a challenge to design TFETs to operate at higher *ON* current, the subthreshold swing can still be maintained under 60 mV/decade over several orders of magnitude of drain current. Two-dimensional heterostructures are quite promising for achieving steep slope devices because of their minimal interface state densities and sharp band edges. Interestingly, 2D semiconductor with finite bandgap (such as TMDs) possess almost all the necessary requirements for TFET applications such as *(i) tunable bandgap and (ii) Fermi-level adjustment through doping* and therefore are promising candidates for realizing steep slope devices.

Given the renewed interest in 2D semiconductors and TMDs and their potential applicability in TFET design, it would be relevant to list some of the work demonstrating BTBT in TMDs and heterostructures and also devices that exhibit a smaller subthreshold slope. Sarkar *et al.* [110] reported the use of Ge/MoS_2 hybrid with atomically thin MoS_2 acting as the channel (as seen in Figure 3.13). Their unique design provides a number of advantages such as *(i) an extremely small barrier width provided by bi-layer MoS_2, whose band bending can be easily altered with a gate voltage, (ii) low EA of Ge and bandgap along with high EA of MoS_2 forms a staggered heterojunction necessary for BTBT, (iii) a vdW heterostructure with a strain-free interface, (iv) exploits the well-established doping principles and procedures for 3D semiconductors to form abrupt doping profiles at source-channel interface with negligible diffusion of dopants across the interface due to vdW gap, (v) finally, the vertical heterostructure with a large overlap area gives multiple channels for tunneling current leading to higher I_{ON} compared to lateral heterojunctions.* With an ultra-low source-drain voltage of 0.1 V, this planar device has demonstrated subthreshold slope of <60 mV/decade over four decades of drain current with a minimum subthreshold value of 3.9 mV/decade. Improvements to drain current can be achieved by treatment of Ge substrate to remain GeO_x layer that acts as an unintentional tunnel barrier whose transmission probability does not significantly improve with applied gate bias. Interestingly, the device represents the first of its kind (i.e., in planar architecture) that

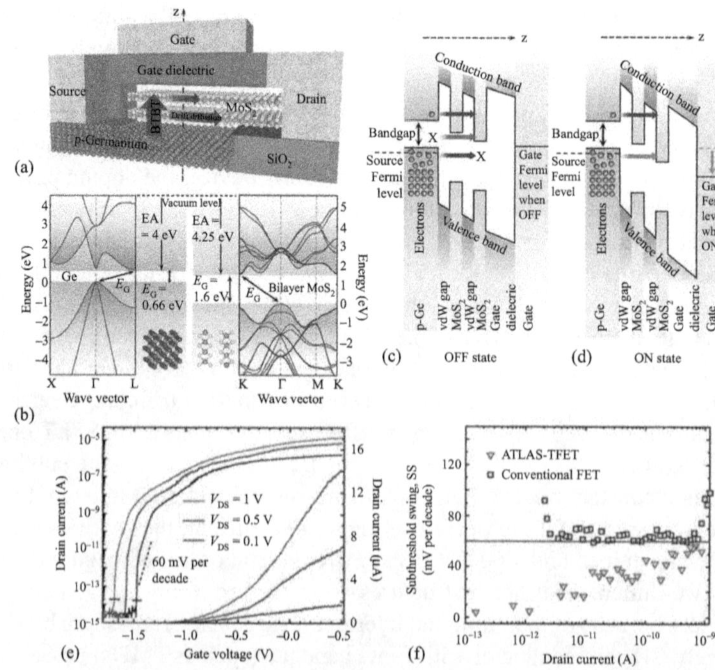

Figure 3.13 *(a) Cross section of ATLAS-TFET with bilayer MoS_2 channel and p-type Ge as the source electrode. The carrier pathway is indicated by the red arrows. The tunneling barrier height is determined by band overlap between Ge and MoS_2, and the width is determined by MoS_2 thickness added to the vdW gap. (b) The band alignment of the two materials and their EA illustrates the formation of staggered heterojunction. Insets show crystalline structure of Ge and MoS_2 along with the band structure shown on either side. (c) and (d) Band diagrams along the vertical dashed line in (a). White regions are forbidden energies with no available DOS. The drain current is perpendicular to the plane of the figure and is not shown. In OFF state electrons in VB of Ge do not have any available states in MoS_2 to tunnel into (denoted by the cross at the end of the black arrow). Arrow with the cross at its tail represents states available in MoS_2 but forbidden in Ge. Electrons in CB of Ge can tunnel into states of MoS_2 due to the availability of states, but this current is very small and negligible (minority carrier). (d) Applying gate voltage lowers CB of MoS_2 below VB of Ge and electrons undergo BTBT leading to abrupt increase in ON-current. (e) Drain current as a function of gate voltage (at different source-drain voltage). Black solid line indicates the cross point where the drain current is same as gate leakage current, and the dashed line is the 60 mV/decade slope. All three measured currents produce an SS slope <60 mV/decade. (f) Comparison of SS slope of conventional and ATLAS MOSFET. ATLAS MOSFET has an SS smaller 60 mV/decade over four orders of drain current. © 2018 Springer Nature. Reprinted, with permission from, Ref [110]*

satisfies International Technology Roadmap for Semiconductors (ITRS) prescription of achieving <60 mV/decade over four decades of drain current (Figure 3.13). Roy *et al.* [111] utilized a MoS_2/WSe_2 heterostructure wherein the carrier concentration and potential of each TMD layer is controlled independently by a symmetric dual gate structure. This device exhibited a unique dual behavior under different bias voltages—Zener tunneling was observed under reverse bias with the device resembling n^+/p^+ diode with a type III band offset (broken gap), whereas under forward bias, the device exhibited rectifying behavior with the band offset at the heterojunction changing to a type II (staggered) gap (Figure 3.14). This behavior is unique to this design as traditional 3D semiconductor heterojunctions cannot exhibit both rectifying and tunneling behavior due to difference in the required doping conditions. The difference in behavior stems from the vdW gap at the junction of the heterostructure that allows the bands of these two materials move relative to each other via gate voltage. Conventional heterostructures do not possess this flexibility as the tunneling interface has covalent bonds that pin the band offset. A modified version of the previous device design, which utilizes $WSe_2/SnSe_2$ heterostructures instead [112] wherein the WSe_2 is the gate-controlled p-type layer while $SnSe_2$ is the degenerate n-type layer, can be employed. Without the use of dual-gate design, the entire device can be controlled with a single gate since $SnSe_2$ has a degenerate n-type behavior and the vdW gap prevents pinning of bands at the interface. At very high negative voltage, the heterojunction has type III alignment, the device behaves as p-TFET with subthreshold slope of 100 mV/decade at room temperature and ON–OFF ratio exceeding 10^3. As the applied voltage becomes less negative, the bands switch to a type-II alignment with the device behaving more like n-MOSFET.

Similarly, in parallel to TMDs, few-layer phosphorene has also been shown to possess the necessary qualities for TFET device and in better performance metrics as compared to TMDs in some avenues. TMD-based TFETs generally possess very low I_{ON} which tends to degrade their operation speed and energy delay product [113]. From the previous section, we can notice that the I_{ON} depends on the transmission probability of the tunnel barrier, which in-turn depends on the effective carrier mass, electric field at the interface and the bandgap of the channel. Although channel materials with small bandgap and small carrier effective mass improve ON-current, it would also increase OFF-current through direct source to drain tunneling and hence pose a limitation on the channel scaling. This complicated interplay of parameters necessitates careful choice of materials. Phosphorene, in this regard, displays a great potential for application due to *(i) atomically thin channel that provides a higher electric field, (ii) smaller effective carrier mass and bandgap to form smaller tunnel barrier and (iii) remains a direct-bandgap semiconductor even in multilayer configuration unlike TMDs, and hence the tunneling probability is unaffected by phonons that usually tend to deteriorate* I_{ON}. Based on these desirable properties, full-band atomistic quantum transport simulation of few-layer phosphorene TFETs show an extremely high ON-current of ~1 mA/μm, which is orders of magnitude higher than those with TMDs, ON–OFF ratio of $\sim 10^6$ for a 15-nm channel with source voltage of 0.5 V. Additionally, the subthreshold slope and energy delay product of these devices are significantly smaller as compared to TMD-based TFETs or conventional CMOS

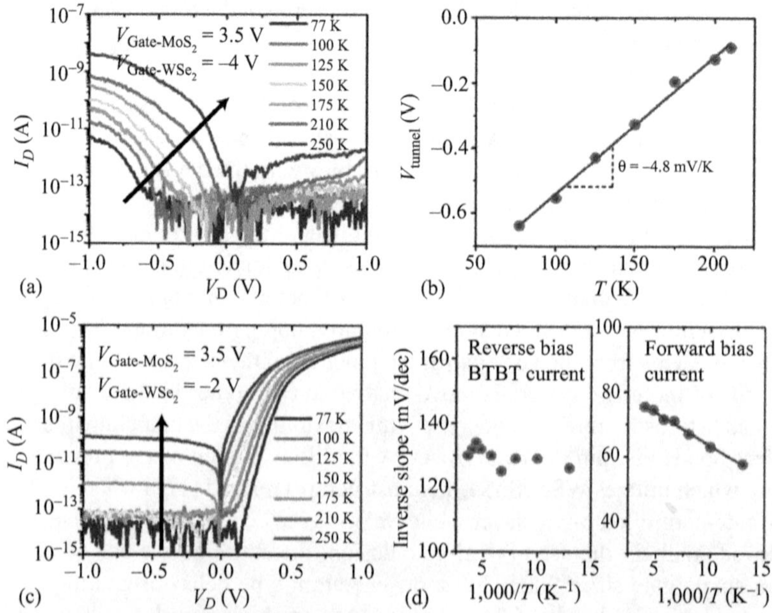

Figure 3.14 *(a) Dependence of tunnel current with drain voltage at different temperatures (The direction of the arrow indicates increasing temperature.). The tunnel current onset voltage shifts positively with temperature—a characteristic of Zener tunneling, i.e., change in bandgap (and band tail) with temperature. (b) Temperature coefficient of tunneling breakdown of the heterostructures is found to be similar to that of Si Zener diode. (c) Temperature dependence of forward rectifying behavior shows decreasing built-in potential with increasing temperature (The direction of the arrow indicates increasing temperature.) and increase in reverse bias current with increasing temperature, (d) inverse slope of tunnel current independent of temperature (left) indicative of Zener tunneling, while the inverse slope in forwards rectifying mode decreases slightly with temperature (right). This weak dependence with temperature in contrast to covalently bonded p–n junctions was explained due to the occurrence of photon or phonon-assisted interlayer majority carrier recombination. © 2018 American Chemical Society. Reprinted, with permission from, Ref [111]*

devices. This serves to highlight the tremendous potential of this material for steep slope tunneling transistors.

3.4.3 *Resonant tunneling devices/diodes*

Given the duality in the nature of behavior of electrons, it is commonplace to observe effects like filtering, diffraction and interference which are traditionally attributed to

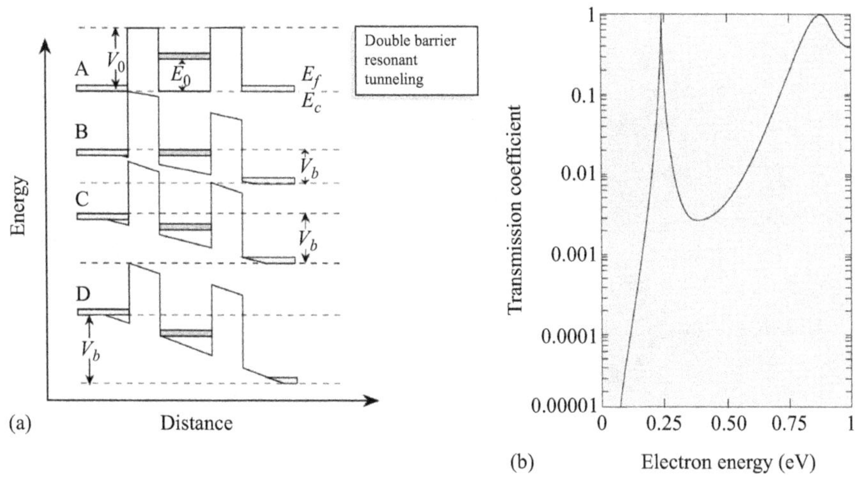

Figure 3.15 (a) Resonant-tunneling operation schematic with the energy band diagram under different bias voltages, (b) transmission coefficient as a function longitudinal electron energy. The sharp peaks represent energies for which transmission coefficient is 1, i.e., no reflection. Reprinted with permission from Springer Nature: Coherent Transport and Mesoscopic Devices by Umesh K. Misra and Japrit Singh. © 2018

electromagnetic waves. For example, filtering of higher energy electrons was seen in TFET devices. The most common configuration of resonant tunneling involves two highly doped regions with quantum wells sandwiched in-between (Figure 3.15(a)). From the structure shown, we can see that at zero bias, from point "*A*," no current flows through the structure as the allowed state in the well is not aligned with the states of the incoming charge carriers. Upon increasing bias, the device goes into "*B*" wherein electrons travel through the barrier as the energy levels on each side lineup and electron current becomes maximum. Upon further increase of bias voltage, i.e., point "*C*," the states on the left and right become misaligned, and the current decreases with increasing voltage (negative resistance region). Further increase in bias voltage results in strong thermionic emission and increases electron current (as indicated by point "*D*"). This structure can be solved analytically by applying Schrodinger's equations in every region. For a complete analytical derivation, the readers can refer to [114]. Tunneling through a double-barrier produces interesting resonance features in the transmission probability function. The sharp peaks in the probability function (Figure 3.15(b)) represent the electron energies for which resonant tunneling through quasi-bound states between the barriers can occur.

Britnell *et al.* [116] first reported the resonant tunneling and negative differential conductivity (NDC) in graphene transistors with h-BN tunneling barriers in-between. Following this work, Mischenko *et al.* [117] also reported graphene/h-BN/graphene vertical heterostructure to form a tunnel transistor with strong NDC that could persist up to room temperature. Attributed to the control of crystallographic orientation between the stacked graphene layers with very high precision (~2°) during device

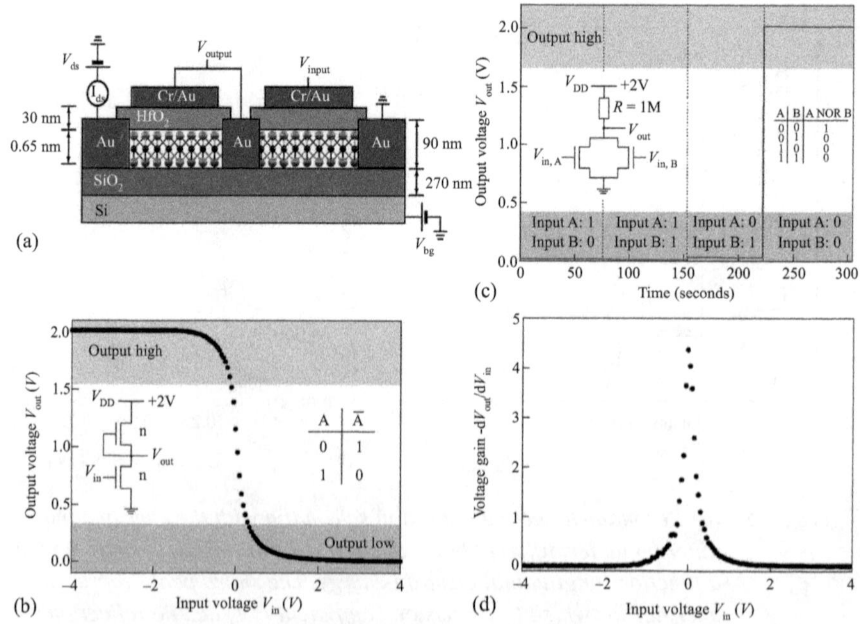

Figure 3.16 Cross-sectional view of monolayer MoS_2 IC in a dual-gate configuration, (b) NOR gate logic implemented by connecting two-MoS_2 transistors in parallel and external 1 MΩ resistive load. The inset shows all different combination of inputs and their corresponding output (V_{out} = 1 only when both inputs are low), (c) inverter operation with two n-type transistors in series that provides a greater than unity gain (~4) (as shown in (d)). © 2018 American Chemical Society. Reprinted, with permission from, Ref [115]

fabrication, resonant tunneling with both momentum and energy conservation could take place and result in a strong NDC response. The authors observed that the orientation between graphene lattice directly influenced the bias point of the resonance. Additionally, the authors devised a high-frequency oscillator circuit that could operate in MHz region (limited only by parasitic capacitances). This tunnel device does not possess the long carrier dwell time commonly observed in standard double-barrier devices which could have favorable implication for applications in THz frequencies. In addition to graphene RTDs, Lin *et al.* [76] reported resonant tunneling and room temperature NDC in TMD/graphene heterostructures. Grown with MOCVD, the MoS_2/WSe_2/graphene stack exhibits NDR with a peak-valley current ratio of 1.9 and 2.2 depending on the order of the stack. These PVCR values are comparable to their reported by in other vdW heterostructures.

The attempts to improve steeper subthreshold swing have primarily focused on utilizing quantum mechanical tunneling for carrier injection instead of thermionic emission. Alternatively, increasing the nonlinearity of the system to improve the steepness of subthreshold swing was also conceived by Gopalakrishnan *et al.* [118] by introducing an amplifier to amplify the ON-state and suppress the OFF-state. The amplification process had the following conditions: (i) amplification arising from a gain mechanism internal to the device and (ii) mechanism should be controllable with gate voltage and be fast enough to switch from ON to OFF state. Impact ionization satisfies these conditions, and a p–n junction pushed into breakdown mode can exhibit extremely steep slopes and also be switched at high frequencies. Such an operation can be achieved with a gated p–i–n junction where the gate voltage modulates the channel length. Simulated behavior of Ge I-MOS produced a subthreshold slope of 5 mV/decade, with an ON-current >1 mA/μm and ON/OFF ratio of $\sim 10^6$ at 400K. Björk *et al.* [119] later demonstrated the subthreshold slope of ~6 mV/decade at room temperature using a Si nanowire using partially surrounded gate. Previous sections have clearly shown the potential of TMDs for low-power devices with high ON/OFF ratio. The applicability of TMDs for I-MOS devices can be considered contingent upon low breakdown field strength which would enable voltage scaling.

3.5 Circuit and system applications

Logic gates are electronically controlled switches such as diodes or transistors that perform Boolean operations and are the fundamental building blocks of digital electronics. The previous chapters were aimed at providing an overview of different aspects of these building blocks themselves—from choosing the appropriate 2D semiconductor material, scalable growth and fabrication techniques for high-density IC manufacturing, controllable doping techniques to alter semiconductor's electrical properties based on requirement, developing heterostructures to increase their functionality to explore new current injection mechanism to enable low-power device that can also allow for voltage scaling and maximize the advantage of miniaturization trend.

2D materials with finite bandgap (~1–2 eV) exhibit high ON–OFF ratio and are well suited for logic operations where clear distinction between the binary states "1" and "0" is imperative along with low standby-power dissipation. MoS_2 transistor-based logic devices were first demonstrated by Radisavljevic *et al.* [115] in 2011. Two n-type MoS_2 monolayer transistors (Figure 3.16(a)) were made to operate as a NOR gate (Figure 3.16(b)) and an inverter (Figure 3.16(c)) with a room-temperature gain >4 (Figure 3.16(d)), thereby affording it the capability to drive the input of next stage. Although these were proof-of-concept devices made from exfoliated MoS_2, these results were the first critical step in utilizing TMDs for performing room-temperature logic operations. Following this, an IC comprising of an inverter, NAND gate, static-RAM and five-stage ring oscillator (Figure 3.17(b)) were fabricated by Wang *et al.* [120] (Figure 3.17(a)), thus making a transition from fabricating standalone logic gates to multi-stage logic circuits that are commonplace in today's ICs. Utilizing both depletion- and enhancement-mode transistors, the devices exhibited ON–OFF ratio of $\sim 10^7$, high ON-current density of ~23 μA/μm and oscillator operation frequency of 1.6 MHz (Figures 3.17(c) and (d)). Interestingly, bi-layer MoS_2 offered several advantages as compared to monolayer, *(i) better gate charge to drain-current*

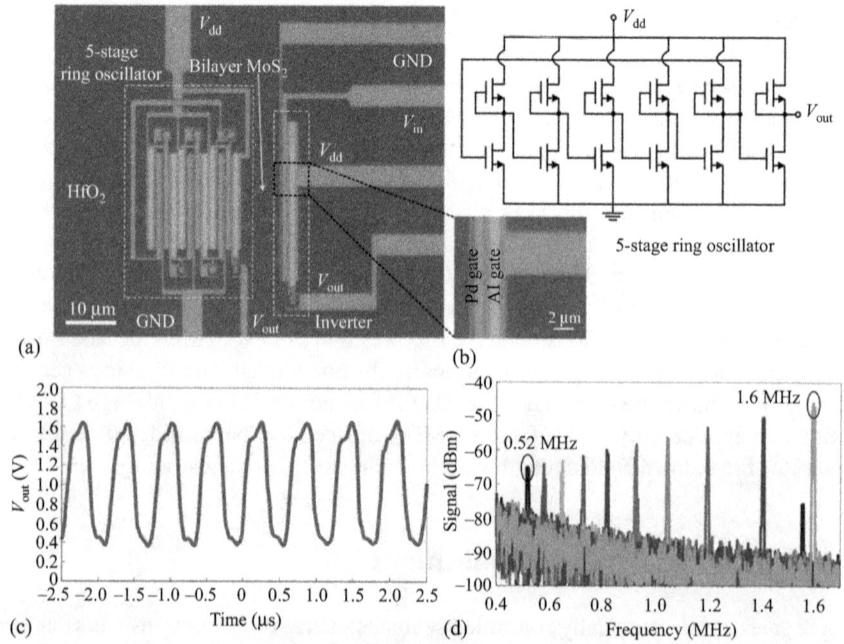

Figure 3.17 (a) Optical image of ring oscillator fabricated with bilayer MoS_2, (b) schematic of ring oscillator circuit with the first five devices forming a positive feedback loop leading to oscillations in its response, (c) output voltage as a function of time oscillates at a fundamental frequency of 1.6 MHz (at $V_{DD} = 2$ V), with a 62.5 ns delay per stage, (d) signal's power spectrum as a function of V_{DD}. From left to right $V_{DD} = 1.15$ V and 1.2–2 V in steps of 0.1 V. As V_{DD} increases, the fundamental oscillation frequency increases from 0.52 to 1.6 MHz. © 2018 American Chemical Society. Reprinted, with permission from, Ref [120]

ratio (decides the speed of device operation) and (ii) improved mobility of ~313 cm^2/V s as compared to ~217 cm^2/V s offered by monolayer. The downside to the bilayer devices was its higher OFF-state power dissipation. Therefore, the choice of the number of layers eventually depends on the final requirement, i.e., high-speed operation or low standby-power dissipation. Interestingly, heterostructures of dissimilar materials have been successfully employed to form inverter, NAND and NOR gates. Laterally stitched graphene–MoS_2 devices have shown the capability to perform logic operations with sharp transitions between logic levels and also drive multiple devices when cascaded due to their larger than unity voltage gain (~7) [121]. Interestingly, non-TMDs such as SnS_2 have been employed to develop logic gates (NOT and NOR) with ON–OFF ratio ~10^7 and with voltage gain exceeding unity (~3.5) [122]. Functionality enhanced field-effect transistors with a polarity control gate form a new class of devices that utilize ambipolar transport in the transistor channel. This device concept in conjunction plays well with the material properties of TMDs in developing ambipolar transport channel. Resta *et al.* [123] have the demonstrated the use of few-layer WSe_2 with a polarity gate that could switch the device operation from an n- to p-type based on the gate bias voltage without the need of any physical doping. This added functionality allows for more compact and less power-consuming circuits. These devices with their dynamic polarity control allow for "*on-the-fly*" polarity change and exhibited an ON/OFF ratio of 10^7 (n-type) and 10^6 (p-type).

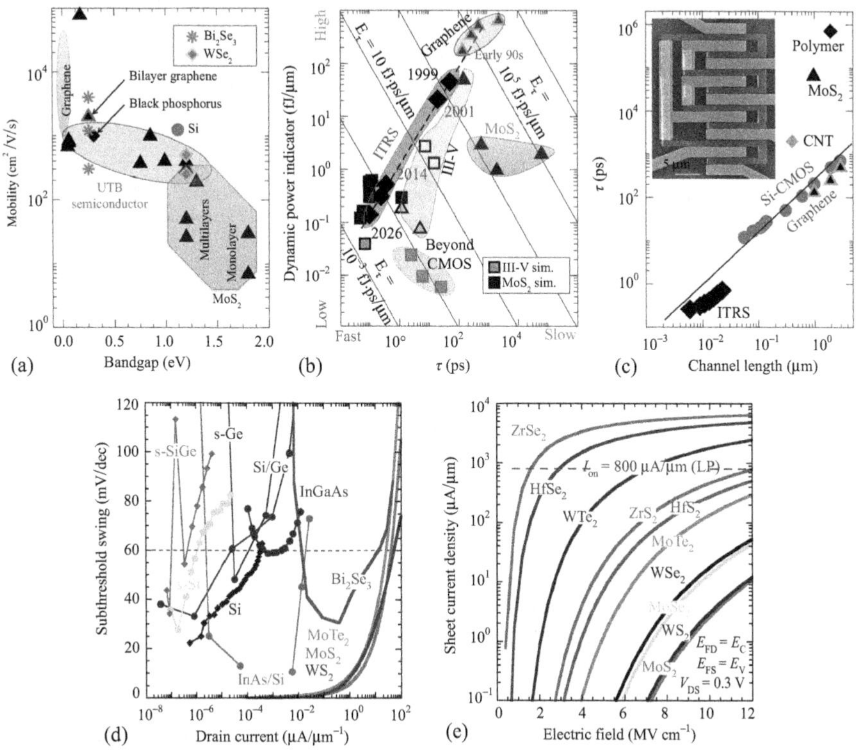

Figure 3.18 (a) Mobility vs bandgap graphene, Si, ultra-thin body semiconductors, BP and MoS_2, (b) dynamic power indicator and τ for different semiconductors along with ITRS stipulation. The oblique lines are DPI $\times$ τ that gives us a figure of merit that gives us a measure of speed and energy consumption of these devices. (c) Shows the switching delay of different material-based devices along with ITRS stipulation. Inset shows an optical image of five-stage ring oscillator. (d) Measured SS as a function of drain current for different TFET devices and (e) simulated tunnel current density as a function of applied electric field of different novel material based TFET devices where the dashed line indicates ON-state. © 2018 Springer Nature. Reprinted, with permission from, Ref [124]

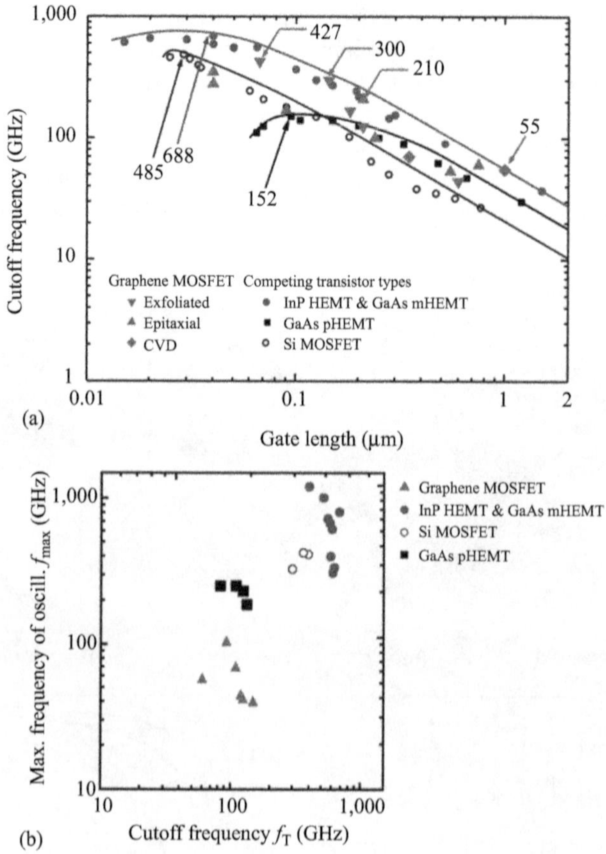

Figure 3.19 *(a) Cutoff frequency as a function of channel length in graphene-based MOSFET devices in comparison to InP and GaAs metamorphic and pseudomorphic HEMTs and Si MOSFET. With long channel length, cutoff frequency is dependent on L^{-1}, and the mobility plays an important role in deciding the maximum cutoff frequency (e.g., Si has the lowest mobility as compared to InP and GaAs and hence has lowest cutoff frequency). At shorter gate lengths, inverse behavior deviates from L^{-1} dependence. External parasitics and short channel effects come into picture and the influence of carrier mobility becomes less pronounced. (b) Relationship between f_{max} and f_T for GFETs and other competing materials for RF FETs. High f_T in graphene does not necessarily translate to higher f_{max} primarily due to poor drain current saturation. © 2018 Cambridge University Press. Reprinted, with permission from, Ref [133]*

3.5.1 RF applications

Devices for high-frequency applications are predominantly SiGe or III–V based with operation frequency being a few hundred GHz. Requirements for increased bandwidth has got researchers working fervently on exploring emerging semiconducting materials or coming up with improved device designs. High carrier mobility, saturation velocity and optimized parasitic capacitances are necessary device parameters for high frequency operation. Important parameters for RF applications are *(i) intrinsic time delay "*τ*" defined as* $\tau = CV_{DD}/I_{ON}$*; gives the time required to switch a logic-gate, (ii) cutoff (*f_T*) frequency where the current gain of the transistor drops to unity and (iii) maximum frequency (*f_{max}*) wherein the unilateral power gain drops to unity.* The maximum operation frequency can be expressed as [124]

$$f_{max} = \frac{f_T}{2\sqrt{((R_G + R_s)/r_d) + 2\pi f_T R_G C_{GD}}}$$

where R_G,R_s and r_d are gate, source and output differential resistance, respectively, and C_{GD} is the gate-drain capacitance. To improve f_{max} series resistance needs to be reduced and r_d has to be maximized (i.e., good current saturation as seen in SiGe or III–V would be necessary). Given these parameter requirements, combined with the fact RF electronics does not need to be switched off unlike regular operation or digital logic, graphene proves to be a promising candidate. Several graphene-based FETs have been reported with high cutoff frequency (>400 GHz) [125,126]. Unfortunately, their maximum oscillation frequency still requires much improvement (currently between 40 and 105 GHz [126–128]) primarily affecting its poor drain-current saturation characterization stemming from its lack of bandgap. Introducing, a bandgap in graphene has been shown to improve its current saturation characteristics and can be potentially employed for high-frequency devices [129,130] but comes at the cost of reduced carrier mobility (Figure 3.18). In the previous sections, we observed that TMDs shows great potential with very low I_{OFF} *and* could be employed to fabricate steep-slope FETs and perform well within ITRS stipulations for low-power devices. Unfortunately, even with their significant current saturation, TMDs have low mobility and saturation and as a result fall far behind graphene, Si and III–V electronics (Figure 3.19) [131]. Only a few experimental reports of high-frequency operation of MoS_2 FETs exist, and their performance is still far-off from their other semiconducting counterparts [132].

3.6 List of abbreviations

AFM—atomic force microscope
Ag—silver
ALD—atomic layer deposition
Ar—argon
Au—gold
Bi_2Se_3—bismuth selenide
Bi_2Te_3—bismuth telluride
BJT—bipolar junction transistor

BLG—bilayer graphene
BP—black phosphorus
BTBT—band to band tunneling
CaF_2—calcium fluoride
CB—conduction band
Cs_2Co_3—cesium carbonate
CVD—chemical vapor deposition
CVT—chemical vapor transport
$(C_2H_5)_2S$—ethylene disulfide
$(C_2H_5)_3B$—triethylboron
$(CH_3)_2Se$—dimethyl selenium
Co—cobalt
Cu—copper
DFT—density functional theory
DOS—density of states
EA—electron affinity
Fe—iron
FET—field effect transistors
Ge—germanium
GHz—gigahertz
GeO_x—germanium oxide
GNR—graphene nanoribbons
Hf—hafnium
HfS_2—hafnium disulfide
HEMT—high mobility electron transistor
HOPG—highly oriented pyrolytic graphite
H_2—hydrogen
H_2O—water
ICs—integrated circuits
I-MOS—impact metal oxide semiconductor
In—indium
IR—infrared
Ir—iridium
ITRS—International Technology Roadmap for Semiconductors
K—potassium
LPCVD—low pressure chemical vapor deposition
MOCVD—metal organic chemical vapor deposition
MOSFET—metal oxide semiconductor field effect transistor
MOVPE—metal organic vapor phase epitaxy
MBE—molecular beam epitaxy
Mo—molybdenum
$MoCl_5$—molybdenum (V) chloride
MoS_2—molybdenum disulfide
$Mo(CO)_6$—molybdenum hexacarbonyl
MoO_3—molybdenum (VI) oxide
$MoTe_2$—molybdenum telluride

NH_3—ammonia
NDC—negative differential conductivity
NDR—negative differential resistance
Nb—niobium
Ni—nickel
N_2—nitrogen
NO_2—nitrogen dioxide
O_2—oxygen
Pd—palladium
P—phosphorus
PH_3—phosphine
p–i–n—p-type-intrinsic-n type
PIII—plasma immersion ion implantation
PLD—pulsed laser deposition
PMMA—poly-methyl methacrylate
Pt—platinum
PTAS—perylene-3,4,9,10-tetracarboxylic acid tetrapotassium salt
RAM—random access memory
RHEED—reflection high energy electron diffraction
RTD—resonant tunneling diodes
Sc—scandium
STM—scanning tunneling microscopy
SB—Schottky barrier
Se—selenium
Si—silicon
S—sulfur
SLG—single layer graphene
S—sculpture
Te—tellurium
THz—terahertz
Ti—titanium
TFET—tunneling field effect transistors
TGA—thermogravimetric analysis
TMDs—transition metal dichalcogenides
UHV—ultra-high vacuum
VB—valence band
vdW—van der Waals
WS_2—tungsten disulfide
WSe_2—tungsten diselenide
$W(CO)_6$—tungsten hexacarbonyl
WO_3—Tungsten (VI) oxide
WF—work function
WKB—Wentzel–Kramers–Brillouin
XRR—X-ray reflectance
Zr—zirconium

References

[1] Geim, Andre K., and Konstantin S. Novoselov. "The rise of graphene." Nature Materials 6.3 (2007): 183.

[2] Bolotin, Kirill I., Sikes, Kenneth J., Jiang, Zhigang, *et al.* "Ultrahigh electron mobility in suspended graphene." Solid State Communications 146.9–10 (2008): 351–355.

[3] Balandin, Alexander A., Ghosh, Suchismita, Bao, Wenzhong, *et al.* "Superior thermal conductivity of single-layer graphene." Nano Letters 8.3 (2008): 902–907.

[4] Murali, Raghunath, Yinxiao Yang, Kevin Brenner, Thomas Beck, and James D. Meindl "Breakdown current density of graphene nanoribbons." Applied Physics Letters 94.24 (2009): 243114.

[5] Ko, Guenwoo, Kim, Hong-Yeol, Ahn, Jaehui, Park, Youngmoo, Lee, Kwan Young, and Kim, Jihyun. "Graphene-based nitrogen dioxide gas sensors." Current Applied Physics 10.4 (2010): 1002–1004.

[6] Schedin, Fredrik, Geim, Andre K., Morozov, Sergei V., *et al.* "Detection of individual gas molecules adsorbed on graphene." Nature Materials 6.9 (2007): 652.

[7] Lu, Chun-Hua, Huang-Hao Yang, Chun-Ling Zhu, Xi Chen, and Guo-Nan Chen, *et al.* "A graphene platform for sensing biomolecules." Angewandte Chemie 121.26 (2009): 4879–4881.

[8] He, Shijiang, Bo Song, Di Li, *et al.* "A graphene nanoprobe for rapid, sensitive, and multicolor fluorescent DNA analysis." Advanced Functional Materials 20.3 (2010): 453–459.

[9] Son, Young-Woo, Marvin L. Cohen, and Steven G. Louie. "Energy gaps in graphene nanoribbons." Physical Review Letters 97.21 (2006): 216803.

[10] Chen, Yen-Chia, Dimas G. De Oteyza, Zahra Pedramrazi, *et al.* "Tuning the band gap of graphene nanoribbons synthesized from molecular precursors." ACS Nano 7.7 (2013): 6123–6128.

[11] Rani, Pooja, and V. K. Jindal. "Designing band gap of graphene by B and N dopant atoms." RSC Advances 3.3 (2013): 802–812.

[12] Denis, Pablo A. "Band gap opening of monolayer and bilayer graphene doped with aluminium, silicon, phosphorus, and sulfur." Chemical Physics Letters 492.4–6 (2010): 251–257.

[13] Fan, Xiaofeng, Zexiang Shen, A. Q. Liu, and Jer-Lai Kuo. "Band gap opening of graphene by doping small boron nitride domains." Nanoscale 4.6 (2012): 2157–2165.

[14] Ni, Zhen Hua, Ting Yu, Yun Hao Lu, Ying Ying Wang, Yuan Ping Feng, and Ze Xiang Shen. "Uniaxial strain on graphene: Raman spectroscopy study and band-gap opening." ACS Nano 2.11 (2008): 2301–2305.

[15] Boukhvalov, D. W., and M. I. Katsnelson. "Tuning the gap in bilayer graphene using chemical functionalization: density functional calculations." Physical Review B 78.8 (2008): 085413.

[16] Zhang, Hang, Bekyarova, Elena, Huang, Jhao-Wun, *et al.* "Aryl functionalization as a route to band gap engineering in single layer graphene devices." Nano Letters 11.10 (2011): 4047–4051.

[17] Novoselov, Kostya S., Andre K. Geim, Sergei V. Morozov, *et al.* "Electric field effect in atomically thin carbon films." Science 306.5696 (2004): 666–669.

[18] Zhang, Yuanbo, Tsung-Ta Tang, Caglar Girit, *et al.* "Direct observation of a widely tunable bandgap in bilayer graphene." Nature 459.7248 (2009): 820.

[19] Xia, Fengnian, Damon B. Farmer, Yu-ming Lin, and Phaedon Avouris, *et al.* "Graphene field-effect transistors with high on/off current ratio and large transport band gap at room temperature." Nano Letters 10.2 (2010): 715–718.

[20] Neto, AH Castro, Francisco Guinea, Nuno MR Peres, Kostya S. Novoselov, and Andre K. Geim, *et al.* "The electronic properties of graphene." Reviews of Modern Physics 81.1 (2009): 109.

[21] Zheng, Binjie, Yuanfu Chen, Zegao Wang, *et al.* "Vertically oriented few-layered HfS2 nanosheets: growth mechanism and optical properties." 2D Materials 3.3 (2016): 035024.

[22] Cho, Suyeon, Sera Kim, Jung Ho Kim, *et al.* "Phase patterning for ohmic homojunction contact in MoTe2." Science 349.6248 (2015): 625–628.

[23] Late, Dattatray J., Bin Liu, HSS Ramakrishna Matte, Vinayak P. Dravid, and C. N. R. Rao, *et al.* "Hysteresis in single-layer MoS2 field effect transistors." ACS Nano 6.6 (2012): 5635–5641.

[24] Cui, Yang, Run Xin, Zhihao Yu, *et al.* "High-performance monolayer WS2 field-effect transistors on high-κ dielectrics." Advanced Materials 27.35 (2015): 5230–5234.

[25] Das, Saptarshi, and Joerg Appenzeller. "WSe2 field effect transistors with enhanced ambipolar characteristics." Applied Physics Letters 103.10 (2013): 103501.

[26] Song, Li, Lijie Ci, Hao Lu, *et al.* "Large scale growth and characterization of atomic hexagonal boron nitride layers." Nano Letters 10.8 (2010): 3209–3215.

[27] Zhang, Haijun, Chao-Xing Liu, Xiao-Liang Qi, Xi Dai, Zhong Fang, and Shou-Cheng Zhang. "Topological insulators in Bi_2Se_3, Bi_2Te_3 and Sb_2Te_3 with a single Dirac cone on the surface." Nature Physics 5.6 (2009): 438.

[28] Tang, Hao, Dong Liang, Richard L.J. Qiu, and Xuan P.A. Gao. "Two-dimensional transport-induced linear magneto-resistance in topological insulator Bi2Se3 nanoribbons." ACS Nano 5.9 (2011): 7510–7516.

[29] Li, Likai, Yijun Yu, Guo Jun Ye, *et al.* "Black phosphorus field-effect transistors." Nature Nanotechnology 9.5 (2014): 372.

[30] Kubota, Yoichi, Kenji Watanabe, Osamu Tsuda, and Takashi Taniguchi. "Deep ultraviolet light-emitting hexagonal boron nitride synthesized at atmospheric pressure." Science 317.5840 (2007): 932–934.

[31] Watanabe, Kenji, Takashi Taniguchi, and Hisao Kanda. "Direct-bandgap properties and evidence for ultraviolet lasing of hexagonal boron nitride single crystal." Nature Materials 3.6 (2004): 404.

[32] Song, Li, Lijie Ci, Hao Lu, *et al.* "Large scale growth and characterization of atomic hexagonal boron nitride layers." Nano Letters 10.8 (2010): 3209–3215.

[33] Britnell, Liam, Gorbachev, Roman, Jalil, Rashid, *et al.* "Field-effect tunneling transistor based on vertical graphene heterostructures." Science 335.6071 (2012): 947–950.

[34] Wang, Jingang, Fengcai Ma, and Mengtao Sun. "Graphene, hexagonal boron nitride, and their heterostructures: properties and applications." RSC Advances 7.27 (2017): 16801–16822.

[35] Topsakal, M., E. Aktürk, and S. Ciraci. "First-principles study of two-and one-dimensional honeycomb structures of boron nitride." Physical Review B 79.11 (2009): 115442.

[36] Castellanos-Gomez, Andres, Vicarelli, Leonardo, Prada, Elsa, *et al.* "Isolation and characterization of few-layer black phosphorus." 2D Materials 1.2 (2014): 025001.

[37] Guo, Gen-Cai, Xiao-Lin Wei, Da Wang, Yanping Luo, and Li-Min Liu, *et al.* "Pristine and defect-containing phosphorene as promising anode materials for rechargeable Li batteries." Journal of Materials Chemistry A 3.21 (2015): 11246–11252.

[38] Kang, Junmo, Deep Jariwala, Christopher R. Ryder, *et al.* "Probing out-of-plane charge transport in black phosphorus with graphene-contacted vertical field-effect transistors." Nano Letters 16 (2016): 2580–2585.

[39] Choi, Seon Jae, Bum-Kyu Kim, Tae-Ho Lee, *et al.* "Electrical and thermoelectric transport by variable range hopping in thin black phosphorus devices." Nano Letters 16.7 (2016): 3969–3975.

[40] Ding, Bangfu, Wei Chen, Zilong Tang, and Junying Zhang, *et al.* "Tuning phosphorene nanoribbon electronic structure through edge oxidization." The Journal of Physical Chemistry C 120.4 (2016): 2149–2158.

[41] Ryder, Christopher R., Joshua D. Wood, Spencer A. Wells, *et al.* "Covalent functionalization and passivation of exfoliated black phosphorus via aryl diazonium chemistry." Nature Chemistry 8.6 (2016): 597.

[42] Koenig, Steven P., Rostislav A. Doganov, Leandro Seixas, *et al.* "Electron doping of ultrathin black phosphorus with Cu Adatoms." Nano Letters 16 (2016): 2145–2151.

[43] Li, Likai, Fangyuan Yang, Guo Jun Ye, *et al.* "Quantum Hall effect in black phosphorus two-dimensional electron system." Nature Nanotechnology 11.7 (2016): 593.

[44] Favron, Alexandre, Etienne Gaufrès, Frédéric Fossard, *et al.* "Photooxidation and quantum confinement effects in exfoliated black phosphorus." Nature Materials 14.8 (2015): 826.

[45] Wang, Gaoxue, William J. Slough, Ravindra Pandey, and Shashi P. Karna, *et al.* "Degradation of phosphorene in air: understanding at atomic level." 2D Materials 3.2 (2016): 025011.

[46] Wood, Joshua D., Spencer A. Wells, Deep Jariwala, *et al.* "Effective passivation of exfoliated black phosphorus transistors against ambient degradation." Nano Letters 14.12 (2014): 6964–6970.

[47] Pei, Jiajie, Xin Gai, Jiong Yang, *et al.* "Producing air-stable monolayers of phosphorene and their defect engineering." Nature Communications 7 (2016): 10450.

[48] Kang, Joohoon, Joshua D. Wood, Spencer A. Wells, *et al.* "Solvent exfoliation of electronic-grade, two-dimensional black phosphorus." ACS Nano 9.4 (2015): 3596–3604.

[49] Hanlon, Damien, Claudia Backes, Evie Doherty, *et al.* "Liquid exfoliation of solvent-stabilized few-layer black phosphorus for applications beyond electronics." Nature Communications 6 (2015): 8563.

[50] Doganov, Rostislav A., Eoin C. T. O'Farrell, Steven P. Koenig, *et al.* "Transport properties of pristine few-layer black phosphorus by van der Waals passivation in an inert atmosphere." Nature Communications 6 (2015): 6647.

[51] Choi, Wonbong, Nitin Choudhary, Gang Hee Han, Juhong Park, Deji Akinwande, and Young Hee Lee, *et al.* "Recent development of two-dimensional transition metal dichalcogenides and their applications." Materials Today 20.3 (2017): 116–130.

[52] Cho, Suyeon, Sera Kim, Jung Ho Kim, *et al.* "Phase patterning for ohmic homojunction contact in MoTe2." Science 349.6248 (2015): 625–628.

[53] Kang, Yimin, Sina Najmaei, Zheng Liu, *et al.* "Plasmonic hot electron induced structural phase transition in a MoS2 monolayer." Advanced Materials 26.37 (2014): 6467–6471.

[54] Najmaei, Sina, Zheng Liu, Wu Zhou, *et al.* "Vapour phase growth and grain boundary structure of molybdenum disulphide atomic layers." Nature Materials 12.8 (2013): 754–759.

[55] Yu, Yifei, Chun Li, Yi Liu, Liqin Su, Yong Zhang, and Linyou Cao, *et al.* "Controlled scalable synthesis of uniform, high-quality monolayer and few-layer MoS2 films." Scientific Reports 3 (2013).

[56] Chang, Yung-Huang, Wenjing Zhang, Yihan Zhu, *et al.* "Monolayer MoSe2 grown by chemical vapor deposition for fast photodetection." ACS Nano 8.8 (2014): 8582–8590.

[57] Wang, Shanshan, Youmin Rong, Ye Fan, *et al.* "Shape evolution of monolayer MoS2 crystals grown by chemical vapor deposition." Chemistry of Materials 26.22 (2014): 6371–6379.

[58] Laskar, Masihhur R., Lu Ma, Santhakumar Kannappan, *et al.* "Large area single crystal (0 0 0 1) oriented MoS2." Applied Physics Letters 102.25 (2013): 252108.

[59] Rhyee, Jong-Soo, Junyeon Kwon, Piyush Dak, *et al.* "High-mobility transistors based on large-area and highly crystalline CVD-grown MoSe2 films on insulating substrates." Advanced Materials 28.12 (2016): 2316–2321.

[60] Kim, Ki Kang, Allen Hsu, Xiaoting Jia, *et al.* "Synthesis of monolayer hexagonal boron nitride on Cu foil using chemical vapor deposition." Nano Letters 12.1 (2011): 161–166.

[61] Shi, Yumeng, Christoph Hamsen, Xiaoting Jia, *et al.* "Synthesis of few-layer hexagonal boron nitride thin film by chemical vapor deposition." Nano Letters 10.10 (2010): 4134–4139.

[62] Kim, Gwangwoo, A-Rang Jang, Hu Young Jeong, Zonghoon Lee, Dae Joon Kang, and Hyeon Suk Shin, "Growth of high-crystalline, single-layer hexagonal boron nitride on recyclable platinum foil." Nano Letters 13.4 (2013): 1834–1839.

[63] Kim, Soo Min, Allen Hsu, Min Ho Park, *et al.* "Synthesis of large-area multilayer hexagonal boron nitride for high material performance." Nature Communications 6 (2015): 8662.

[64] Orofeo, Carlo M., Orofeo, Carlo M., Satoru Suzuki, Hiroyuki Kageshima, and Hiroki Hibino, "Growth and low-energy electron microscopy characterization of monolayer hexagonal boron nitride on epitaxial cobalt." Nano Research 6.5 (2013): 335–347.

[65] Park, Ji-Hoon, Jin Cheol Park, Seok Joon Yun, *et al.* "Large-area monolayer hexagonal boron nitride on Pt foil." ACS Nano 8.8 (2014): 8520–8528.

[66] Ismach, Ariel, Harry Chou, Domingo A. Ferrer, *et al.* "Toward the controlled synthesis of hexagonal boron nitride films." ACS Nano 6.7 (2012): 6378–6385.

[67] Han, Jaehyun, Jun-Young Lee, Heemin Kwon, and Jong-Souk Yeo, "Synthesis of wafer-scale hexagonal boron nitride monolayers free of aminoborane nanoparticles by chemical vapor deposition." Nanotechnology 25.14 (2014): 145604.

[68] Tay, Roland, Yingjie, Hongling Li, Siu Hon Tsang, *et al.* "Trimethylamine borane: a new single-source precursor for monolayer h-BN single crystals and h-BCN thin films." Chemistry of Materials 28.7 (2016): 2180–2190.

[69] Chang, Ren-Jie, Xiaochen Wang, Shanshan Wang, *et al.* "Growth of large single-crystalline monolayer hexagonal boron nitride by oxide-assisted chemical vapor deposition." Chemistry of Materials 29.15 (2017): 6252–6260.

[70] Lu, Guangyuan, Tianru Wu, Qinghong Yuan, *et al.* "Synthesis of large single-crystal hexagonal boron nitride grains on Cu–Ni alloy." arXiv preprint arXiv: 1503.02804 (2015).

[71] Smith, Joshua B., Daniel Hagaman, and Hai-Feng Ji. "Growth of 2D black phosphorus film from chemical vapor deposition." Nanotechnology 27.21 (2016): 215602.

[72] Kang, Kibum, Saien Xie, Lujie Huang, *et al.* "High-mobility three-atom-thick semiconducting films with wafer-scale homogeneity." Nature 520.7549 (2015): 656.

[73] Eichfeld, Sarah M., Lorraine Hossain, Yu-Chuan Lin, *et al.* "Highly scalable, atomically thin WSe2 grown via metal–organic chemical vapor deposition." ACS Nano 9.2 (2015): 2080–2087.

[74] Lin, Yu-Chuan, Ram Krishna Ghosh, Rafik Addou, *et al.* "Atomically thin resonant tunnel diodes built from synthetic van der Waals heterostructures." Nature Communications 6 (2015): 7311.

[75] Kobayashi, Y., T. Akasaka, and T. Makimoto. "Hexagonal boron nitride grown by MOVPE." Journal of Crystal Growth 310.23 (2008): 5048–5052.

[76] Li, Jing, Dahal, Rajendra, Majety, Sashikanth, Lin, Jingyu, and Jiang, Hongxing. "Hexagonal boron nitride epitaxial layers as neutron detector materials." Nuclear Instruments and Methods in Physics Research Section A: Accelerators, Spectrometers, Detectors and Associated Equipment 654.1 (2011): 417–420.

[77] Radisavljevic, Branimir and Andras Kis, "Mobility engineering and a metal–insulator transition in monolayer MoS 2." Nature Materials 12.9 (2013): 815.

[78] Liu, Han, Mengwei Si, Sina Najmaei, *et al.* "Statistical study of deep submicron dual-gated field-effect transistors on monolayer chemical vapor deposition molybdenum disulfide films." Nano Letters 13.6 (2013): 2640–2646.

[79] Paduano, Qing S., Michael Snure, James Bondy, and Timothy W. C. Zens, "Self-terminating growth in hexagonal boron nitride by metal organic chemical vapor deposition." Applied Physics Express 7.7 (2014): 071004.

[80] Kim, Dong Y., Yeong, Nam Han, Hokyeong Jeong, "Pressure-dependent growth of wafer-scale few-layer h-BN by metal–organic chemical vapor deposition." Crystal Growth & Design 17.5 (2017): 2569–2575.

[81] Arthur, John R. "Molecular beam epitaxy." Surface Science 500.1–3 (2002): 189–217.

[82] Zhang, Yi, Tay-Rong Chang, Bo Zhou, *et al.* "Direct observation of the transition from indirect to direct bandgap in atomically thin epitaxial MoSe2." Nature Nanotechnology 9.2 (2014): 111–115.

[83] Vishwanath, Suresh, Xinyu Liu, Sergei Rouvimov, *et al.* "Comprehensive structural and optical characterization of MBE grown MoSe2 on graphite, CaF2 and graphene." 2D Materials 2.2 (2015): 024007.

[84] Barton, Adam T., Yue, Ruoyu, Anwar, Sarkar R.M., *et al.* "Transition metal dichalcogenide and hexagonal boron nitride heterostructures grown by molecular beam epitaxy." Microelectronic Engineering 147 (2015): 306–309.

[85] Laskar, Masihhur R., Digbijoy N. Nath, Lu Ma, *et al.* "p-Type doping of MoS2 thin films using Nb." Applied Physics Letters 104.9 (2014): 092104.

[86] Fang, Hui, Mahmut Tosun, Gyungseon Seol, *et al.* "Degenerate n-doping of few-layer transition metal dichalcogenides by potassium." Nano Letters 13.5 (2013): 1991–1995.

[87] Lin, Jia Dan, Cheng Han, Fei Wang, *et al.* "Electron-doping-enhanced trion formation in monolayer molybdenum disulfide functionalized with cesium carbonate." ACS Nano 8.5 (2014): 5323–5329.

[88] Nipane, Ankur, Debjani Karmakar, Naveen Kaushik, Shruti Karande, and Saurabh Lodha, "Few-layer MoS2 p-type devices enabled by selective doping using low energy phosphorus implantation." ACS Nano 10.2 (2016): 2128–2137.

[89] Fang, Hui, Steven Chuang, Ting Chia Chang, Kuniharu Takei, Toshitake Takahashi, and Ali Javey, "High-performance single layered WSe2 p-FETs with chemically doped contacts." Nano Letters 12.7 (2012): 3788–3792.

[90] Sun, Minglei, Wencheng Tang, Qingqiang Ren, Sa-ke Wang, Jin Yu, and Yanhui Du, "A first-principles study of light non-metallic atom

substituted blue phosphorene." Applied Surface Science 356 (2015): 110–114.

[91] Yang, Lifang, Wenbo Mi, and Xiaocha Wang. "Tailoring magnetism of black phosphorene doped with B, C, N, O, F, S and Se atom: a DFT calculation." Journal of Alloys and Compounds 662 (2016): 528–533.

[92] Kong, Long-Juan, Guang-Hua Liu, and Yu-Jun Zhang. "Tuning the electronic and optical properties of phosphorene by transition-metal and nonmetallic atom co-doping." RSC Advances 6.13 (2016): 10919–10929.

[93] He, Yuanyuan, Feifei Xia, Zhibin Shao, Jianwei Zhao, and Jiansheng Jie, "Surface charge transfer doping of monolayer phosphorene via molecular adsorption." The Journal of Physical Chemistry Letters 6.23 (2015): 4701–4710.

[94] Xiang, Du, Cheng Han, Jing Wu, *et al.* "Surface transfer doping induced effective modulation on ambipolar characteristics of few-layer black phosphorus." Nature Communications 6 (2015): 6485.

[95] Koenig, Steven P., Rostislav A. Doganov, Leandro Seixas, *et al.* "Electron doping of ultrathin black phosphorus with Cu adatoms." Nano Letters 16.4 (2016): 2145–2151.

[96] Yu, Xuechao, Shengli Zhang, Haibo Zeng, and Qi Jie Wang, "Lateral black phosphorene P–N junctions formed via chemical doping for high performance near-infrared photodetector." Nano Energy 25 (2016): 34–41.

[97] Ryder, Christopher R., Joshua D. Wood, Spencer A. Wells, *et al.* "Covalent functionalization and passivation of exfoliated black phosphorus via aryl diazonium chemistry." Nature Chemistry 8.6 (2016): 597.

[98] Das, Saptarshi, Hong-Yan Chen, Ashish Verma Penumatcha, and Joerg Appenzeller, "High performance multilayer MoS2 transistors with scandium contacts." Nano Letters 13.1 (2012): 100–105.

[99] Liu, Wei, Jiahao Kang, Deblina Sarkar, Yasin Khatami, Debdeep Jena, and Kaustav Banerjee, *et al.* "Role of metal contacts in designing high-performance monolayer n-type WSe2 field effect transistors." Nano Letters 13.5 (2013): 1983–1990.

[100] Kang, Jiahao, Deblina Sarkar, Wei Liu, Debdeep Jena, and Kaustav Banerjee, "A computational study of metal-contacts to beyond-graphene 2D semiconductor materials." Electron Devices Meeting (IEDM), 2012 IEEE International. IEEE, 2012.

[101] Popov, Igor, Gotthard Seifert, and David Tománek. "Designing electrical contacts to MoS_2 monolayers: a computational study." Physical Review Letters 108.15 (2012): 156802.

[102] Leong, Wei Sun, Xin Luo, Yida Li, Khoong Hong Khoo, Su Ying Quek, and John T. L. Thong, *et al.* "Low resistance metal contacts to MoS2 devices with nickel-etched-graphene electrodes." ACS Nano 9.1 (2014): 869–877.

[103] Kappera, Rajesh, Damien Voiry, Sibel Ebru Yalcin, *et al.* "Phase-engineered low-resistance contacts for ultrathin MoS_2 transistors." Nature Materials 13.12 (2014): 1128.

[104] Ling, Xi, Yuxuan Lin, Qiong Ma, *et al.* "Parallel stitching of 2D materials." Advanced Materials 28.12 (2016): 2322–2329.

[105] Jena, Debdeep. "Tunneling transistors based on graphene and 2-D crystals." Proceedings of the IEEE 101.7 (2013): 1585–1602.

[106] Sze, S. M. and K. K. Ng, Physics of Semiconductor Devices, 3rd ed. New York, NY, USA: Wiley-Interscience, 2006.

[107] Bardeen, John. "Tunnelling from a many-particle point of view." Physical Review Letters 6.2 (1961): 57.

[108] Ionescu, Adrian M., and Heike Riel. "Tunnel field-effect transistors as energy-efficient electronic switches." Nature 479.7373 (2011): 329.

[109] Chang, Leland, David J. Frank, Robert K. Montoye, *et al.* "Practical strategies for power-efficient computing technologies." Proceedings of the IEEE 98.2 (2010): 215–236.

[110] Sarkar, Deblina, Xuejun Xie, Wei Liu, *et al.* "A subthermionic tunnel field-effect transistor with an atomically thin channel." Nature 526.7571 (2015): 91.

[111] Roy, Tania, Mahmut Tosun, Xi Cao, *et al.* "Dual-gated MoS2/WSe2 van der Waals tunnel diodes and transistors." ACS Nano 9.2 (2015): 2071–2079.

[112] Roy, Tania, Mahmut Tosun, Mark Hettick, Geun Ho Ahn, Chenming Hu, and Ali Javey, "2D–2D tunneling field-effect transistors using WSe2/SnSe2 heterostructures." Applied Physics Letters 108.8 (2016): 083111.

[113] Ameen, Tarek A., Hesameddin Ilatikhameneh, Gerhard Klimeck, and Rajib Rahman, "Few-layer phosphorene: an ideal 2D material for tunnel transistors." Scientific reports 6 (2016): 28515.

[114] Mishra, Umesh, and Jasprit Singh. Semiconductor Device Physics and Design. Netherlands: Springer Science & Business Media, 2007.

[115] Radisavljevic, Branimir, Michael Brian Whitwick, and Andras Kis. "Integrated circuits and logic operations based on single-layer MoS2." ACS Nano 5.12 (2011): 9934–9938.

[116] Britnell, Liam, Gorbachev, Roman V., Geim, Andre K., *et al.* "Resonant tunneling and negative differential conductance in graphene transistors." Nature Communications 4 (2013): 1794.

[117] Mishchenko, Artem, Tu, J.S., Cao, Yang, *et al.* "Twist-controlled resonant tunneling in graphene/boron nitride/graphene heterostructures." Nature Nanotechnology 9.10 (2014): 808.

[118] Gopalakrishnan, Kailash, Peter B. Griffin, and James D. Plummer. "I-MOS: A novel semiconductor device with a subthreshold slope lower than kT/q." Electron Devices Meeting, 2002. IEDM'02. International. IEEE, 2002.

[119] Björk, Mikael T., Hayden, Oliver, Schmid, Heinz, Riel, Heike, and Riess, Walter. "Vertical surround-gated silicon nanowire impact ionization field-effect transistors." Applied Physics Letters 90.14 (2007): 142110.

[120] Wang, Han, Lili Yu, Yi-Hsien Lee, *et al.* "Integrated circuits based on bilayer MoS2 transistors." Nano Letters 12.9 (2012): 4674–4680.

[121] Ling, Xi, Yuxuan Lin, Qiong Ma, *et al.* "Parallel stitching of 2D materials." Advanced Materials 28.12 (2016): 2322–2329.

[122] Song, Haisheng S., Li, Songlin, Gao, Liang, *et al.* "High-performance top-gated monolayer SnS_2 field-effect transistors and their integrated logic circuits." Nanoscale 5.20 (2013): 9666–9670.

[123] Resta, Giovanni V., Surajit Sutar, Yashwanth Balaji, *et al.* "Polarity control in WSe_2 double-gate transistors." Scientific Reports 6 (2016): 29448.

[124] Fiori, Gianluca, Francesco Bonaccorso, Giuseppe Iannaccone, *et al.* "Electronics based on two-dimensional materials." Nature Nanotechnology 9.10 (2014): 768.

[125] Cheng, Rui, Jingwei Bai, Lei Liao, *et al.* "High-frequency self-aligned graphene transistors with transferred gate stacks." Proceedings of the National Academy of Sciences 109.29 (2012): 11588–11592.

[126] Yu, Chenxue, He, Zhezhao, Li, Jia, *et al.* "Quasi-free-standing bilayer epitaxial graphene field-effect transistors on 4H-SiC (0001) substrates." Applied Physics Letters 108.1 (2016): 013102.

[127] Wu, Yanqing, Keith A. Jenkins, Alberto Valdes-Garcia, *et al.* "State-of-the-art graphene high-frequency electronics." Nano Letters 12.6 (2012): 3062–3067.

[128] Guo, Zelei, Rui Dong, Partha Sarathi Chakraborty, *et al.* "Record maximum oscillation frequency in C-face epitaxial graphene transistors." Nano Letters 13.3 (2013): 942–947.

[129] Szafranek, Bart N., Gianluca Fiori, Daniel Schall, Daniel Neumaier, and Heinrich Kurz, *et al.* "Current saturation and voltage gain in bilayer graphene field effect transistors." Nano Letters 12.3 (2012): 1324–1328.

[130] Fiori, G. and G. Iannaccone. "Insights on radio frequency bilayer graphene FETs." IEEE Int. Electron Dev. Meeting, 2012, 17.3.1–17.3.4.

[131] Fiori, Gianluca, Bartholomäus N. Szafranek, Giuseppe Iannaccone, and Daniel Neumaier, "Velocity saturation in few-layer MoS_2 transistor." Applied Physics Letters 103.23 (2013): 233509.

[132] Cheng, Rui, Shan Jiang, Yu Chen, *et al.* "Few-layer molybdenum disulfide transistors and circuits for high-speed flexible electronics." Nature Communications 5 (2014): 5143.

[133] Lemme, Max C., Lain-Jong Li, Tomás Palacios, and Frank Schwierz, "Two-dimensional materials for electronic applications." MRS Bulletin 39.8 (2014): 711–718.

Chapter 4
WSe_2 polarity-controllable devices

Giovanni V. Resta[1], *Iuliana P. Radu*[2], *Giovanni De Micheli*[1], *and Pierre-Emmanuel Gaillardon*[3]

This chapter is dedicated to polarity-controllable devices fabricated with two dimensional (2D) semiconducting tungsten diselenide (WSe_2). The chapter is organized as follows: first, a general introduction on 2D materials is presented, followed by a section dedicated to summarize the state of the art in the growth of 2D materials. The concept of ambipolarity is then introduced, and the main experimental results on WSe_2 ambipolar devices are presented. We transition then to the core part of the chapter describing recent advances in polarity-controllable transistors fabricated with ambipolar WSe2. We then focus on quantum transport simulations carried out to assess the performances of the devices at ultra-scaled gate lengths. We conclude with a summary, highlighting the main concept presented in the chapter and with a brief outlook on the future challenges in the field.

4.1 Two-dimensional transition metal dichalcogenides

2D materials of the *transition metal dichalcogenide* (TMDC) family [1] are layered compounds of general formula MX_2, where M is a transition metal of group IV, V or VI, such as molybdenum (Mo), tungsten (W) , hafnium (Hf), etc., and X is a chalcogen such as sulfur (S), selenium (Se) or tellurium (Te). Bulk TMDC crystals are formed by vertical stacking of 2D layers, separated by around 6.5 Å and held together by weak Van der Waals interactions, in a similar fashion as graphite [2]. Among TMDC materials, the ones formed by group VI metals (Mo and W) show a semiconducting behavior and have exhibited excellent electrical properties [3–5].

[1]Department of Electrical Engineering, Integrated System Laboratory (LSI), École Polytechnique Fédérale de Lausanne (EPFL), Switzerland
[2]IMEC, Leuven, Belgium
[3]Laboratory of NanoIntegrated Systems (LNIS), Department of Electrical and Computer Engineering, University of Utah, USA

The peculiar vertical-layered structure of these materials allows for the growth or exfoliation of few and even monolayer films that have been shown to provide almost perfect electrostatic control of the transistor channel, making them robust to short-channel effects and well suited for beyond-silicon low-power logic applications [3–5].

4.1.1 *Synthesis of two-dimensional materials*

Until now, demonstration of novel devices based on 2D films has been carried out mainly thanks to the exfoliation technique originally developed for graphene [2]. Although this top-down technique allows for the production of high-quality thin flakes, with dimensions as big as tens of microns, this method is not scalable, does not allow any systematic control of the flake size and thickness and makes the fabrication much more challenging. In order to enable the creation of a reliable technology based on 2D-TMDCs, that would go beyond the simple demonstration of few working devices, it is essential to develop bottom-up methods for the productions of large-area, defect-free, atomically thin films with uniform electrical, optical and structural properties. Current methods for the growth of TMDCs materials are based on *chemical vapor deposition* (CVD) [4,6–14], Van der Waals epitaxy [15,16] and *molecular-organic CVD* (MOCVD) [17]. These techniques have shown promising results for mono(few)-layer MoS_2 [4,6,7,9,10,16,17], WSe_2 [12–14,18], WS_2 [8,17] and for the creation of 2D heterostructures [11,12] that researchers are exploring for the realization of efficient tunnel *field-effect transistors* (FETs). Basic operational circuits [4] have been demonstrated on monolayer CVD grown MoS_2, and single devices have shown gigahertz radio frequency performances [9] combined with high-current densities, exceeding 200 $\mu A/\mu m$ and intrinsic low-field mobility up to 55 $cm^2/V/s$. Multilayer WSe_2 grown by CVD using graphene as seed material has shown *p*-type mobility higher than 80 $cm^2/V/s$, when using oxygen as doping material [14]. Recently, uniform growth of monolayer MoS_2 and WS_2 has been demonstrated [17] on a 4-in. wafer, using MOCVD, with a resulting device yield of 99%. This process has the downside of being very time-consuming (26 h needed to grow a uniform 4-in. monolayer). Moreover, several of these growth techniques often use template substrates (such as sapphire, gold, graphite, etc.) that are not suitable for the realization of electronic devices. Thus, an efficient transfer technique for the semiconducting thin-film is necessary. The ability of transferring the grown 2D material would completely decouple the material growth (which usually requires high temperature) from the device realization and will play a key role in enabling 3D integration on the back-of-the-line *complementary metal-oxide-semiconductor* (CMOS) process [19], where low-temperature processes are needed ($<300\,°C$).

4.1.2 *Value of ambipolarity*

Ambipolarity refers to the capability of a semiconducting material to conduct both type of charge carriers, i.e., electrons and holes. Ambipolarity is usually considered a drawback in conventional CMOS electronics, where the building blocks are unipolar, *n*- or *p*-type, doped transistors with ohmic contacts. Physical doping is introduced in silicon with ion implantation during the fabrication process, and irreversibly sets the polarity of the transistors according to the dopant atoms used (Arsenic (As) for *n*-type doping and phosphorus (P) for *p*-type doping). Physical doping of 2D materials is currently an open challenge, and the realization of complementary transistors has only been demonstrated with the use of chemical doping and different metal contacts [20,21] or using two different semiconducting materials for *n*- and *p*-type devices [19]. When no physical or chemical doping is introduced, contact to an intrinsic 2D-semiconductor usually results in the creation of Schottky barriers at source and drain. If the Fermi level of the metal aligns with the Fermi level of the semiconductor in such a way that the Schottky barriers created are similar for electrons and holes (mid-gap contact), both can be injected in the channel, thus achieving ambipolar conduction. Electrostatic doping can then be used to modulate the Schottky barriers (with a polarity gate (PG)) and achieve polarity-controllable behavior of the device [22]. Thus, a symmetric ambipolar behavior must be regarded as the starting point for the realization of polarity-controllable devices. However, in novel 2D materials, ambipolarity does not always come easy. For example, in the most studied 2D-TMDCs, MoS_2, ambipolar conduction is blocked by a partial Fermi-level pinning [23] present at the contacts interface. One way that ambipolar conduction has been achieved on few-layers MoS_2 is by decoupling the semiconductor channel from the metal contacts using h-BN tunneling contacts, thus preventing the Fermi-level pinning [24] . However, this technique greatly complicates the fabrication process and causes an undesired increase in the contact resistance. Moreover, the higher bandgap reported for MoS_2, with respect to other 2D materials such as tungsten diselenide (WSe_2), zirconium selenide ($ZrSe_2$) and hafnium selenide ($HfSe_2$) [25], also makes it less appealing for polarity-controllable applications, as it would produce higher Schottky barriers, thus limiting carrier injection at the contacts [26]. In contrast, hole conduction has been demonstrated on WSe_2, and high mobilities have been reported [27,28]. Reports of *n*-type behavior on WSe_2 [29] have also shown high mobilities and current densities for electrons. Ambipolar conduction has been reported [30–32], and a stable complementary technology has been demonstrated [21]. The remainder experimental section of this chapter will focus on WSe_2, which to date appears to be the most suitable 2D material for demonstration of polarity-controllable devices.

4.2 Polarity-controllable devices on WSe_2

WSe_2 flakes, prepared by mechanical exfoliation, were used in the experimental demonstrations. The flakes were exfoliated on 20 nm SiO_2 substrate, analyzed with *atomic force microscopy* (AFM) to extract the flake thickness and then transferred to a target substrate, where buried metal gates had previously been patterned. A schematic illustration of the dry transfer process is shown in Figure 4.1. Thick ($\sim$10 μm) spin-coated *poly-methyl methacrylate* (PMMA) was used as a transferring agent. After selecting the flake for transfer (1), PMMA was spin-coated on the sample and annealed at 165 °C on a hot plate (2). PMMA was then diced around the flake using a micro-engraver and, upon release of the WSe_2/PMMA stack, picked up using a microneedle (3). The WSe_2/PMMA stack was then transferred to the target substrate and aligned

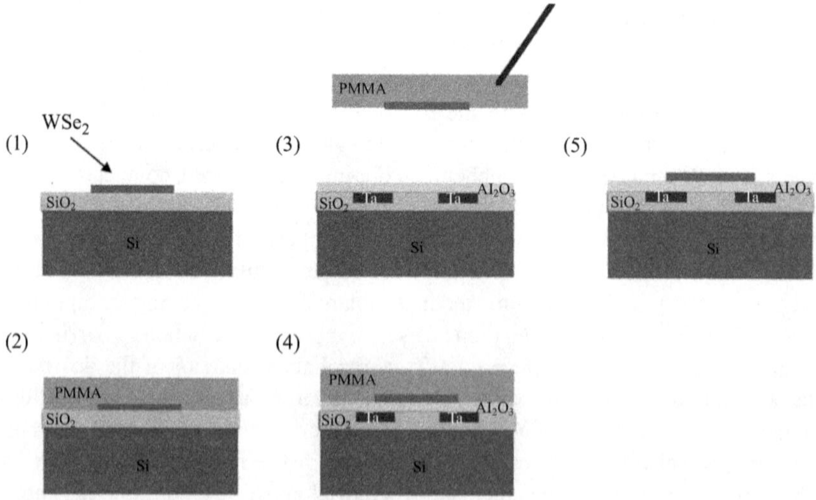

Figure 4.1 Flake-transfer procedure. The process enables the transfer and accurate positioning of the 2D flakes. © 2016 Nature publishing group. Adapted, with permission, from Reference [22]

with respect to the buried program gate by a manual pick-and-drop process. Adhesion of the WSe_2/PMMA stack was assured by 2 min hot-plate annealing at 190 °C (4). Finally, PMMA was dissolved using dichloromethane, and the sample was cleaned with a hot acetone bath (~12 h) to ensure the absence of PMMA residues (5).

Figure 4.2 shows the AFM topography of the flake (a), with its thickness extracted across the cutline (b). Thanks to the transfer procedure described above, the flake was placed in between two buried metal lines, and metal contacts (2 nm Ti/50 nm Pd) were evaporated (the final device is shown in Figure 4.2(c)). The buried metal line acted at the contact interface as PG and thus controlled the selection of the charge carriers preferably injected in the channel. The bulk silicon wafer was used as the *control gate* (CG) to modulate a potential barrier in the central region of the channel, controlling the ON/OFF state of the device (see Figure 4.2(d)).

As mentioned in Section 4.1.2, it is essential to achieve a symmetric ambipolar behavior to fully exploit polarity control. In the experimental results presented in [22], Ti/Pd-contacted WSe_2 FETs on SiO_2 dielectric substrate showed a considerably higher electron current with respect to the hole current, when measured after contact liftoff and without any additional treatment (Figure 4.3(a)). To improve the ambipolar behavior and increase the symmetry between the *n*- and *p*-conduction branches, a contact-annealing step was performed (200 °C for 10 h in an open tube furnace). The effect of contact annealing was found to be reproducible, and for the particular device presented in Figure 4.3(a), a 10× increase in hole ON-current (I_{ON}) and a 4× decrease in OFF-current (I_{OFF}) was achieved, while the electron ON current only showed a minor gain. The increase in both *n*- and *p*-type currents points to an improved physical contacts between the Ti/Pd contacts and the semiconducting channel, possibly

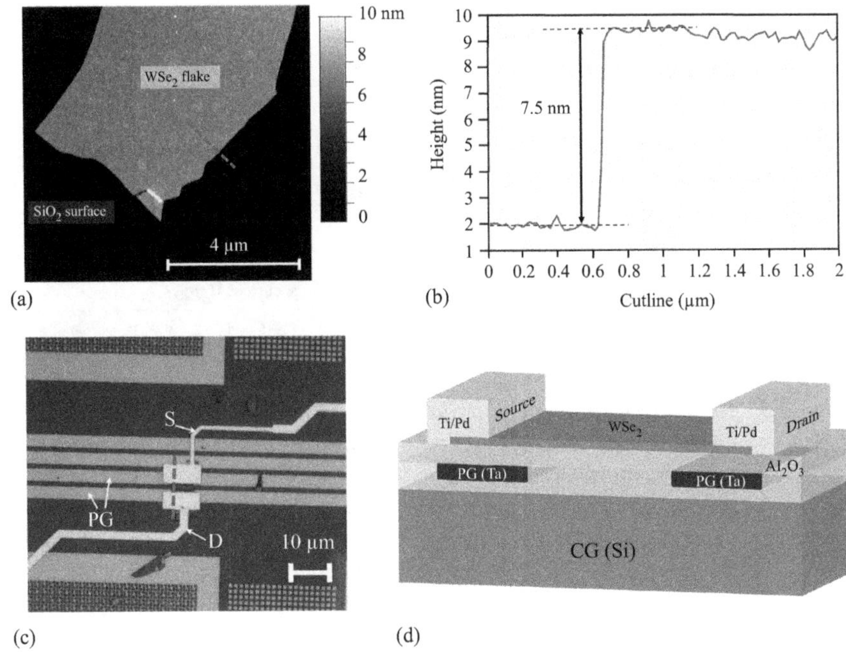

Figure 4.2 *WSe_2 flake properties and device fabrication. (a) AFM topography of the exfoliated flake before the transfer process. (b) Height profile across the cutline showed in (a). The extracted flake thickness is 7.5 nm. (c) Optical micro image of the fabricated device. (d) 3D schematic cross sections across the cutline in (c). © 2016 Nature publishing group. Adapted, with permission, from Reference [22]*

coupled with an increased mobility of charge carriers due to desorption of impurities from the WSe_2 flake. Annealing was also performed on the devices realized after transferring the flakes to the substrate with the buried PG. The full back-gate transfer characteristics were measured leaving the PG floating; thus, no control of the Schottky barriers was used. Thanks to the annealing procedure a pronounced symmetry, see Figure 4.3(b), was achieved which allowed for the realization of both *n*- and *p*-type devices without the addition of any physical doping. The measured I_{OFF} was below the pA range (100 fA), with I_{ON} of 4 μA for electrons and 0.25 μA for holes.

The modulation of the Schottky barriers at drain and source was tested by applying an independent bias to the PG, enabling the selection of the type of carriers (electrons or holes) that are favorably injected in the channel. For negative voltage values of the PG (Figure 4.4(a)), electron injection in the channel was completely suppressed and local *p*-type electrostatic doping in the contact regions was introduced. The bias applied to the PG induced positive charges at the interface and caused the effective Schottky barrier height to decrease (thinning of tunneling barrier and thus increased tunneling probability for holes). For the lowest applied voltage

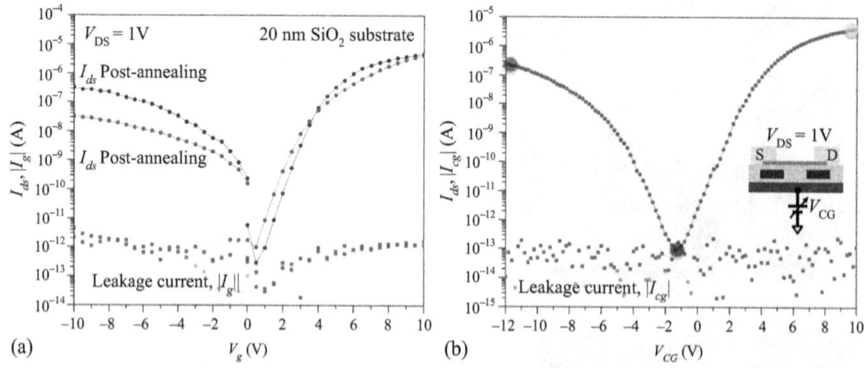

Figure 4.3 Characterization of ambipolar behavior. (a) Effect of contact annealing on the transfer characteristics of the device. In this case, the full transfer characteristics were measured with two separate V_g sweeps, thus creating the discontinuity at $V_g = 0$ V. (b) Ambipolar transfer characteristics of the device presented in Figure 4.2, measured with the PG left floating. © 2016 Nature publishing group. Adapted, with permission, from Reference [22]

(−12 V in Figure 4.4(a)), the ON/OFF-current levels were restored to the previous values extracted from the back-gate measurement with the program gate floating, and *p*-type behavior of the device was achieved. In a similar fashion, when applying a positive voltage to the PG, electrons were preferably injected in the channel, and the hole current was suppressed for all negative voltages applied on the CG (Figure 4.4(b)). Again for the highest PG applied voltage (10 V in Figure 4.4(b)), the I_{ON} and I_{OFF} levels matched with the ones extracted in Figure 4.3(a). I_{ON}/I_{OFF} ratios of 10^7 for *n*-type operation and of 10^6 for *p*-type operation were achieved, indicating optimal electrostatic control of the channel for both carriers and the potential for low-power applications, thanks to the low OFF-current measured in both configurations.

The switching properties of the device were further characterized by sweeping the voltage bias on the PG while fixing the bias of the CG (Figure 4.5(a)). In this configuration, the polarity of the transistor was changed during each sweep showing the ability of the device to transit from *p*- to *n*-type behavior, or vice-versa, in the same measurement, thus demonstrating on-the-fly polarity transition. Notice that for $V_{CG} = 0$ V, the device showed an ambipolar OFF-state behavior, conducting either electrons or holes according to the bias of the PG. The device presented operated as a comparison driven switch, comparing the voltage values applied to the PG and CG. When loaded with a resistor on one terminal and grounded on the other, the device implements a pseudo (because not full swing) exclusive OR (XOR) function (Figure 4.5(b)). The transistor is not conducting when the PG and CG are biased with opposite polarities, which in Figure 4.5(b) are marked as 01 and 10 cases, and thus no current would be flowing in the device leaving the output to 1 (high). When the values of PG and CG have the same logic value, both high or both low, then the transistor

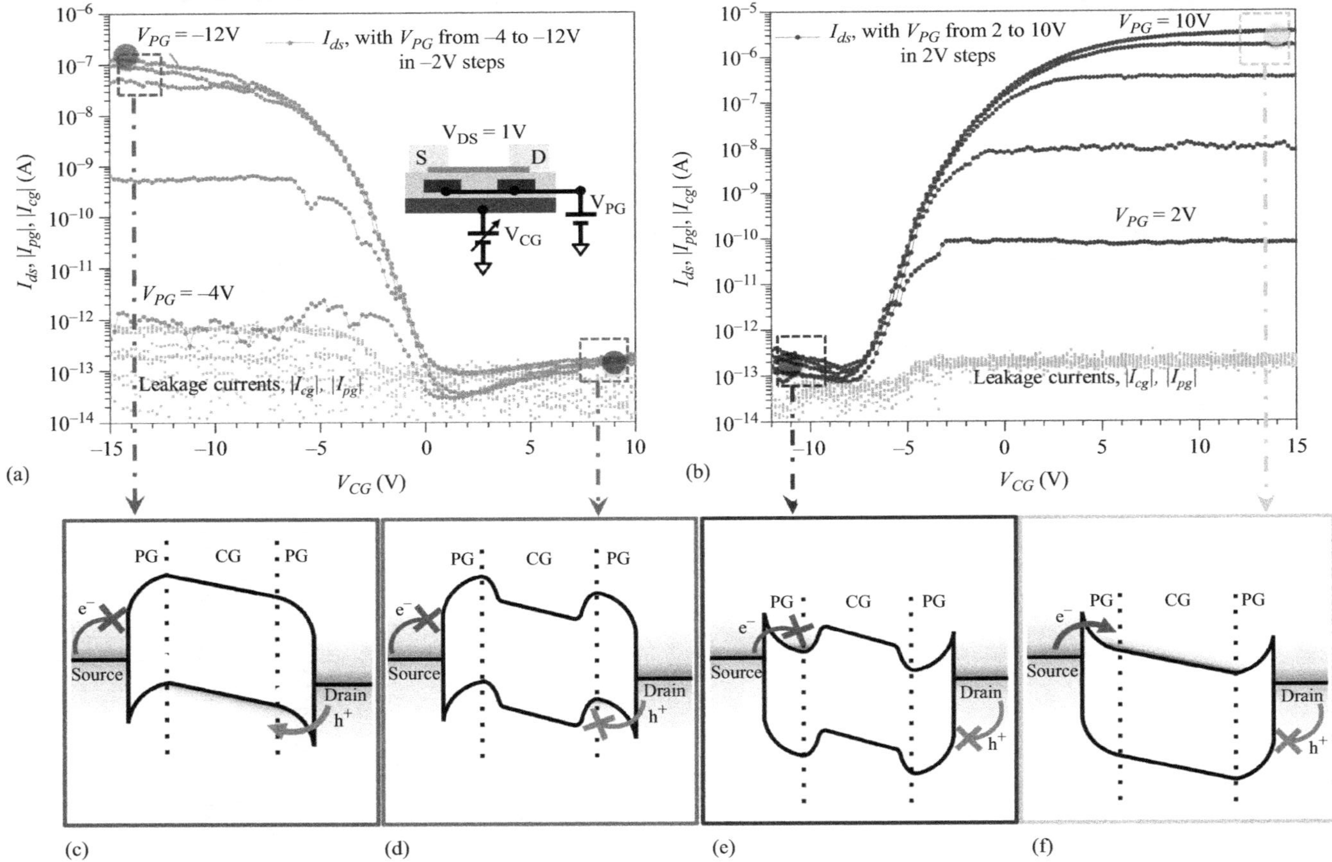

Figure 4.4 *Polarity-controllable behavior. (a) p-Type transfer characteristics obtained sweeping the bias on the control gate, for negative biases of the polarity gate. (b) n-Type transfer characteristics obtained sweeping the bias on the control gate, for positive biases of the polarity gate. (c–f) Schematic band diagrams of the four region of operation (ON-p-type, OFF-p-type, ON-n-type, OFF-p-type). © 2016 Nature publishing group. Adapted, with permission, from Reference [22]*

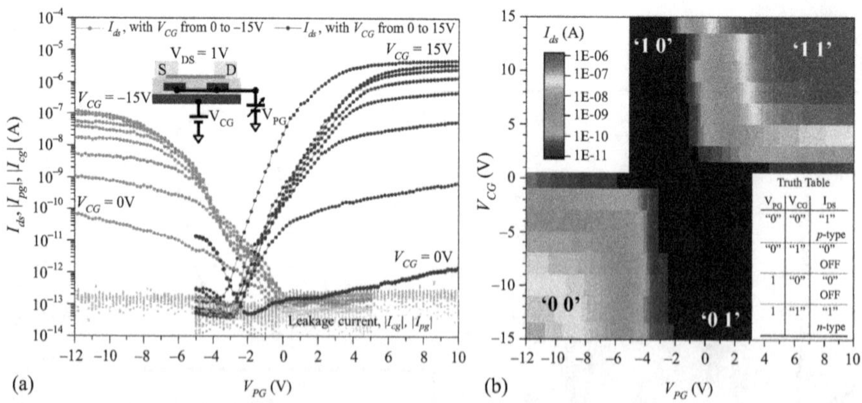

Figure 4.5 Polarity change "on the fly" and intrinsic-XOR behavior. (a) Transfer characteristics obtained sweeping the bias of the polarity gate at fixed biases of the control gate. (b) 2D current map showing the complete switching properties of the device. © 2016 Nature publishing group. Adapted, with permission, from Reference [22]

is conducting and the output of the logic gate would be 0 (low) (Figure 4.5(b) inset). The comparison-driven switching property of the device can be exploited at higher circuit level as it gives the possibility to realize logic gates (e.g., XOR, majority gates, as well as other circuit primitives) with fewer transistors than possible in conventional CMOS technology.

4.3 Quantum transport simulations

This section focuses on quantum transport simulations of 2D polarity-controllable devices. Given the limitations of current fabrication technologies on 2D materials (both related to material synthesis, doping and oxide growth on top of 2D materials), simulations of ultra-scaled devices are essentials to evaluate the real potential of new materials and new device structures. Due to the natural 2D structure of TMDC materials and the ultra-scaled length of the channels that are being investigated (<10 nm), it has become essential to use simulator tools that include quantum effects and perform quantum transport simulations, rather than using the regular drift-diffusion-based tools. In order to study the properties of ultra-scaled 2D polarity-controllable devices, the conduction and valence band of 2D semiconducting channels were modeled with a two-band tight-binding Hamiltonian, that was created from the material properties presented in Table 4.1. The hopping parameter t_{hop}, to be used by the open-source quantum-transport code NanoTCAD Vides [33,34] in the NEGF simulations, is given by [35]

$$|t_{hop}|^2 = \frac{2\hbar E_G}{3a^2 m_R^*} \tag{4.1}$$

Table 4.1 Material properties used to construct the effective mass Hamiltonian

Material property	1L WSe_2	2L WSe_2	2L MX_2
a (Å)	3.28	3.28	3.70
E_G (eV)	1.5	1.1	0.8
ϕ_{SB} (eV)	0.75	0.55	0.4
m_e	0.33	0.33	0.3
m_h	0.45	0.45	0.4

where a is the lattice constant, E_G is the energy bandgap, m_R^* is the reduced effective mass and $\hbar$ is the reduced Planck constant. Here, the material parameters such as lattice constant, effective masses and bandgaps are taken from literature [36–40] and reported in Table 4.1. This approach has been widely used to project performance of nanoscale transistors based on Si, III–V [41] and now 2D materials [40,42,43]. Further, to model Schottky contacts, the Hamiltonian was extended at the contacts for the zero-bandgap metal, and Dirichlet boundary conditions were applied. The model was extended to bilayer 2D materials by adding an interlayer hopping parameter in the effective-mass Hamiltonian, to account for coupling between the two layers [44]. This model provides a good trade-off between accuracy and computational time which is crucial in advanced device design with exotic materials for future technology nodes.

Figure 4.6 shows the 3D schematic structures of the simulated devices with top-gate (TG) and double-gate (DG) geometry (Figure 4.6(a) and (b), respectively). In the TG configuration, HfO_2 ($\kappa = 25$, *equivalent-oxide-thickness* (EOT) = 0.47 nm) was used as top dielectric, while SiO_2 ($\kappa = 3.9$, EOT = 30 nm) was considered as bottom dielectric. For the DG geometry, HfO_2 ($\kappa = 25$, EOT = 0.47 nm) was used for top and bottom gate dielectrics. The switching properties of the device were studied performing self-consistent ballistic simulations, iteratively solving Poisson and Schrödinger equations (within the NEGF formalism), with NanoTCAD Vides. No doping was introduced at source, and drain contacts for both gate geometries and mid-gap SB contacts were assumed, to have symmetric characteristics for the two polarities. The device performances were evaluated at different gate lengths, while keeping the same length for both the CG and PG gates ($L_{CG} = L_{PG}$) and the length ungated channel region (L_{OPEN}), separating PG and CG, fixed to $L_{CG}/2$, as shown in Figure 4.6. Thus, in the remainder of this section, we will refer to L_G as the length of each gated segment. No underlap, between the polarity gates and the source and drain contacts, was used the simulations, in order to provide the most efficient modulation of the Schottky barrier. The voltage bias on the PG (V_{PG}) was fixed, thus setting the device polarity, and the switching properties of the transistor, as a function of the CG voltage (V_{CG}), were studied.

The operation principle of the device is shown in Figure 4.7 with the help of the band-diagrams extracted from the simulations on monolayer (1L) WSe_2 at $L_G = 8$ nm. The PG controls the device polarity by tuning the effective Schottky barriers height (ϕ_{SB}) at source and drain (n-type behavior at $V_{PG} = 1$ V in Figure 4.7(a) and p-type

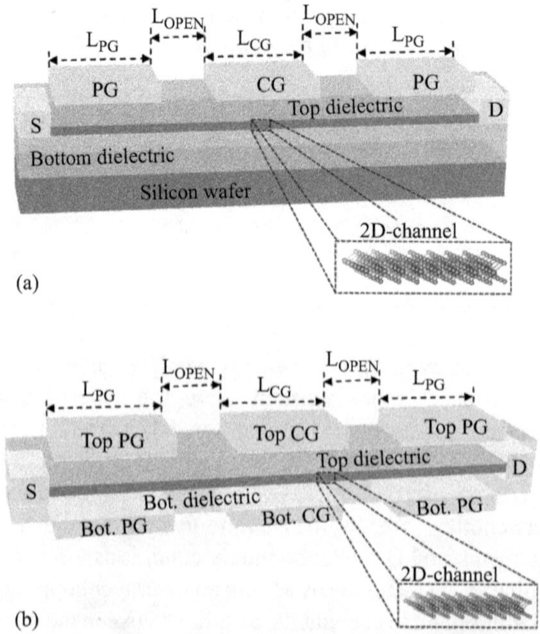

Figure 4.6 3D schematic structure of the simulated devices. (a) Top-gate (TG) device structure. (b) Double-gate (DG) device structure. The semiconducting channel is highlighted, showing the atomic structure of the 2D material. © 2017 Nature publishing group. Adapted, with permission, from Reference [26]

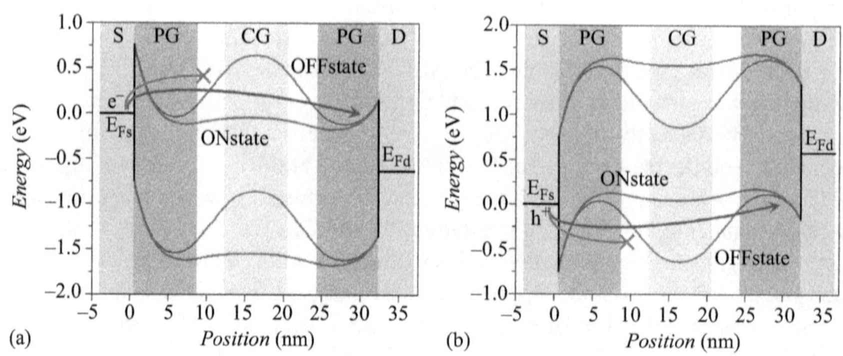

Figure 4.7 Band bending diagrams extracted from the simulation of monolayer WSe_2 at $L_G = 8$ nm. (a) n-Type operation for $V_{PG} = 1$ V. (b) p-Type operation for $V_{PG} = -1$ V. © 2017 Nature publishing group. Adapted, with permission, from Reference [26]

behavior at $V_{PG} = -1$ V in Figure 4.7(b)), while the CG determines the ON/OFF state of the FET by controlling the potential barrier in the central region of the channel. For *n*-type behavior, see Figure 4.7(a), in the OFF state ($V_{CG} = 0$ V), the potential barrier created by the CG blocks electron conduction. When the CG is biased at $V_{CG} = 0.8$ V, the barrier is removed and electrons flow from source to drain. For *p*-type behavior, see Figure 4.7(b), in the OFF state ($V_{CG} = 0$ V), the potential barrier created by the CG blocks hole conduction. When the CG is biased at $V_{CG} = -0.8$ V, the barrier is removed and holes flow from source to drain. The simulations showed the feasibility of polarity control at ultra-scaled dimensions, down to 4 nm gate lengths, when direct tunneling through the CG potential barrier begins to considerably degrade the device OFF state.

Figure 4.8 shows the transfer characteristics for the case of 1L-WSe_2 semiconducting channel, for both TG and DG geometries. 1L-WSe_2 was modeled with 1.5 eV bandgap, and the Schottky-barrier height was set to 0.75 eV ($\phi_{SB} = 0.75$ eV) for both charge carriers, simulating mid-gap Schottky contacts. It can be seen that, thanks to the high semiconducting bandgap, 1L-WSe_2 provides an excellent control of the device OFF state but also compromises the ON-state behavior of the transistor. In fact, the Schottky-barrier height at the source and drain contacts prevents efficient injection of charge carriers, with ON currents only reaching few μA/μm in the DG geometry.

To increase the ON current of the devices, bilayer (2L) WSe_2 was studied as a channel material. Bilayer WSe_2 has a reduced bandgap of 1.1 eV, which will reduce the height of the Schottky barrier at the contacts. Moreover, the increased mobile charge density, provided by the additional layer, is predicted to improve the device ON-state. Figure 4.9 shows the simulated transfer characteristics of 2L-WSe_2 FETs for both polarities and gate geometries, at different gate lengths. The Schottky-barrier height (ϕ_{SB}) was set to 0.55 eV for both charge carriers, simulating mid-gap Schottky contacts.

As a result of the decrease in Schottky-barrier height at the contact interface ($\phi_{SB} = 0.55$ eV), the ON-currents are increased by around two orders of magnitude. With the lowering of the semiconducting bandgap, the potential barrier created by the CG in the OFF state of the device is also decreased, deteriorating the device OFF current (I_{OFF}) that is increased by around three orders of magnitude. Nevertheless, the transfer characteristics, presented in Figure 4.9, show that even at the shortest gate length simulated ($L_G = 4$ nm), I_{OFF} is lower than 10^{-3} μA/μm, providing $I_{ON}/I_{OFF} > 10^6$. The use of a DG geometry benefits the electrostatic control of the gates over the channel and eliminates the charge screening effect between the layers that occurs in the TG structure. The improvement in the device electrostatics, given by the DG configuration, is shown in Figure 4.10 where the I_{OFF} and I_{ON} (Figure 4.10(a) and (b), respectively), extracted from the transfer characteristics of *n*-type devices with TG and DG structures, are compared. The I_{OFF} were extracted from the simulations at $V_{CG} = -0.2$ V and $V_{PG} = 1$ V. It was found that, until $L_G = 5$ nm, the OFF current in the DG configuration is consistently one order of magnitude lower with respect to the TG geometry, while the I_{ON}, extracted from the simulations at $V_{CG} = 0.8$ V and $V_{PG} = 1$ V, showed an average 2× improvement. For $L_G = 4$ nm,

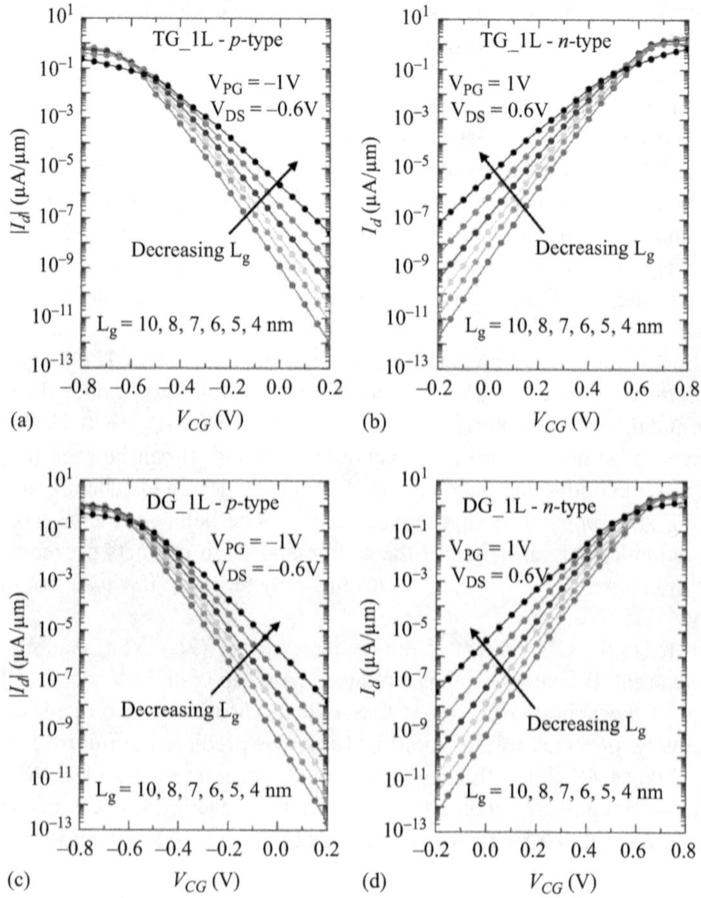

Figure 4.8 Simulated transfer characteristics for 1L-WSe$_2$ polarity-controllable FETs at different gate lengths. (a and b) Transfer characteristics of p- and n-type FET with top-gate geometry. (c and d) Transfer characteristics of p- and n-type FET with double-gate geometry. For both geometries, the length of each gated segment is varied from 10 nm down to 4 nm. © 2017 Nature publishing group. Adapted, with permission, from Reference [26]

the potential barrier created by the CG starts to becomes thin enough to have tunneling effects, deteriorating the OFF state of the device and thus lowering the positive impact of the DG. Similar results can be found for the *p*-type characteristics simulated on the same device.

Further analysis of the transfer characteristics is presented in Figure 4.11, where the effects of the ultra-scaled gate lengths were analyzed in relation to the *subthreshold slope* (SS) and the *drain-induced barrier lowering* (DIBL). The SS was evaluated as the average slope of the transfer characteristics in the subthreshold regime (from

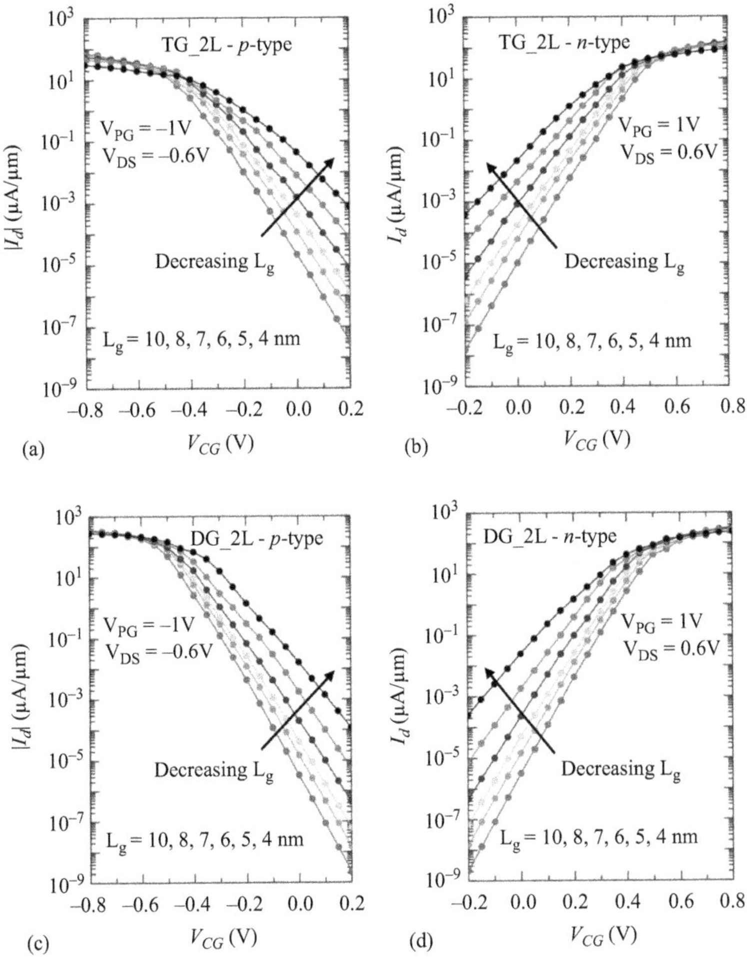

Figure 4.9 Simulated transfer characteristics for 2L-WSe_2 polarity-controllable FETs at different gate lengths. (a and b) Transfer characteristics of p- and n-type FET with top-gate geometry. (c and d) Transfer characteristics of p- and n-type FET with double-gate geometry. For both geometries, the length of each gated segment is varied from 10 nm down to 4 nm. © 2017 Nature publishing group. Adapted, with permission, from Reference [26]

−0.2 to 0.2 V_{CG}) for both *p*- and *n*-type operation mode (Figure 4.11(a) and (b), respectively). It was shown that the SS greatly benefits from the double-gate geometry, which is able to mitigate the detrimental effect of increased channel thickness for the bilayer device. The DIBL was calculated as the variation of threshold voltage (V_{th}) of the device divided by the variation of applied V_{DS} ($DIBL = \Delta V_{th}/\Delta V_{DS}$ [mV/V]). A variation of threshold voltage of 25 mV was extracted at the end of the subthreshold

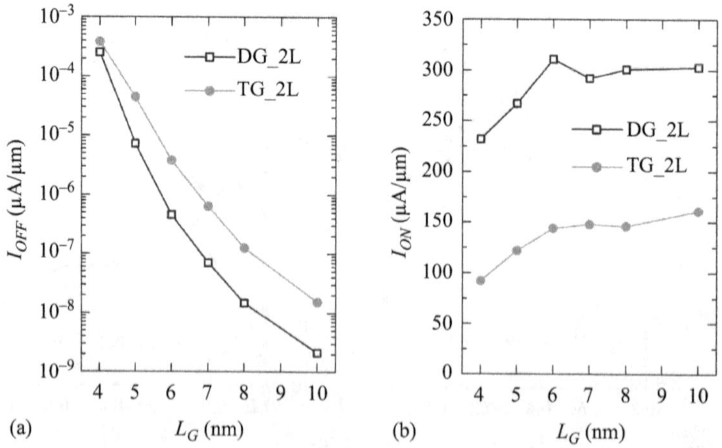

Figure 4.10 Benefits of double-gate geometry in 2L-WSe_2 n-type FETs. (a) Comparison between top- and double-gate OFF current (I_{OFF}). (b) Comparison between top- and double-gate ON current (I_{ON}). © 2017 Nature publishing group. Adapted, with permission, from Reference [26]

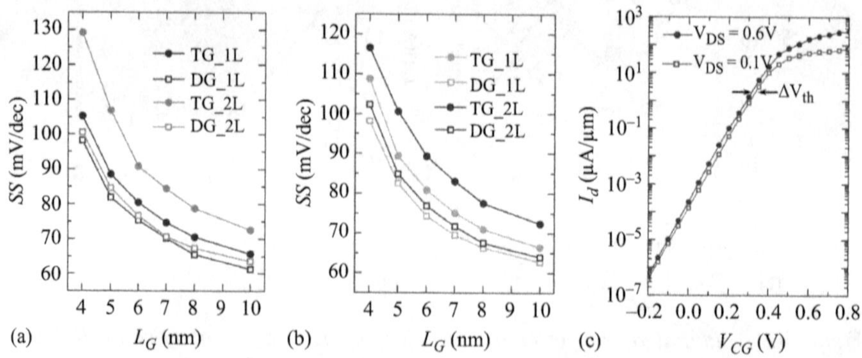

Figure 4.11 Evaluation of subthreshold slope and DIBL. (a and b) Subthreshold slope extracted from the transfer characteristics in p- (a) and n-type operation mode (b) for both TG and DG geometry with 1L- and 2L-WSe_2. (c) Evaluation of DIBL effect at $L_G = 6$ nm for DG n-type operation. © 2017 Nature publishing group. Adapted, with permission, from Reference [26]

regime, $V_{CG} = 0.3$ V, as the lateral shift between the transfer characteristics simulated at $V_{DS} = 0.1$ and 0.6 V (see Figure 4.11(c)). The computed DIBL was in the order of 50 mV/V for $L_G = 6$ nm, showing excellent immunity to DIBL effects. The observed immunity to DIBL is an added benefit of the SB polarity-controllable FETs, as the drain voltage drop in the channel is concentrated at the Schottky junction at drain.

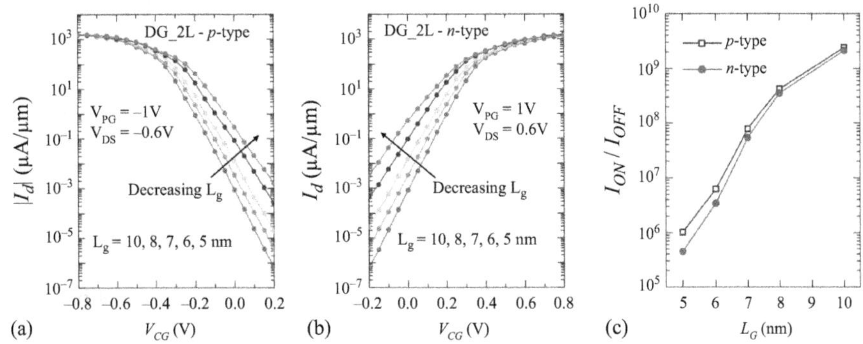

Figure 4.12 Analysis of performances for DG polarity-controllable device with 2L-MX_2 material. (a) Transfer characteristics of p-type FET with double-gate geometry. (b) Transfer characteristics of n-type FET with double-gate geometry. The length of each gated segment was varied from 10 nm down to 5 nm. (c) I_{ON}/I_{OFF} for both p- and n-type operation mode. © 2017 Nature publishing group. Adapted, with permission, from Reference [26]

The change in V_{DS} does not affect the height of the potential barrier created by the CG, which is ultimately responsible for the lowering of the threshold voltage of the device.

These analyses showed that the double-gate geometry provides the best electrostatic control and enhances the performances of the device by lowering the I_{OFF}, while improving the I_{ON} and SS. Nevertheless, the I_{ON} reachable with 2L-WSe_2, in both *n*- and *p*-type operation mode, is still much lower that what is achievable with 2D-doped devices [40]. The Schottky-barrier height at source and drain ($\phi_{SB} = 0.55$ eV) does not allow to have efficient tunneling at the contact interface. In order to reduce the impact of the Schottky-barriers on the device, 2D materials with a lower semiconducting bandgap were studied. Theoretical calculations [36–38] have shown that in the family of 2D-TMDCs, several materials, such as ZrS_2, HfS_2, $HfSe_2$, etc., have a bandgap of around 0.7–0.9 eV and could prove to be well suited for application in polarity-controllable FETs. For many of these materials, experimental results are very limited [25,45–48], and even in the theoretical ab-initio calculations, there are discrepancies in the computed material properties [36–38]. There are especially great variations in the reported value of the semiconducting bandgap, depending on the functional used in the ab-initio simulations. Based on these theoretical analyses, a 2D material was modeled, according to the properties presented in Table 4.1, and its potential application as a semiconducting channel in polarity-controllable devices was studied. A 2L-MX_2 material with an increased lattice constant, a lower bandgap and similar effective masses with respect to WSe_2 (as predicted for ZrS_2, HfS_2, $HfSe_2$) was considered. The MX_2 material was modeled with a 0.8 eV bandgap, and the Schottky-barrier height (ϕ_{SB}) was set to 0.4 eV for both charge carriers, simulating mid-gap Schottky contacts. Figure 4.12(a) and (b) shows the transfer characteristics at different L_G for a DG geometry for both *p*- and *n*-type polarities, while the device performances in terms of I_{ON}/I_{OFF} ratios are presented in Figure 4.12(c). The lower Schottky-barrier height at source and drain ($\phi_{SB} = 0.4$ eV) allowed for a greater number of carriers to be injected in the channel, increasing the I_{ON} to 1.5 mA/μm, while keeping I_{OFF} well below 10^{-2} μA/μm down to $L_G = 5$ nm. For both polarities, it was shown that $I_{ON}/I_{OFF} > 10^5$ down to $L_G = 5$ nm. The lower I_{ON}/I_{OFF} ratios for *n*-type behavior shown in Figure 4.12(c) were caused by the lower effective mass of electrons, which increases the transmission probability of carriers over the potential barrier created by the CG, thus increasing the I_{OFF}. These performances are comparable to the ones predicted, using similar quantum transport simulations based on NEGF, for conventional 2D-doped devices [40,42,43] and thus show a feasible scaling path for polarity-controllable technologies applied to 2D semiconductors.

4.4 Summary

This chapter was dedicated to polarity-controllable devices fabricated with semiconducting 2D WSe_2. We aimed at giving a broad overlook of the field, and of the state-of-the-art of 2D materials growth. We focused on experimental devices realized with WSe_2 and considered novel 2D materials (such as ZrS_2, HfS_2, $HfSe_2$, etc.) in quantum transport simulations. Some of the key aspects that have been elucidated in this chapter are as follows:

(1) Undoped Schottky-barrier (SB) FETs for the conduction of both types of charge carriers (ambipolar behavior).
(2) The ambipolar behavior provides an added degree of freedom for the realization of doping-free or lightly doped devices.
(3) Novel 2D materials can be adapted to this technology as Schottky contacts are easily created (realizing Ohmic contacts to 2D materials is still a great challenge).
(4) Polarity-controllable devices fabricated on WSe_2 achieved high ON/OFF ratios ($> 10^6$) for both polarities on the same device, indicating optimal electrostatic control and the potential for low-power applications.
(5) Quantum transport simulations have shown that novel 2D materials (such as ZrS_2, HfS_2, $HfSe_2$, etc.), thanks to their lower bandgap, allow for a greater number of carriers to be injected in the channel, achieving I_{ON} of ~1.5 mA/μm, while keeping I_{OFF} well below 10^{-2} μA/μm down to $L_G = 5$ nm.

References

[1] Wilson J, Yoffe A. The transition metal dichalcogenides discussion and interpretation of the observed optical, electrical and structural properties. Advances in Physics. 1969;18(73):193–335.
[2] Novoselov K, Geim AK, Morozov S, *et al.* Two-dimensional gas of massless Dirac fermions in graphene. Nature. 2005;438(7065):197–200.

[3] Jariwala D, Sangwan VK, Lauhon LJ, *et al.* Emerging device applications for semiconducting two-dimensional transition metal dichalcogenides. ACS Nano. 2014;8(2):1102–1120.

[4] Wang QH, Kalantar-Zadeh K, Kis A, *et al.* Electronics and optoelectronics of two-dimensional transition metal dichalcogenides. Nature Nanotechnology. 2012;7(11):699–712.

[5] Fiori G, Bonaccorso F, Iannaccone G, *et al.* Electronics based on two-dimensional materials. Nature Nanotechnology. 2014;9(10):768–779.

[6] Yu L, Lee YH, Ling X, *et al.* Graphene/MoS_2 hybrid technology for large-scale two-dimensional electronics. Nano Letters. 2014;14(6):3055–3063.

[7] Lee YH, Yu L, Wang H, *et al.* Synthesis and transfer of single-layer transition metal disulfides on diverse surfaces. Nano Letters. 2013;13(4):1852–1857.

[8] Gao Y, Liu Z, Sun DM, *et al.* Large-area synthesis of high-quality and uniform monolayer WS_2 on reusable Au foils. Nature Communications. 2015;6:8569.

[9] Sanne A, Ghosh R, Rai A, *et al.* Radio frequency transistors and circuits based on CVD MoS_2. Nano Letters. 2015;15(8):5039–5045.

[10] Smithe KK, English CD, Suryavanshi SV, *et al.* High mobility in monolayer MoS_2 devices grown by chemical vapor deposition. In: Device Research Conference (DRC), 2015 73rd Annual. IEEE; 2015. p. 239–240.

[11] Gong Y, Lei S, Ye G, *et al.* Two-step growth of two-dimensional WSe_2/$MoSe_2$ heterostructures. Nano Letters. 2015;15(9):6135–6141.

[12] Li MY, Shi Y, Cheng CC, *et al.* Epitaxial growth of a monolayer WSe_2-MoS_2 lateral *p–n* junction with an atomically sharp interface. Science. 2015;349(6247):524–528.

[13] Gao Y, Hong YL, Yin LC, *et al.* Ultrafast growth of high-quality monolayer WSe_2 on Au. Advanced Materials. 2017;29(29):1700990.

[14] Tang HL, Chiu MH, Tseng CC, *et al.* Multilayer graphene–WSe_2 heterostructures for WSe_2 transistors. ACS Nano. 2017;11(12):12817–12823.

[15] Geim AK, Grigorieva IV. Van der Waals heterostructures. Nature. 2013; 499(7459):419–425.

[16] Shi Y, Zhou W, Lu AY, *et al.* Van der Waals epitaxy of MoS_2 layers using graphene as growth templates. Nano Letters. 2012;12(6):2784–2791.

[17] Kang K, Xie S, Huang L, *et al.* High-mobility three-atom-thick semiconducting films with wafer-scale homogeneity. Nature. 2015;520(7549):656–660.

[18] Zhou H, Wang C, Shaw JC, *et al.* Large area growth and electrical properties of *p*-type WSe_2 atomic layers. Nano Letters. 2014;15(1):709–713.

[19] Sachid AB, Tosun M, Desai SB, *et al.* Monolithic 3D CMOS using layered semiconductors. Advanced Materials. 2016;28(13):2547–2554.

[20] Tosun M, Chuang S, Fang H, *et al.* High-gain inverters based on WSe_2 complementary field-effect transistors. ACS Nano. 2014;8(5):4948–4953.

[21] Yu L, Zubair A, Santos EJ, *et al.* High-performance WSe_2 complementary metal oxide semiconductor technology and integrated circuits. Nano Letters. 2015;15(8):4928–4934.

[22] Resta GV, Sutar S, Balaji Y, *et al.* Polarity control in WSe_2 double-gate transistors. Scientific Reports 6, 29448 (2016).

[23] Gong C, Colombo L, Wallace RM, *et al.* The unusual mechanism of partial Fermi level pinning at metal–MoS_2 interfaces. Nano Letters. 2014;14(4): 1714–1720.

[24] Li XX, Fan ZQ, Liu PZ, *et al.* Gate-controlled reversible rectifying behaviour in tunnel contacted atomically-thin MoS_2 transistor. Nature Communications. 2017;8(1):970.

[25] Mleczko MJ, Zhang C, Lee HR, *et al.* $HfSe_2$ and $ZrSe_2$: Two-dimensional semiconductors with native high-κ oxides. Science Advances. 2017;3(8):e1700481.

[26] Resta GV, Agarwal T, Lin D, *et al.* Scaling trends and performance evaluation of 2-dimensional polarity-controllable FETs. Scientific Reports 7, 45556 (2017).

[27] Podzorov V, Gershenson M, Kloc C, *et al.* High-mobility field-effect transistors based on transition metal dichalcogenides. Applied Physics Letters. 2004;84(17):3301–3303.

[28] Movva HC, Rai A, Kang S, *et al.* High-mobility holes in dual-gated WSe_2 field-effect transistors. ACS Nano. 2015;9(10):10402–10410.

[29] Liu W, Kang J, Sarkar D, *et al.* Role of metal contacts in designing high-performance monolayer *n*-type WSe_2 field effect transistors. Nano Letters. 2013;13(5):1983–1990.

[30] Das S, Appenzeller J. WSe_2 field effect transistors with enhanced ambipolar characteristics. Applied Physics Letters. 2013;103(10):103501.

[31] Das S, Demarteau M, Roelofs A. Ambipolar phosphorene field effect transistor. ACS Nano. 2014;8(11):11730–11738.

[32] Chuang HJ, Tan X, Ghimire NJ, *et al.* High mobility WSe_2 *p*-and *n*-type field-effect transistors contacted by highly doped graphene for low-resistance contacts. Nano Letters. 2014;14(6):3594–3601.

[33] Fiori G, Iannaccone G. NanoTCAD ViDES. 2008. http://vides.nanotcad.com/vides/

[34] Bruzzone S, Iannaccone G, Marzari N, *et al.* An open-source multiscale framework for the simulation of nanoscale devices. IEEE Transactions on Electron Devices. 2014;61(1):48–53.

[35] Agarwal T, Radu I, Raghavan P, *et al.* Effect of material parameters on two-dimensional materials based TFETs: An energy-delay perspective. In: Solid-State Device Research Conference (ESSDERC), 2016 46th European. IEEE; 2016. p. 47–50.

[36] Zhang W, Huang Z, Zhang W, *et al.* Two-dimensional semiconductors with possible high room temperature mobility. Nano Research. 2014;7(12):1731–1737.

[37] Gong C, Zhang H, Wang W, *et al.* Band alignment of two-dimensional transition metal dichalcogenides: Application in tunnel field effect transistors. Applied Physics Letters. 2013;103(5):053513.

[38] Rasmussen FA, Thygesen KS. Computational 2D materials database: Electronic structure of transition-metal dichalcogenides and oxides. The Journal of Physical Chemistry C. 2015;119(23):13169–13183.

[39] Zhao W, Ribeiro RM, Toh M, *et al.* Origin of indirect optical transitions in few-layer MoS_2, WS_2, and WSe_2. Nano Letters. 2013;13(11):5627–5634.

[40] Cao W, Kang J, Sarkar D, *et al.* 2D semiconductor FETs—Projections and design for sub-10 nm VLSI. IEEE Transactions on Electron Devices. 2015; 62(11):3459–3469.

[41] Park SH, Liu Y, Kharche N, *et al.* Performance comparisons of III–V and strained-Si in planar FETs and nonplanar FinFETs at ultrashort gate length (12 nm). IEEE Transactions on Electron Devices. 2012;59(8):2107–2114.

[42] Yoon Y, Ganapathi K, Salahuddin S. How good can monolayer MoS_2 transistors be? Nano Letters. 2011;11(9):3768–3773.

[43] Mishra V, Smith S, Ganapathi K, *et al.* Dependence of intrinsic performance of transition metal dichalcogenide transistors on materials and number of layers at the 5 nm channel-length limit. In: Electron Devices Meeting (IEDM), 2013 IEEE International. IEEE; 2013. p. 5–6.

[44] Fiori G, Iannaccone G. Performance analysis of graphene bilayer transistors through tight-binding simulations. In: Computational Electronics, 2009. IWCE'09. 13th International Workshop on. IEEE; 2009. p. 1–4.

[45] Li L, Fang X, Zhai T, *et al.* Electrical transport and high-performance photoconductivity in individual ZrS_2 nanobelts. Advanced Materials. 2010;22(37): 4151–4156.

[46] Kang M, Rathi S, Lee I, *et al.* Electrical characterization of multilayer $HfSe_2$ field-effect transistors on SiO_2 substrate. Applied Physics Letters. 2015;106(14):143108.

[47] Xu K, Wang Z, Wang F, *et al.* Ultrasensitive phototransistors based on few-layered HfS_2. Advanced Materials. 2015;27(47):7881–7887.

[48] Kanazawa T, Amemiya T, Ishikawa A, *et al.* Few-layer HfS_2 transistors. Scientific Reports. 2016;6:22277.

Chapter 5
Carrier type control of MX_2 type 2D materials for functionality-enhanced transistors

Shu Nakaharai[1]

5.1 MX_2 materials

5.1.1 2D materials for FEDs

Two-dimensional (2D) materials, such as graphene, phosphorene, MX_2 type materials of transition metal dichalcogenides (TMDCs) and some other oxides or chalcogenides, have attracted much attention due to their atomically thin sheet form with atomically flat interfaces and thickness [1–6]. While graphene lacks energy band gap, some other MX_2 type TMDC semiconductors have moderate size band gaps, and in addition, some of these semiconductors exhibit carrier mobility comparable to silicon. The atomically flat and smooth surfaces are the major reason that these 2D materials have been expected to replace silicon on insulator channel layers for the advancement beyond the physical size limitation in the transistor size scaling caused by the short channel effects. And, what's more, some of these 2D semiconductors have been revealed to have favorable characteristics of ambipolar carrier injection from Schottky junctions, which is a crucial feature for the FED technology [7,8]. In this chapter, how TMDC materials are applied to the channel materials in FED is introduced.

5.1.2 Crystalline structure and electric characteristics of TMDCs

In the layered structure, TMDC materials, atoms of transition metal and chalcogen are covalently bonded to form a transition metal monoatomic layer sandwiched by chalcogen monoatomic layers, and these sheets are stacked by weak van der Waals force to each other (Figure 5.1). Due to this stacked structure of weakly bonded monoatomic sheets, the TMDC monolayers can be easily exfoliated by sticky tape similarly with graphene. The unit cell configuration in a TMDC atomic sheet has mainly two polytypes, one is trigonal prism type in which chalcogen atoms form a prism shape (Figure 5.2). The monolayers of this polytype stack periodically with two monolayers, then it is called 2H phase, exhibiting semiconducting carrier conduction.

[1]Quantum Device Engineering Group, WPI Center for Materials Nanoarchitechtonics, National Institute for Materials Science, Japan

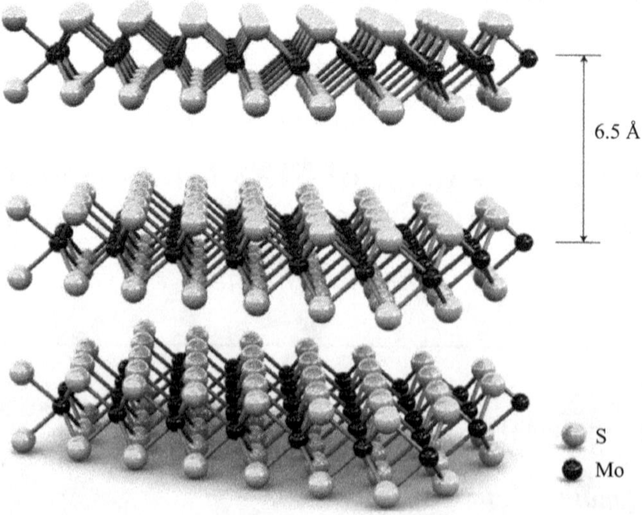

Figure 5.1 Crystalline structure of MoS_2. 2D sheets are weakly stacked by von der Waals force [6]

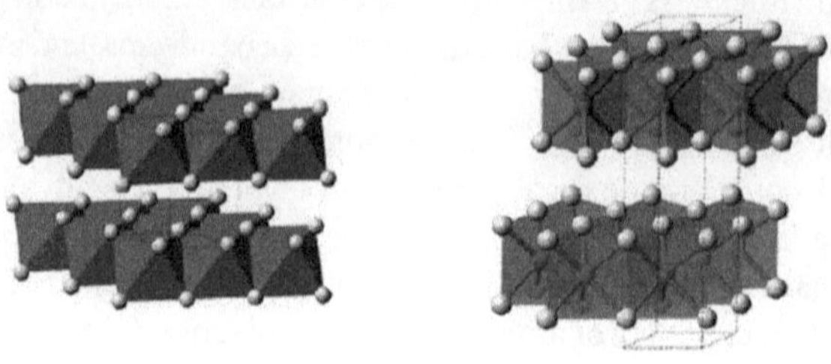

Figure 5.2 Typical polytypes of crystalline structure of MX_2 type materials, 1T and 2H [5]

Another polytype of 2D layers of TMDCs is the octahedral type in which chalcogen atoms form an octahedron with a transition metal atom at the center. In this case, since each layer stack in the same order, these 2D layers are called 1T or 1T′ type. Commonly, the 1T (or 1T′) crystals behave as metal or semi-metal.

In the 2H type TMDC semiconductors, the band gap is the direct transition type in the case of monolayer, and indirect in multilayers, and the band gap values are the largest in monolayer cases. For example, in the case of 2H-MoS_2, the band gap is 1.8 eV for monolayer and 1.2 eV for a bulk. Carrier mobility in these TMDC semiconductors strongly depends on the number of layers. In the case of thin layers (less than 5–10 layers), mobility is degraded seriously by scattering charged impurity or difficulty in Charrier injection from the contacts. The highest mobility is commonly obtained when the layer number is around ten [9]. The main origin of scattering is considered to be the charged impurities or polar optical phonon (Figure 5.3) depending

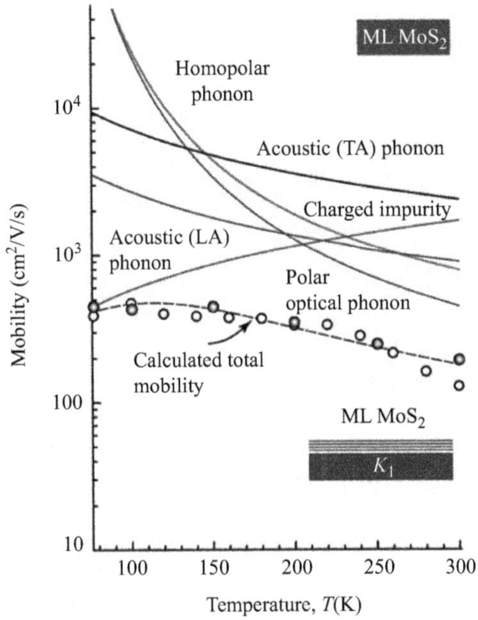

Figure 5.3 Temperature dependence of carrier mobility in MoS_2 transistors, along with the factors that limit the mobility [10]

on the temperature range [10]. Due to the screening effect, high-k gate dielectric layers for both sides of a TMDC layer are quite effective for reducing the degradation of carrier mobility.

5.2 MX_2 materials in transistors

5.2.1 Need for 2D materials channel in advanced FETs

Semiconductors of 2D materials such as TMDCs have attracted much attention for future channel materials of the ultimate thin-body field effect transistors (FETs) overcoming the problems in the thinned conventional 3D semiconductor crystals such as silicon, germanium or III–V compounds. In the conventional 3D crystals of semiconductors, shrinking of the transistor size was hindered by so-called short channel effects, which leads to losing of gate control over the channel due to the space charge distribution in the depth direction around junctions. For overcoming these issues, silicon on insulator, germanium on insulator or III–V on insulator structures have been developed, and for this reason, the fully depleted Fin-FET structure has been adopted. As the scaling goes further, the fin width is expected to be thinner until the regime of volume inversion. However, the thinning of the channel layer will inevitably be faced by the limit as the body thickness becomes lower than about 4 nm, where carrier mobility is seriously degraded due to the interface roughness and the body thickness fluctuation [11]. 2D materials such as TMDCs are promising for the expectation of overcoming this problem, because of their surface flatness and the thickness homogeneity in the atomic level. In addition, since the atoms in the TMDC semiconductors are bonded covalently to each other within the monolayer, there are in principle no dangling bonds on the surface. This could also work favorably for keeping performance of device when they are thinned down about a couple of nanometers level.

5.2.2 Carrier doping

On the other hand, there are many issues to be addressed in TMDC for device applications. One of them is the problem in the carrier doping for both electrons and holes especially in the local doping with high spatial resolution. Ion implantation is commonly used for the conventional CMOS fabrication, but it is barely applicable to TMDCs due to the extremely thin layer thickness, and therefore, many methods have been developed for TMDCs so far. One of the typical doping methods is by charge transfer from surface dopants attached, for example, molecules of NO_x [12] or F4TCNQ [13] for hole doping, and some suboxides of AlO_x or TiO_x for electron dopants. It was also demonstrated that self-limiting layer-by-layer oxidation of the top-single-atomic layer of WSe_2 works as a surface hole doping layer [14]. This is a promising method for local hole doping in TMDC channel with a fine spatial resolution. A substitutional doping of Nb atoms for Mo in MoS_2 atomic layer has been proved to work as n-type doping [15]. Here, the Nb doping was done during the crystal growth process, and therefore, it is not applicable to the local doping for CMOS or functionality-enhanced transistors. Another idea for carrier doping is electrostatic doping via capacitive coupling with gate electrodes. In this case, local control of carrier doping is possible, and also, the carrier type can be changed dynamically during runtime, which is the essential feature for the FED technology. Ionic liquid gating is also another method for electrostatic doping, in which a huge capacitance is induced between ionic liquid and TMDC channel interface, resulting in a large carrier density in the channel that any solid gates cannot access due to the limitation of electrostatic breakdown.

5.2.3 Carrier injection via Schottky junctions

Another issue in TMDC applications to FETs is carrier injection from Schottky junctions. Similar to typical semiconductors, the junction between a TMDC semiconductor and a metal forms a Schottky barrier, and Fermi level pinning effect causes the difficulty in controlling the polarity of the junction. Fermi level pinning occurs generally in Schottky junctions in which the Fermi level of the metal comes to a certain level in the band gap of the semiconductor and it moves only little even if the work function of the metal is largely changed. This is a fatal problem for realization of the polarity-controllable transistors, because both electrons and holes should be injected into an intrinsic semiconductor channel. In a junction with a high Schottky barrier, carriers are injected via tunneling through the barrier only if the barrier is thin enough. Electrostatic control of the carrier density in semiconductors induces thinning of the Schottky barrier at the junction, which enhances the injection of both electrons and holes by tunneling. However, it is still difficult to make carriers tunnel through the barriers when the barrier is very high. Therefore, depinning of Fermi level at the junction is still a critical point in the polarity-controllable transistors.

5.2.4 *Fermi level pinning*

In the case of MoS_2 with solid gates, for example, the Fermi level of metals comes near to the conduction band edge and only electrons can be injected from contact metals with a wide range of work functions from Sc (3.5 eV) to Pt (5.9 eV), while the estimated Schottky barrier heights for these metals were 30 and 230 meV, respectively, as shown in Figure 5.4 [9]. In other case of TMDCs, such as WSe_2, the Fermi levels of metals at Schottky junction come commonly around the midgap or slightly close to the valence or conduction band edge, depending on the crystal growth conditions.

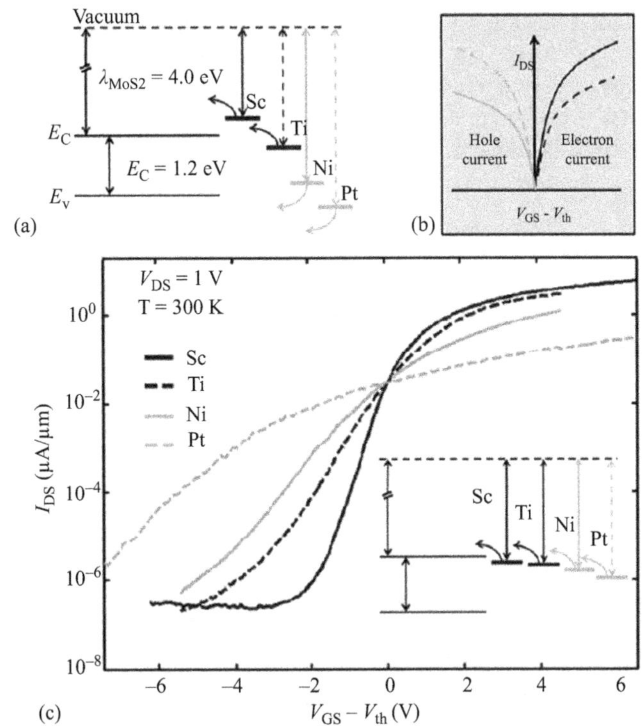

Figure 5.4 (a) Expected relation between work function of the metals and the bands of MoS_2, (b) Expected transfer curves of devices with contacts of different metals. N-type is expected in Sc and Ti contact devices, and p-type is expected in Ni and Pt and (c) Experimentally obtained transfer curves of devices with different contact metals, in which all devices work as n-type FET. The inset is the actual energy levels of metals at the junction [9]

Ambipolar conduction is possible in some of these materials by solid-gate operations, in which the carrier injection is based on tunneling of thinned Schottky barrier by electrostatic doping for both electrons and holes. In the following sections, it will be reviewed on the importance and some techniques on the ambipolar conductance in TMDC materials.

5.3 Polarity controllable transistors on $MoTe_2$

5.3.1 *Ambipolar channels for polarity controllable transistors*

For the realization of polarity-controllable transistors, ambipolarity is one of the most crucial requirements, and for this reason, TMDC materials of WSe_2 [16] and $MoTe_2$ [17] have been investigated due to their excellent ambipolarity. In these cases, transistor polarity is determined mainly by the property of the carrier injection into the channel, as well as the electrostatic doping of an intrinsic (or lightly doped) semiconducting channel. The transistor operation should be regarded as the gated Schottky-diode device operation, in which one of the gate electrodes is used for thinning the Schottky barrier, and hence, the thin-film structure of TMDCs is favorable for inducing band bending by electrostatic doping, compared to the bulk or thick films. In this section, we focus on $MoTe_2$ channel devices.

5.3.2 *$MoTe_2$ channel*

The first example of polarity-controllable transistor on TMDC was fabricated on alpha-phase $MoTe_2$ [17] in which an excellent ambipolarity had been reported [8]. The ambipolar conduction in a $MoTe_2$ device is shown in Figure 5.5, in which

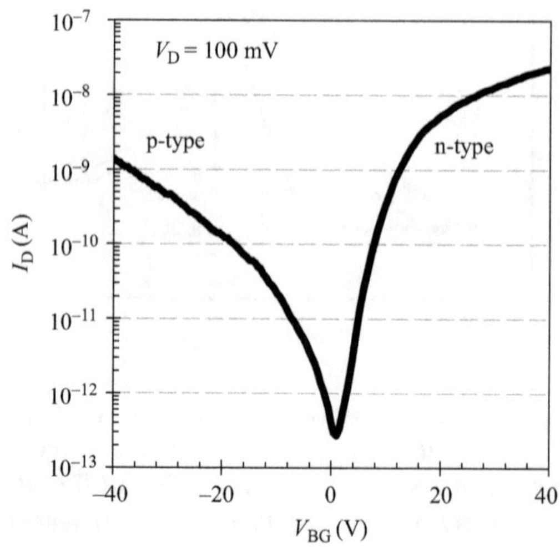

Figure 5.5 Ambipolar conduction in a $MoTe_2$ device with Ti contacts at a drain bias of 100 mV. It works as a ambipolar device, but the current level of hole conduction branch is lower than that of electron branch [17]

Ti was used for the contact metal and the gate bias was applied through a 285-nm-thick SiO_2 dielectric layer from a silicon substrate back gate. Here, few layer $MoTe_2$ flakes were mechanically exfoliated from a single crystal by sticky tape and deposited on the SiO_2 surface. As shown in Figure 5.5, electron conduction (n-type) is one order of magnitude higher than the hole conduction (p-type), implying that the Fermi level of Ti at the Schottky junction is pinned near the conduction band edge. Detailed estimation of Schottky barrier height in $MoTe_2$ will be discussed later. Figure 5.6(a) and (b) shows one of the fabricated polarity-controllable devices, which has a single top gate with 30-nm-thick SiO_2 as a gate dielectric layer added to the device. By a back gate biasing, a high carrier density is induced in the access regions between source/drain contacts and the edge of the top gate, and at the same time, carrier injection from the source/drain contacts is also enhanced by thinning the barrier width of the Schottky barrier.

5.3.3 Single top gate device on $MoTe_2$

Figure 5.6(c)–(f) shows the polarity-controlled operation of the fabricated device with a +100 mV drain bias at room temperature in vacuum, as well as the schematics of the band diagrams in both p- and n-type modes. When the back gate bias was negative (Figure 5.6(c) and (e)), the transistor turned "ON" by $V_{tg} < V_{th}$ as a p-type transistor, and turned "OFF" by $V_{tg} > V_{th}$ by stopping the hole conduction at a potential barrier induced at the center of the transistor. On the other hand, by applying a positive gate bias (Figure 5.6(d) and (f)), the transistor worked as an n-type one, by raising up ($V_{tg} < V_{th}$) or pushing down ($V_{tg} > V_{th}$) the potential barrier at the top-gated region. In this polarity-controlled transistor operation, the imbalance of the ON-currents of p- and n-modes reflects the ambipolar characteristics of the $MoTe_2$ channel shown in Figure 5.5, corresponding with the difference of the Schottky barrier heights for electrons and holes. Also, the threshold voltage, V_{th}, in the transistor operation in both modes shifted in the minus direction, indicating the electron-doped channel. The mobility of electrons was also degraded to be 0.14 cm^2/V s, which is one order of magnitude smaller than that without the top gate (3.0 cm^2/V s). These doping and degradation of mobility are considered to be induced by the SiO_2 dielectric layer which was evaporated by e-gum, in which a high density of the interface states was estimated to be in the order of 10^{13} cm^{-2}.

5.3.4 Dual top-gate device on $MoTe_2$

The second example of the polarity-controlled $MoTe_2$ transistors is a dual-gated one illustrated in Figure 5.7(a) and (b). In this case, there are two top gates added on the device with 30-nm-thick-SiO_2-dielectric layer with a 100 nm gap in between. In this device concept, drain current is modulated by both top gates of source and drain parts instead of back gating. These gate biasing control not only the channel potential but also the Schottky junction configurations of the source and drain contacts; that is, the device could be understood as a combination of two gated junctions. The device structure is the same as that of the previous single top-gated one except for the top-gate structure, and the device operation principle is illustrated in Figure 5.7(c) and (d) for p- and n-FET modes, respectively. Here, the drain terminal is positively biased. When

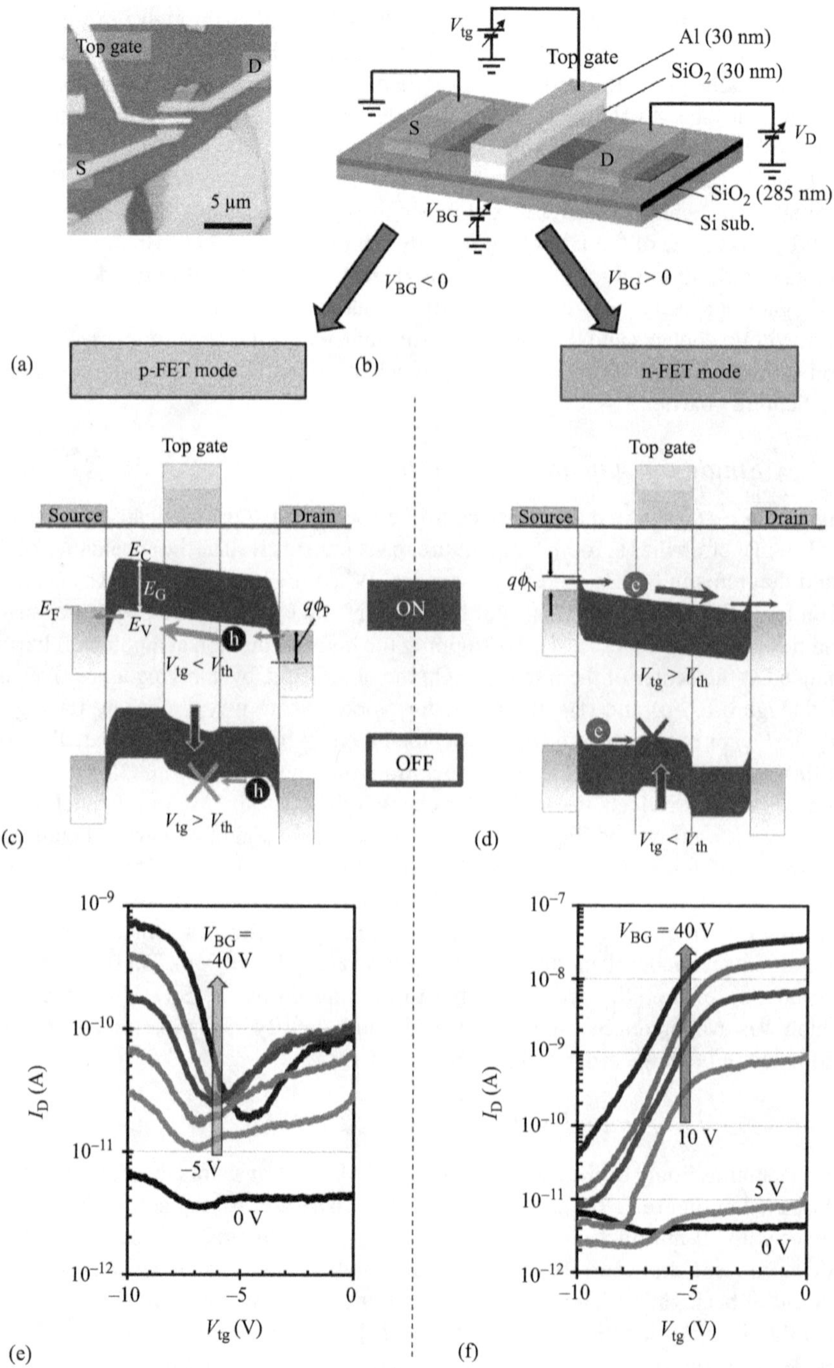

Figure 5.6 (a) Optical micrograph of a fabricated device with single top gate. (b) Schematic of the transistor structure. (c) Schematic band diagram in the p-FET mode with $V_{BG} < 0$. (d) Schematic band diagram in the n-FET mode with $V_{BG} > 0$. (e) Transistor operation in the p-FET mode with $V_{BG} < 0$. (f) Transistor operation in the n-FET mode with $V_{BG} > 0$ [17]

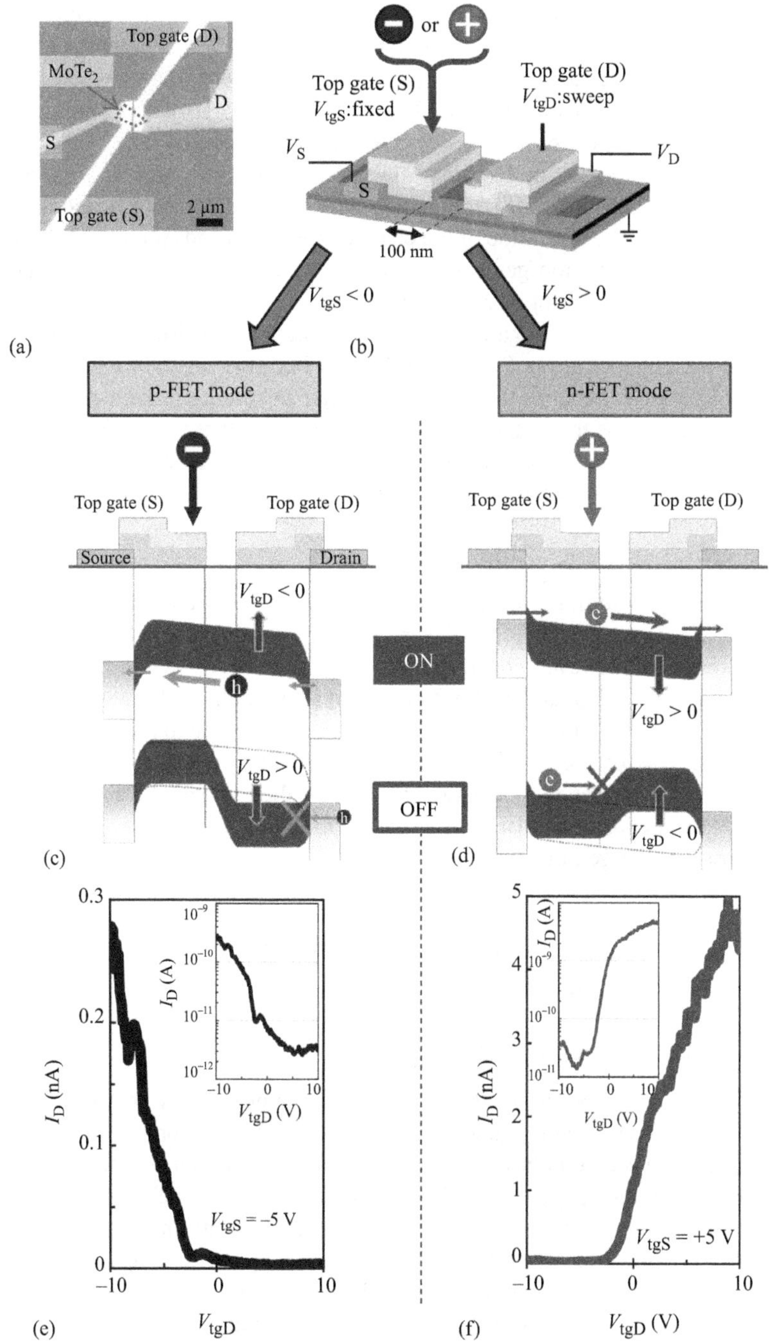

Figure 5.7 *(a) Optical micrograph of a fabricated device with dual top gates. (b) Schematic of the transistor structure. (c) Schematic band diagram in the p-FET mode with $V_{BG} < 0$. (d) Schematic band diagram in the n-FET mode with $V_{BG} > 0$. (e) Transistor operation in the p-FET mode with fixed $V_{tgS} < 0$. The inset is a logarithmic plot. (f) Transistor operation in the n-FET mode with fixed $V_{tgS} > 0$. The inset is a logarithmic plot [17]*

one of the top gates, for example the source side gate, is negatively biased ($V_{tgS} < 0$), then the drain current flows only if the top gate of the other side is also negatively biased ($V_{tgD} < 0$) and holes are not injected from the drain contact when $V_{tgD} > 0$. As a result, this device works as a p-FET by a gating operation of V_{tgD} (Figure 5.7(e)). On the other hand, when the source side gate is positively biased ($V_{tgS} > 0$), electrons are injected from the source junction, and then the device works as an n-FET for the gating operation of V_{tgD} (Figure 5.7(f)). The drain current is pinched off by the potential barrier in the drain side of the channel. One of the advantages of this device concept is that the two top gates are symmetric, and consequently the function of these top gates can be swapped each other.

On the other hand, the gate biasing work not only for lowering the potential barrier of the channel at the Schottky junctions but also for thinning the width of the Schottky barrier because the dominant carrier injection mechanism is based on tunneling of electrons through the barrier. This suggests that the ON/OFF switching of the device requires a larger gate voltage than the case of carrier injection over the barrier, resulting in a drawback of larger subthreshold slope (S-factor).

5.3.5 Issues in top-gate dielectrics

In both cases of single and dual top-gate structures, the top surface of the channel layer must be covered with gate dielectric layer, and in the above cases, e-beam evaporation of SiO_2 was used. However, covering the $MoTe_2$ channel by SiO_2 seriously degrade the on-currents in both p- and n-FET modes, and also the channel is negatively doped, resulting in a V_{th} shifted in the negative voltage direction due to the interface charge traps. In the case of high-k dielectric like Al_2O_3 on TMDCs, on the other hand, channels are heavily doped by negative charges by the top-gate dielectric deposition by atomic layer deposition (ALD), while carrier mobility is enhanced by the screening of scattering by charged impurities.

5.4 Enhanced ambipolarity by Schottky junction engineering in $MoTe_2$

5.4.1 Schottky junctions in $MoTe_2$

In this section, one possible method of enhancing ambipolarity in TMDCs will be discussed based on a Schottky junction engineering of hybrid contact on $MoTe_2$ [18]. As discussed above, ambipolarity is crucial for improving the performance of polarity-controllable transistors, and hence, the lack of ideal ambipolar channel has been the bottleneck. One of the most serious origins of the difficulty in ambipolarity is the Fermi level pinning effect. Because the sum of Schottky barrier heights for electrons and holes at a junction is fixed to be the band gap value of the semiconductor, at least one type of carriers (electron or hole) has to be injected by tunneling, resulting in imbalance of the on-current level of the p- and n-branches. One of the methods to avoid this problem is to use n-type contact in the source side and p-type contact in the drain side, but still this requires solving the Fermi level pinning effect. Recently, one possible solution to this problem has been proposed by adopting $MoTe_2$ as the channel based on Schottky junction engineering, and it will be reviewed in this section.

In contrast to other TMDC materials, it has been revealed experimentally that $MoTe_2$ is able to mitigate the Fermi level pinning for some kind of metals, and therefore, the polarity of the Schottky junction can be changed simply by selecting the contact metals [18]. Figure 5.8 shows the transistor operations of $MoTe_2$ devices (Figure 5.8(a)) with different contact metals for p-type (Pt) (Figure 5.8(b)) and n-type (Ti, Ni) (Figure 5.8(c) and (d)). In the cases of Au and Pd contacts, the devices exhibited ambipolar behavior. Here, a $MoTe_2$ channel was exfoliated onto a 285-nm-thick SiO_2 layer on a silicon substrate, and the $MoTe_2$ channel was attached with various metal contacts. Fabricated transistors were operated with a fixed drain bias, and the drain current was modulated by the back gate bias. Ambipolar devices exhibited a balanced ON-current which are lower than the ON-currents of p- or n-type ones, indicating that the Schottky barriers for both electrons and holes are larger than the p-type barrier in the Pt contact device or n-type barriers in the Ti and Ni devices.

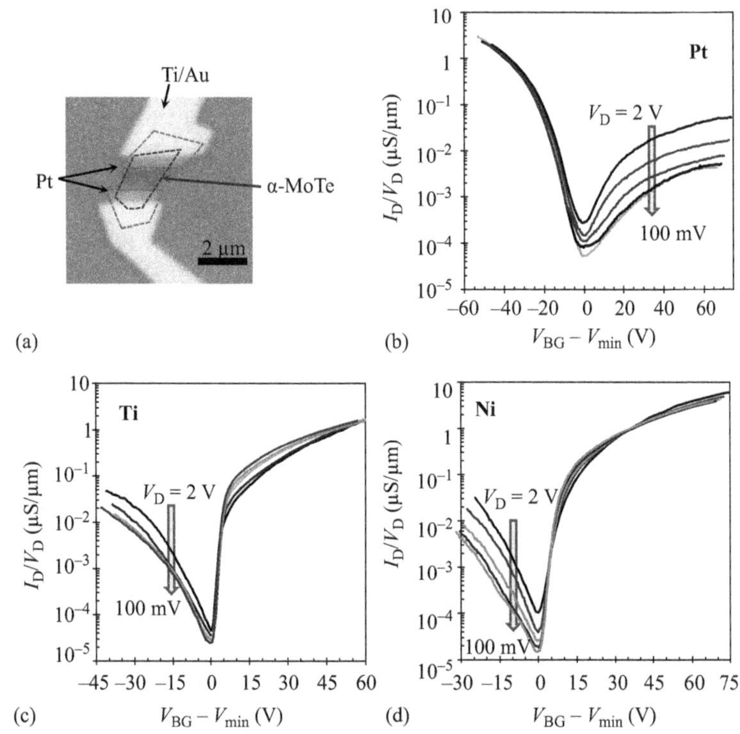

Figure 5.8 (a) Optical image of one of the fabricated devices with Pt contacts. (b) P-type transistor operation in a Pt contact device with different drain biases. N-type transistor operation. N-type FET operation was observed in Ti contact (c) and Ni contact (d) devices with the same measurement configurations [18]

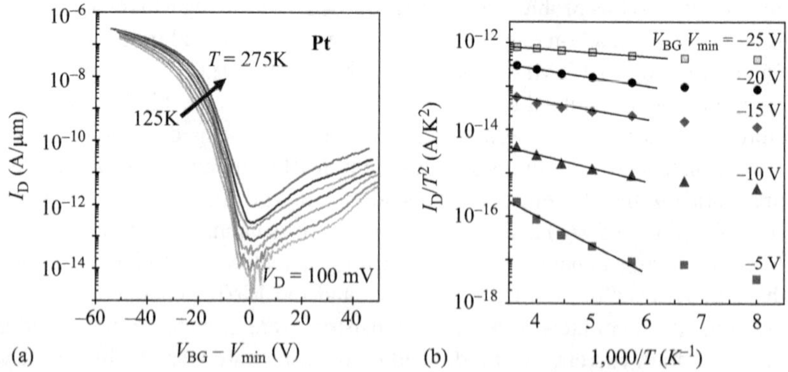

Figure 5.9 (a) Temperature dependence of the drain current in the Pt contact device. (b) I_D/T^2 plotted against 1000/T at different back gate bias, from which the injection barrier height is extracted [18]

In Figure 5.8(b)–(d), drain bias dependence of conductivity $G = I/V$ is presented. Here, in the case of Ti, drain bias dependence of conductivity is much weaker in the on state than the off state, indicating nearly Ohmic contact for electrons and tunneling injection of holes through a Schottky barrier which is thinned by drain bias.

5.4.2 Barrier heights of $MoTe_2$ Schottky junctions

For p- and n-type transistors, Schottky barrier height for electrons (measured from the conduction band edge) and holes (measured from the valence band edge) was estimated according to Das *et al.* by the flat band condition of the Schottky junction band alignment [9]. Figure 5.9 shows the temperature dependence of I–V curves (Figure 5.9(a)) and a back gate bias dependence of injection barrier (Figure 5.9(b)), which is evaluated from temperature dependence of current at each gate bias value according to the thermionic emission regime (5.1).

$$I_D = A^* T^\alpha \exp\left(-\frac{q\Phi_B}{k_B T}\right)\left\{1 - \exp\left(-\frac{qV_D}{k_B T}\right)\right\} \tag{5.1}$$

where q is the elementary charge ($q > 0$ for electrons), T is the temperature, A^* is Richardson's constant, k_B is Boltzmann's constant and Φ_B is the injection barrier potential measured from the Fermi level. Here, α should be 2 or 3/2 depending on the dimensionality of the channel, and here we take 2 according to [9]. As illustrated in Figure 5.10, the injection barrier at the flat band condition corresponds to the Schottky barrier. The obtained values of the Schottky barrier heights are 50 meV for Ti (from the conduction band edge), 100 meV for Ni (from the conduction band edge) and 40 meV for Pt (from the valence band edge). In the cases of Au and Pd, Fermi level is considered to be around midgap of the semiconductor band gap [9]. As illustrated in Figure 5.11, the Fermi level at the Schottky junction varies accordingly to the work functions of the metal, implying that the Fermi level pinning is very week in $MoTe_2$,

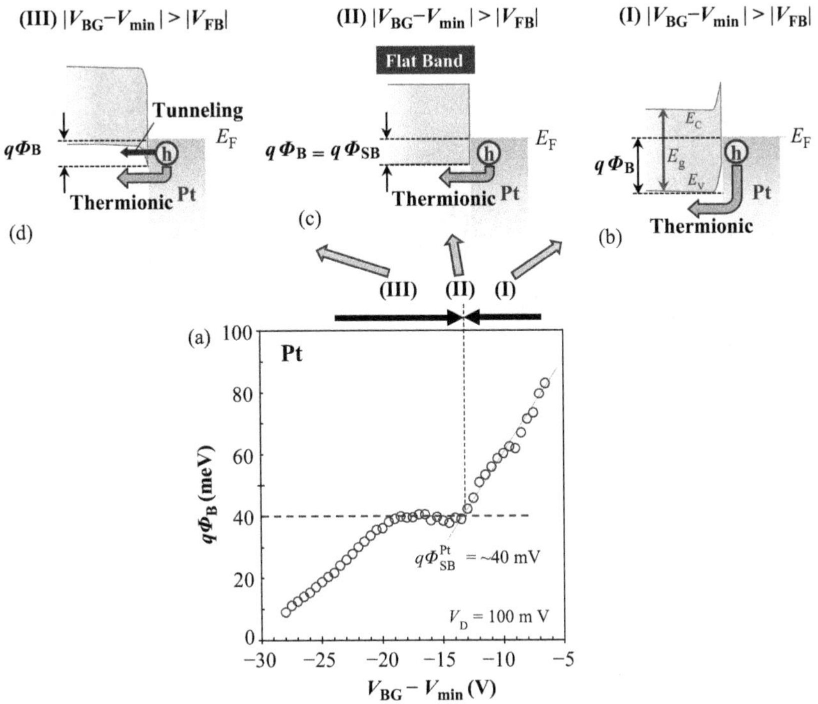

Figure 5.10 (a) The gate bias dependence of the carrier injection barrier in the Pt contact device. The Schottky barrier height of 40 meV was extracted from the flat band condition indicated in the figure. (b) The band configuration of the Schottky junction at the drain contact when the back-gate bias is lower than V_{FB}, which is labeled (I) in (a). (c) The band configuration at the flat band condition, which is labeled (II) in (a). (d) The band configuration when the back-gate bias exceeds V_{FB}, which is labeled (III) in (a) [18]

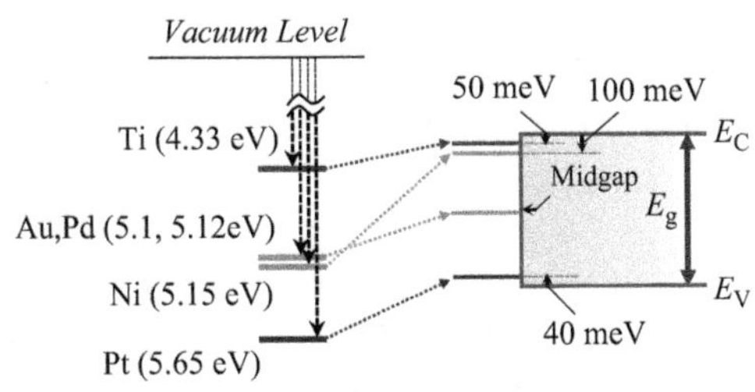

Figure 5.11 Schematic illustration of the band alignment in a Schottky junction of the metal/$MoTe_2$ interface, which suggests the weak Fermi level pinning effect in the metal/$MoTe_2$ Schottky junction [18]

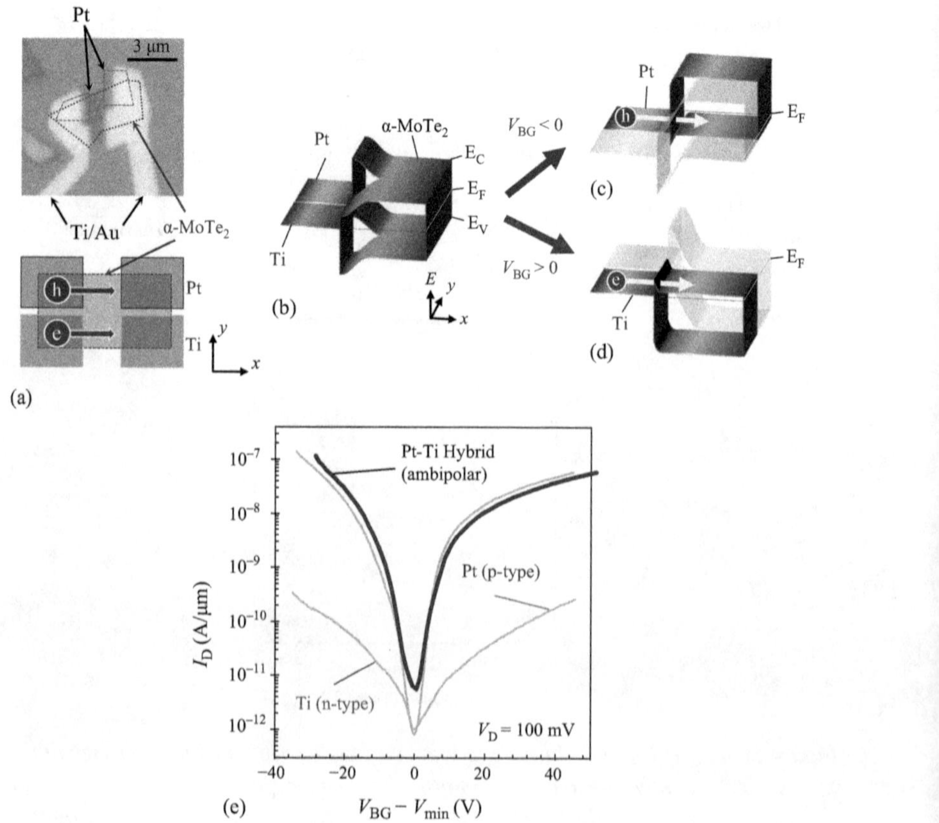

Figure 5.12 (a) Optical micrograph of the fabricated Pt–Ti hybrid contact device. This structure is also illustrated schematically in the lower picture. (b) Band diagram of the Schottky junction of the source contact of the Pt–Ti hybrid device. (c) Holes are injected from Pt in the negative gate bias case. In this case, the device behaves as a p-FET. (d) Electrons are injected from Ti in the positive gate bias case. In this case, the device behaves as an n-FET. (e) Back-gate modulation of drain currents in the Pt contact (p-type), Ti contact (n-type), and Pt–Ti hybrid contact (ambipolar) devices [18]

except for Ni. In the case of Ni, it is supposed to be ambipolar considering the work function which is close to that of Au or Pd; however, as shown in Figure 5.11, it is apparently n-type. One possible explanation should be that, in the cases of highly reactive metals like Ti or Ni, Fermi level pinning effect is still strong and the pinning level is pinned below the conduction band edge just similarly to MoS_2. On the other hand, in the cases of less reactive metals of Au, Pd and Pt, Fermi level pinning effect induced by the interface layer should be weaker than Ti or Ni, resulting in Fermi levels of metals close to the valence band edge.

5.4.3 Enhanced ambipolarity in $MoTe_2$

A polarity-controllable transistor can be formed by taking n-type and p-type contacts for source and drain sides, respectively, since electrons are injected from source and holes are injected from drain. However, in this device configuration, there remains a large barrier for extracting carriers from the channel. And also, the direction of current flowing will be fixed to one direction, which will induce some limitation in circuit design. This limitation in the fixed current direction should be avoided by adopting an idea of "hybrid contact" on a $MoTe_2$ channel, as explained below. For a bidirectional current flow, both electrons and holes must be injected from both of source and drain under a field effect control of the junctions from the gate electrodes. A hybrid contact configuration illustrated in Figure 5.12(a) is the solution for fulfilling this requirement, in which both p-type and n-type contacts of Pt and Ti, respectively, are placed in parallel on the both sides of source and drain. The band alignment at the junction of the hybrid contact is illustrated in Figure 5.12(b)–(d). In this hybrid contact regime, electrons are injected from Ti contact when the junction is negatively biased from the gate, and holes are injected from Pt contact when the junction is negatively biased. Figure 5.12(a) shows the picture of the device, in which a few layer $MoTe_2$ flake is placed on a 285-nm-thick SiO_2 layer, and contacts of Pt and Ti are attached in parallel. The I_D–V_{BG} curve of this device is shown in Figure 5.12(e), where I–V curves of Ti contact and Pt contact devices are superimposed on it. A thick solid line is the I–V curve of the Pt–Ti hybrid contact one. I–V curve of this device follows the on-currents of Pt in the p-branch and Ti in the n-branch. These results indicate that electrons are injected from Ti contact with positive gate bias, while holes are injected from Pt contact with negative gate bias, suggesting that the hybrid contact worked as expected. Thus, an ideal ambipolar transistor is enabled by a combination of ideal p-type and n-type contacts based on the idea of hybrid contacts. An ideal polarity-controllable transistor can be realized only on a perfect ambipolar transistor, and the idea of this hybrid contact device is one of the crucial techniques for it.

5.5 Conclusions

This chapter highlights the application of MX_2 type 2D materials to the polarity-controllable transistors, especially focused on the polarity control of carriers injected into TMDC materials. One of the bottlenecks in realization of polarity-controllable transistors consists in carrier injection of both electrons and holes into intrinsic semiconductor channel through Schottky junctions, and therefore, a novel method overcoming this bottleneck had been desired. Here, it was reviewed that the new kind of semiconducting materials of TMDCs has a promising feature for this purpose, and also, $MoTe_2$, which is one of the TMDC family, has a great potential for polarity controllable transistors for its weak Fermi level pinning effect.

References

[1] A. K. Geim, I. V. Grigorieva, Nature 499, p. 420 (2013).
[2] D. Jariwala, V. K. Sangwan, L. J. Lauhon, T. J. Marks, M. C. Hersam, ACS Nano 8, p. 1102 (2014).
[3] G. Fiori, F. Bonaccorso, G. Iannaccone, *et al.*, Nat. Nano 9, p. 768 (2014).
[4] Q. H. Wang, K. Kalantar-Zadeh, A. Kis, J. N. Coleman, M. S. Strano, Nat. Nano 7, p. 699 (2012).
[5] S. Z. Butler, S. M. Hollen, L. Cao, *et al.*, ACS Nano 7, p. 2898 (2013).
[6] B. Radisavljevic, A. Radenovic, J. Brivio, V. Giacometti, A. Kis, Nat. Nanotechnol. 6, p. 147 (2011).
[7] V. Podzorova, M. E. Gershenson, Ch. Kloc, R. Zeis, E. Bucher, Appl. Phys. Lett. 84, p. 3301 (2004).
[8] Y.-F. Lin, Y. Xu, S. T. Wang, *et al.*, Adv. Mater. 26, p. 3263 (2014).
[9] S. Das, H.-Y. Chen, A. V. Penumatcha, J. Appenzeller, Nano Lett. 13, p. 100 (2013).
[10] S. Kim, A. Konar, W.-S. Hwang, *et al.*, Nat. Commun. 3, p. 1011 (2012).
[11] K. Uchida, S. Takagi, Appl. Phys. Lett. 82, p. 2916 (2003).
[12] H. Fang, M. Tosun, G. Seol, *et al.* Nano Lett. 12, p. 3788 (2012).
[13] H. Fang, M. Tosun, G. Seol, *et al.* Nano Lett. 13, p. 1991 (2013).
[14] M. Yamamoto, S. Nakaharai, K. Ueno, and K. Tsukagoshi, Nano Lett. 16, p. 2720 (2016).
[15] S. Das, M. Demarteau, A. Roelofs, Appl. Phys. Lett. 106, p. 173506 (2015).
[16] G. V. Resta, S. Sutar, Y. Balaji, *et al.*, Sci. Rep. 6, p. 29448 (2016).
[17] S. Nakaharai, M. Yamamoto, K. Ueno, Y.-F. Lin, S.-L. Li, K. Tsukagoshi, ACS Nano 9, p. 5976 (2015).
[18] S. Nakaharai, M. Yamamoto, K. Ueno, K. Tsukagoshi, ACS Appl. Mat. Int. 8, p. 14732 (2016).

Chapter 6

Three-independent gate FET's super steep subthreshold slope

Jorge Romero-González[1]
and Pierre-Emmanuel Gaillardon[1]

6.1 Introduction

Successful scaling of *complementary metal-oxide-semiconductor* (CMOS) *field-effect transistors* (FETs) has led to remarkable improvements in the switching speed, density, and power of highly integrated circuits. As Gordon Moore predicted in 1965, the number of transistors that can fit within a single chip will continue to exponentially increase [1]. However, recent transistor generations have strayed from ideal scaling due to reliability and leakage concerns. The *International Technology Roadmap for Semiconductors* (ITRS) [2] developed scaling projections that show alarming chip power consumption below the 5-nm technology node. Current and future designs have to be power limited—bringing about the creation of dark silicon where a fraction of a chip is turned off to meet power constraints. Using ITRS's scaling factors and Intel's technological data [3], current predictions exceed 50% dark silicon in the 7-nm chip technology node [4,5]—no longer scaling the nominal supply voltage without impacting the performance.

In 1974, Dennard *et al.* observed that power consumption per chip area stayed constant with Moore's law. As the technology node advanced, the feature size, supply voltage (V_{DD}), threshold voltage (V_T), and OFF-current (I_{OFF}) all scaled at the same rate. Under the same area consideration, Dennard scaling model [6] predicted the benefits of increasing transistor count as power density remained constant. The trend of decreasing feature size over new generations continued to scale. However, the ideal scaling model did not take into consideration the physical limitations of MOSFETs such as the impact of subthreshold leakage on the overall chip power. At advanced technology nodes, reducing V_T results in an exponential increase in leakage current—coming from a MOSFET's minimum *subthreshold slope* (SS) value of $\approx$60 mV/dec at room temperature. This meant that the threshold voltage V_T stopped scaling at approximately 0.6 V to conserve leakage power and consequently placed a lower bound on V_{DD}.

[1] Department of Electrical and Computer Engineering, University of Utah, USA

The fundamental limitation of ~60 mV/dec, coming from the thermionic (temperature dependent) injection of electrons, in conventional MOSFETs, is a major obstacle in increasing the performance of low-power electronics. However, there are alternative technologies that are capable of breaking the ~60 mV/dec limit. These devices, commonly known as steep slope devices, utilize different switching mechanisms that are not limited by the thermal SS limit. *Tunnel FETs* (TFETs) [7] are promising transistors that utilize quantum-mechanical band-to-band tunneling. These devices offer an improved ON-current/OFF-current (I_{ON}/I_{OFF}) ratio and reduced power consumption compared to a standard MOSFET. However, their large band-to-band tunneling can produce small ON-current densities—leading to large switching delays. Lower bandgap channel materials can improve the performance of TFETs [8]. *Negative-Capacitance FETs* (NCFETs) [9] use the negative capacitance of a ferroelectric layer to provide a gate voltage amplification—leading to steeper SS capabilities. This mechanism occurs when the slope of the polarization versus electric field is negative (under stable configuration). However, these devices have experimentally demonstrated hysteresis behavior in circuit implementations [10]. *Impact-ionization MOS* (IMOS) [11] are great candidates due to their super-steep SS capabilities based on an impact ionization (II) switching mechanism. The main drawback of IMOS transistors is the large supply voltage, negative bias requirement to sustain the breakdown process, and reliability issue of hot carriers. Device structures with better electrostatic control over the channel and lower bandgap materials do help scale the voltage down. *Nano-electromechanical FETs* (NEMFETs) [12,13] are also promising candidates for low-power applications. They utilize the ultra-abrupt movement of a mechanical beam to achieve steep SS while enhancing the I_{ON}/I_{OFF} ratio. They are able to suppress short channel effects in the OFF-state mode and remove gate leakage current in the ON-state. However, the switching speed, rather than being limited by the intrinsic delay, is limited by the speed of the mechanical gate movement [13]. While it is possible to scale the supply voltage, integration of NEMFET technology into advanced silicon CMOS will come as a challenge.

As an attempt to replace CMOS in low-power applications, the *functionality-enhanced devices* (FEDs) [14–16] extend the functional diversity of a single transistor. In particular, the *three-independent-gate FET* (TIGFET) [16] technology utilizes three MOS gate terminals between metallic source and drain regions—replacing the highly doped S/D regions in MOSFETs with Schottky barriers. This design allows for dynamic control of the polarity [15] and dynamic control of the threshold voltage [16]—allowing for novel architecture designs. While having promising benchmarking performances, the technology also offers steep SS capabilities. The principle switching mechanism consists of a positive feedback induced by weak II to allow for very steep transitions—experimentally demonstrated steep SS of 6 mV/dec over five decades of drain current [17].

In this chapter, the steep SS capabilities of TIGFET technology are studied. In Section 6.2, the principles of SS in MOSFETs are examined. In Section 6.3, a brief background of TIGFET technology is added. In Section 6.4, the working principle of TIGFETs is discussed. In Section 6.5, the structure and fabrication techniques are listed. The experimental SS dependency on voltage and temperature is given in

Section 6.6. Device simulations are used to show SS dependency on device dimensions and voltage in Section 6.7, and the chapter is concluded in Section 6.8.

6.2 Subthreshold slope

In a MOSFET, the subthreshold region occurs when the gate bias (V_G) is below the threshold voltage (V_T) and the channel surface is in weak inversion. The slope of drain current ($\log_{10}$ (I_d)) versus gate bias, in this mode, shows how fast the transistor can be turned on. This figure of merit is known as the SS and is heavily dependent on the switching mechanism of the device. In MOSFETs, the switching process involves the thermionic (temperature-dependent) injection of electrons over a potential barrier—leading to a fundamental SS limit of ≈60 mv/dec at room temperature [18]. A lower bound on SS is particularly important since it places a scaling limit on the supply voltage—reducing the performance for low-power applications in advanced technology nodes.

The subthreshold slope, *SS*, is defined by how much V_G is required to change I_d by one order of magnitude (expressed in mV/dec) or simply $\partial V_G/\partial(\log_{10} I_d)$. The relationship can be expanded into two terms (*n* and *m*) as shown in (6.1).

$$SS = \underbrace{\frac{\partial V_G}{\partial \psi_S}}_{n} \cdot \underbrace{\frac{\partial \psi_S}{\partial(\log_{10} I_d)}}_{m} \cong \underbrace{\left(1 + \frac{C_d}{C_{ox}}\right)}_{n} \cdot \underbrace{\ln(10)\frac{\kappa T}{q}}_{m}, \tag{6.1}$$

$$SS_{min} = 1 \cdot \frac{\kappa T}{q} \ln(10) \cong 60 \text{ mV/dec}\Big|T = 300 \text{ K}, \tag{6.2}$$

where ψ_S is the surface potential, $\kappa T/q$ is the thermal voltage, C_d is the depletion capacitance, and C_{ox} is the oxide capacitance. The term n $(1 + C_d/C_{ox})$ is commonly referred as the transistor's body factor, while the term n $(\ln(10)\kappa T/q)$ refers to the thermal excitation of carriers.

In order to reach the minimum theoretical SS (SS_{min}) of MOSFETs, a body factor (*n*) of 1 must be achieved, as shown in (6.2). To further decrease SS_{min} of any device, the body factor may be reduced below 1, or the injection factor (*m*) may be reduced below $\ln(10)\kappa T/q$. As an example, the switching mechanism of quantum-mechanical band-to-band tunneling in TFETs [7] reduce the *m* factor—enabling lower SS. The negative capacitance within NCFET's ferroelectric layer [9] or instability points between the electrical and mechanical forces in NEMFETs [12] scale down the *n* factor to obtain very steep SS. Furthermore, TIGFET technology lessens the *n* factor by allowing a positive feedback induced by weak II to occur—leading to a very steep subthreshold slope of 6 mV/dec over five decades of drain current [17]. These devices are conceptually plotted in Figure 6.1 along with a theoretical SuperFET that may enable a reduction in the *m* and *n* factors to achieve steep SS.

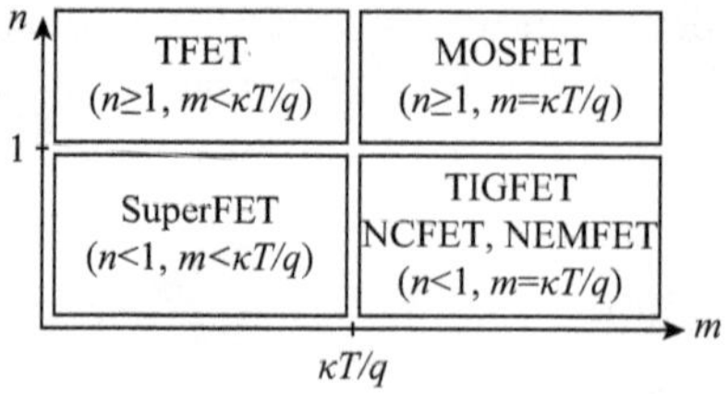

Figure 6.1 Conceptual illustration of switching capabilities for the various technologies. A device may achieve sub-thermal SS if n is less than 1 or if m is less than ln(10)κT/q

6.3 TIGFET background

Introduced in 2005, the carbon nanotube FETs [19] with dual gates pioneered reconfigurable technology as an attempt to eliminate ambipolar characteristics and achieve pure n- and p-type devices. By exploiting this ambipolarity technology with silicon nanowires [14,15], *double-independent gate FETs* (DIGFETs) increased the functionality of a single transistor—giving rise to FEDs that allowed for more complex and novel circuit architectures. As a final extension of FED technology, the TIGFET introduced dynamic control of the threshold voltage [16] and dynamic control of the SS [17]. We refer the reader to Chapter 10 "BCB Benchmarking for Three-Independent-Gate Field Effect Transistors" for a detailed analysis of TIGFET's circuit architecture and how they compare to other Beyond-CMOS devices.

As illustrated in Figure 6.2, a typical TIGFET device has three independent MOS gate contacts on a semiconductor channel between metallic source and drain regions. The *control gate* (CG) regulates the channel conduction while the *polarity gate at source* (PGS) and the *polarity gate at drain* (PGD) modulate the Schottky barrier to respectively allow electrons or holes to flow into the channel. As illustrated in Figure 6.2, TIGFET technology can be implemented with multiple silicon nanowires [14,16]. However, there are multiple other structures that sync well with TIGFET technology, such as silicon fin structures [17] and two-dimensional (2D) materials [20–23].

The MOSFETs' primary source of resistance comes from contacting the doped source/drain regions with metal interconnects as well as the increasing parasitic resistance contribution between the channel and S/D regions (extension resistance). The latter is particularly important as the cross section of the channel decreases using state-of-the-art nanowire and fin structures in current node technologies. That said, in TIGFET transistors, traditional ohmic contact resistances are no longer a primary concern of current limitation. Instead, TIGFET transistors primary resistance contribution comes from the metal–semiconductor junction's specific contact resistance. This contact resistance (different from a traditional ohmic contact resistance) is dependent on the barrier height, temperature, Richardson constant, and planks constant [24].

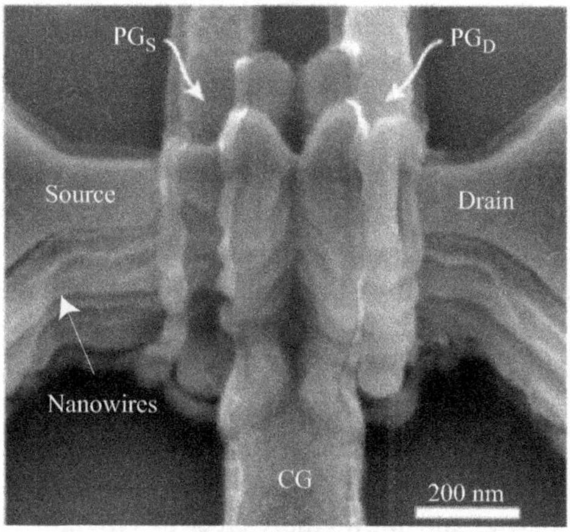

Figure 6.2 Scanning electron microscope image of a vertically stacked silicon-nanowire TIGFET device [16]

6.4 TIGFET working principle

Most TIGFET devices operate along the same set of principles. As shown in Figure 6.3, the thickness of the Schottky barriers (at the source/drain contacts) is modulated by the potential of PGS and PGD—allowing for different types of carriers to flow. Electrons are injected through the barrier at the source side when PGS is high ("1"), and holes are injected through the barrier at the drain side when PGD is low ("0"). The potential at CG is used to regulate the current by modulating a potential barrier in the center of the channel. In the n-type configuration, the device switches between the OFF-state and the ON-state when the potential at CG changes from low ("0") to high ("1"), as shown in Figure 6.3(a). In the p-type configuration, the polarity gates are set low, while the potential at CG transitions between high ("1") and low ("0") to turn the device ON, as shown in Figure 6.3(b).

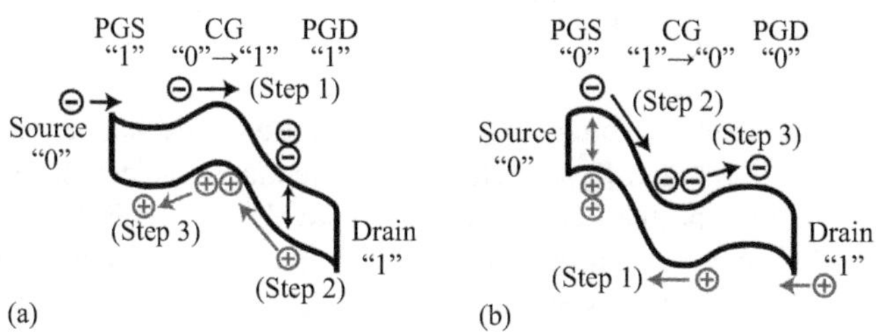

Figure 6.3 The conceptual switching mechanism of TIGFET technology to achieve steep SS by the help of impact ionization induced by positive feedback in the (a) n-type and (b) p-type configuration

The steep subthreshold swing capability of TIGFET transistors is illustrated by the various steps in Figure 6.3(a) and (b). Let us first discuss the n-type configuration, as shown in Figure 6.3(a). The TIGFET device is in the standard OFF-state configuration where PGS is set high ("1"), CG is set low ("0"), and PGD is set high ("1"). In this state, the Schottky barrier on the source side is lowered, and this enables electrons with sufficient energy to pass through but are stopped by the potential barrier in the center of the device. Now, once the CG voltage starts increasing, and if the electrons can gather enough energy, then a weak II occurs (step 1) and electron/hole pairs are generated. Some of the holes that are generated become trapped under the CG region (step 2). This accumulation of holes creates a positive potential at this location and helps the energy bands fall down faster which, in turn, creates more electrons for II, forming a positive feedback [25]. At this point, the energy bands near the PGS region also fall due to the increasing number of generated holes in this region (step 3). This lowers the Schottky barrier and allows more electrons to tunnel through. Step 3 also helps maintain the potential well between PGS and CG for the accumulation in the CG region. The I_{ON}/I_{OFF} ratio is improved since this mechanism occurs until the transistor is completely on [17]. The same technique can be transposed with the p-type configuration, as shown in Figure 6.3(b), but with holes as the majority carriers.

The behavior has been shown before in floating-body silicon-on-insulator (SOI) MOSFETs [26] except that TIGFETs allows for additional feedback. The correct expression of the subthreshold channel current (I_d) and the corresponding minimum SS (SS_{min}) are as follows:

$$I_D = I_0 \exp\left(\frac{qV_g}{n\kappa T}\right) \tag{6.3}$$

$$SS_{min} = n \cdot \frac{\kappa T}{q} \ln 10 \tag{6.4}$$

where

$$n = \frac{1 + r}{1 + r((d(V_{B,CG} - V_{B,PGS}))/dV_{CG})} \quad \text{and} \quad r = \frac{C_d}{C_{ox}}, \tag{6.5}$$

where $\kappa T/q$ is the thermal voltage, C_d is the depletion capacitance, and C_{ox} is the oxide capacitance. $V_{B,CG}$ and $V_{B,PGS}$ refer to the potential in the center of the channel below the CG region and the PGS region, respectively. The body factor expression in (6.5) now takes into account the effect on the channel's charge when carriers are generated by II. This equation is best described as follows. First, assuming there is no II present, a linear change in V_{CG} will cause a linear change in the barrier height in the center of the channel. This leads to a unity n factor and the SS_{min} approaches the thermal limit. Now, if weak II does occur near the CG–PGD region, a nonlinear effect in the potential difference between $V_{B,CG}$ and $V_{B,PGS}$ is seen and it becomes possible to have $d(V_{B,CG} - V_{B,PGS})/dV_{CG} > 1$—leading to $n < 1$ and SS_{min} below thermal limit. This behavior is written in (6.6) as a function of V_{CG}.

$$\frac{d(V_{B,CG} - V_{B,PGS})}{dV_{CG}} \simeq \frac{\eta}{n} \cdot \frac{1}{1 + (I_{gt}/I_{gi})}$$

$$\propto \frac{\eta}{n} \underbrace{\exp\left[\frac{q}{\kappa T}\left(\frac{V_{CG}}{n} - \frac{V_{B,CG} - V_{B,PGS}}{\eta}\right)\right]}_{I_{gi}} \cdot \underbrace{\exp\left(-\frac{\beta_i l}{V_{DS}}\right)}_{I_{gt}} \quad (6.6)$$

where I_{gi} is the II current (dependency on V_{CG} and $V_{B,CG} - V_{B,PGS}$) and I_{gt} is the OFF-state leakage current (dependency on V_{DS}). β_i is a constant and l is a structural parameter. η is an ideality factor between the region controlled by the CG and PGS regions.

These set of equations allow us to predict how the SS will behave for varying potential voltages and temperature for a TIGFET device. For instance, in the OFF-state leakage current, if V_{DS} increases linearly, the $d(V_{B,CG} - V_{B,PGS})/dV_{CG}$ term will increase. On the other hand, the II current is exponentially dependent on the difference between V_{CG} and $V_{B,CG} - V_{B,PGS}$. As expected, this term will increase when $V_{B,CG} - V_{B,PGS}$ becomes smaller than V_{gs} (with the ideality factors taken into account). In other words, the $d(V_{B,CG} - V_{B,PGS})/dV_{CG}$ term will increase if the rate of change of V_{CG} decreases the $V_{B,CG} - V_{B,PGS}$ factor. In the next section, a silicon-based finFET device with TIGFET technology is fabricated, and its switching characteristics show an SS_{min} of 3.4 mV/dec and an average SS of 6.0 mV/dec over five decades of current swing [17].

6.5 Structure and fabrication of TIGFETs

The capability of TIGFET technology to show SS below MOSFET's thermal limit has been shown experimentally using a silicon-based fin channel [17]. In this section, we present the details of the device structure and its fabrication steps.

6.5.1 Device structure and fabrication

The capability to achieve steep-slope behavior in FED devices comes from the voltage isolation between the CG and the polarity gate terminals. The realization of a TIGFET with its two polarity gates connected together is known as the DIGFET. The DIGFET device structure under consideration uses a silicon-based fin structure with gate at the CG and *Schottky barrier bias* (SBB) at both polarity gates as shown in Figure 6.4. The same working principle is applied here—the gate controls the potential barrier in the center of the channel while the SBB regions modulate the Schottky barrier thicknesses.

The process flow to fabricate the proposed fin-based structure [17] is shown in Figure 6.5(a). Alignment markers are patterned on an SOI substrate with lightly *p*-doped 340 nm silicon layer and a 2 μm *buried oxide* (BOX) layer. E beam lithography is used to etch an 800 nm long fin with 340 nm height and 50, 60, and 70 nm width channels as shown in Figure 6.5(b). Next, an oxidation step is used to form

Figure 6.4 Sketch of a TIGFET device with a fin-based silicon channel between nickel silicide source and drain pillars

a 15 nm layer of SiO_2 surrounding the fin. Polysilicon deposition, PR coating, and UV exposure form the SBB regions. After a second oxidation, polysilicon deposition, PR coating, and UV exposure, the gate is self-aligned to the SBB regions. Silicon nitride (SiNi) spacers are added to isolate the SBB region from the gate electrode. To create the contacts at the source and drain regions, a 20-nm nickel layer is deposited by sputtering and annealing (20'@200°C + 20'@300°C + 20'@400°C) to produce NiSi. NiSi is used to form ideal midgap work functions between the metallic S/D and the silicon channel. Selective wet etching in hot piranha solution is used to remove unreacted nickel.

The SEM image of the final structure is shown in Figure 6.5(c). The gate length (L_{gate}) is 200 nm, the height of the fin (H_{fin}) is 340 nm, and the width of the fin (W_{fin}) is 40, 50, or 60 nm. The SiO_2 thickness (t_{ox}) at the SBB region is 15 nm, and at the central gate is 30 nm.

6.6 SS dependency in TIGFETs

In this section, the voltage-dependent and temperature-dependent SS characteristics of a FED device are examined with experimental results and theoretical equations. The same fin-based TIGFET device will be used as discussed in Section 6.5.

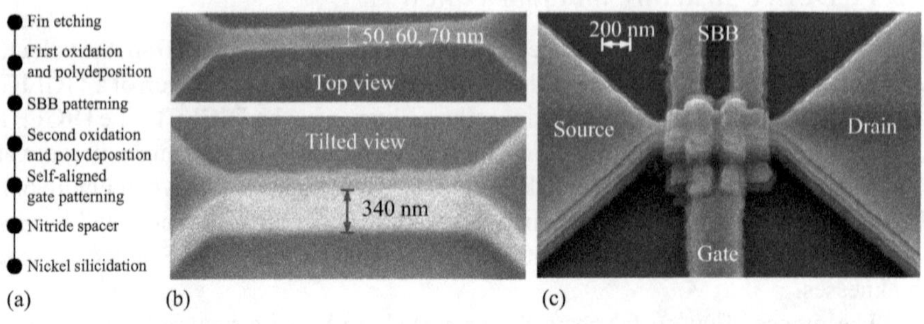

Figure 6.5 Fabrication of the Si-based fin TIGFET device for the super steep SS experimentation [17]: (a) Process flow. (b) SEM image of the fin. (c) SEM image of the final device. Channel length between source and drain is ~600 nm and the final W_{fin} are 40, 50, and 60 nm considering the silicon consumed during oxidation

6.6.1 Experimental dependency on voltage

The measured results of the 40 nm W_{fin} n-type device features V_{DS} and V_{sbb} sweeping, as shown in Figure 6.6. The first plot (Figure 6.6(a)) shows the drain current (I_d) versus gate voltage (V_g) with SBB bias (V_{sbb}) set to 5 V while sweeping V_{DS} from 1 to 5 V. The corresponding SS_{min} is displayed for each sweep. The results show that the TIGFET device performed below the thermal limit of MOSFETs (60 mV/dec) for V_{DS} values between 2 and 5 V and reached a minimum SS value of 3.4 mV/dec using the largest value of V_{DS} (5 V). The inset of the first figure shows no hysteresis as V_{CG} is swept forward and backward. The second figure (Figure 6.6(b)) plotted the corresponding SS values versus I_d for $V_{DS} = 3$, 4, and 5 V. The dashed line shows the thermal limit of 60 mV/dec. This relationship helps to show the ranges of I_d that enable SS_{min} below the thermal limit of MOSFETs. The preferable curve would be a tub-like image with a flat bottom such as the case when V_{DS} is set to 5 V. This curve shows ~3.4 mV/dec sweep for five decades of current. The final figure (Figure 6.6(c)) shows the I–V plot when the SBB potential is set to 3, 4, and 5 V at a constant V_{DS} value of 3 V. The I_{ON}/I_{OFF} ratio favorably increases and the SS favorably decreases when SBB potential increases. The SS_{min} metric transitions between 60 mV/dec when $V_{sbb} = 3$ V to 44 mV/dec when $V_{sbb} = 5$ V. Note that, while the SS_{min} does fall below the thermal limit of MOSFETs for the range of voltages, it does not fall to the impressive 3.4 mV/dec when $V_{DS} = 5$ V. In other words, the V_{DS} potential, rather than V_{sbb}, affects the electrostatic fields substantially within the device. Our previous theoretical calculations agree with the voltage dependency measurements.

Next, the n-type and p-type current symmetry characteristic of a TIGFET device is evaluated in Figure 6.7. In this experiment, the I–V (I_d versus V_g) curves are plotted for the n-type case when V_{sbb} is set to 4 V and for the p-type case when V_{sbb} is set to −2 V. The n-type device shows $> 10 \times 10^6$ I_{ON}/I_{OFF} ratio with minimum SS of 38 mV/dec, while the p-type device shows $> 10 \times 10^7$ I_{ON}/I_{OFF} ratio and a minimum SS of 75 mV/dec. These results show that the switching transitions are still comparable with the polarity control within a single TIGFET transistor.

6.6.2 Origin of voltage dependency

The results obtained from the I–V measurements, as presented in Section 6.6.1, show that as V_{DS} or V_{sbb} increases, the SS_{min} decreases. Here, two methods are used to describe why SS_{min} might be affected by the gate voltages.

6.6.2.1 Body effect

The subthreshold region of a transistor is said to be linearly dependent on the body effect factor. For TIGFET transistors, this factor is dependent on how fast the body potential changes (between the inner regions of gate and SBB near source) as the potential at gate is swept and changes the device between the ON-state and OFF-state. As shown in (6.6), the $d(V_{B,CG} - V_{B,PGS})/dV_{CG}$ term is dependent on V_{DS} and on $V_{B,CG} - V_{B,PGS}$. The extracted $dV_{B,CG} - V_{B,PGS}/dV_{CG}$ versus V_{DS} plot for the n-type device is shown in Figure 6.8(a). As expected, as V_{DS} increases, $V_{B,CG} - V_{B,PGS}$ increases—leading to a decrease in SS_{min}. Note that the rate of change is highest when

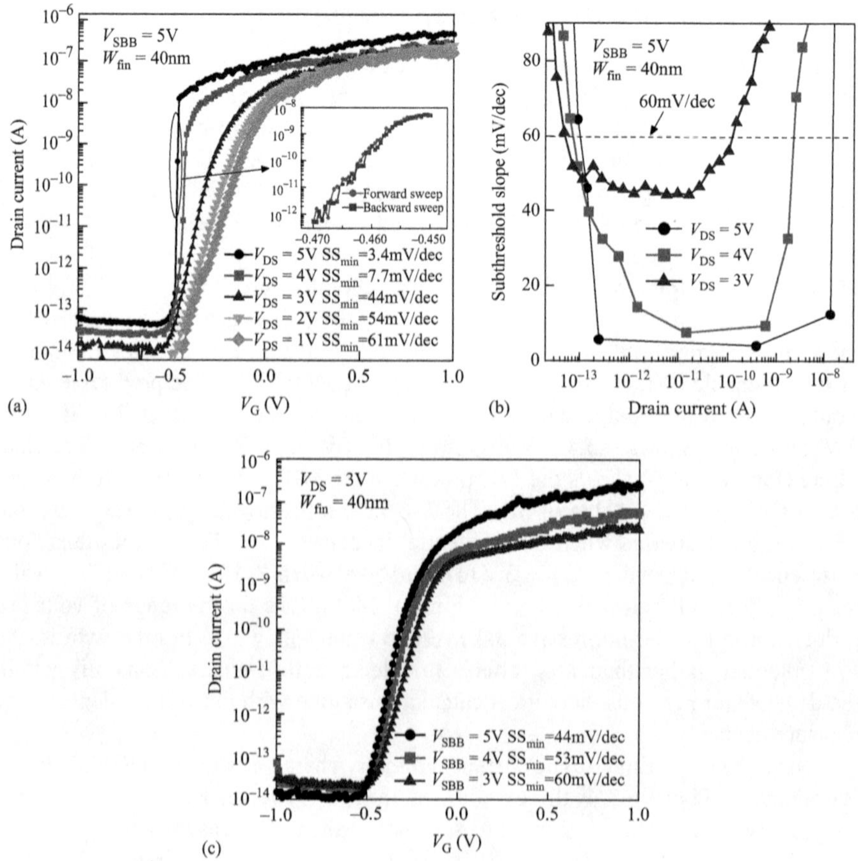

Figure 6.6 Measured results of the fin-based TIGFET at room temperature [17]. (a) I_d versus V_g sweeping V_{DS}. Inset: double sweep showing negligible hysteresis. Minimal SS= 3.4 mV/dec for $V_{DS} = 5$ V. (b) SS versus I_d sweeping V_{DS}. (c) I_d versus V_{CG} sweeping V_{sbb}

V_{DS} is between 3 and 4 V. This is shown experimentally since there is a considerable jump in the SS_{min} (between 44 and 7.7 mV/dec). A high V_{DS}, in other words, means stronger fields within the channel—leading to an increased rate of II and a lower SS_{min}. This will be further examined in Section 6.7.1.

6.6.2.2 Activation energy

Another method of extracting the steep SS is through the activation energy (known as the barrier height of the Schottky barriers) E_a [28]. The Arrhenius plot was used to extract E_a within different operation points. In a MOSFET, the thermal SS limit occurs when the gate voltage has complete control over the channel barrier. Similarly in TIGFETs, $dE_a/dV_g = 1$ means that any change in the gate voltage corresponds to

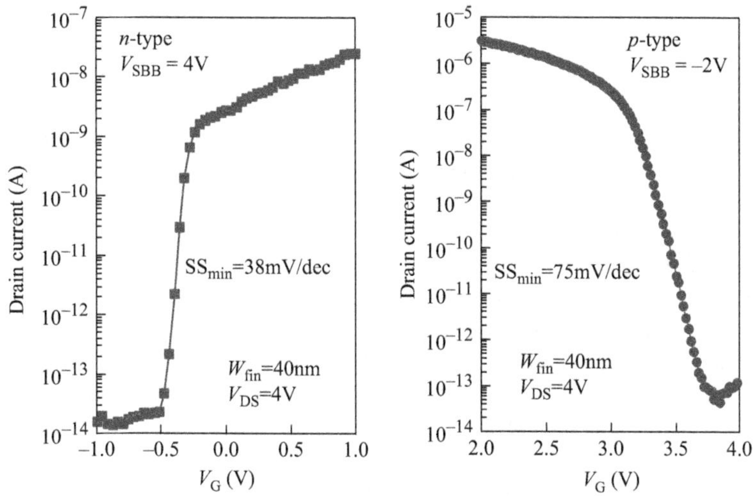

Figure 6.7 Measured (a) n-type and (b) p-type I–V curves for the same fin-based TIGFET device at room temperature [17]

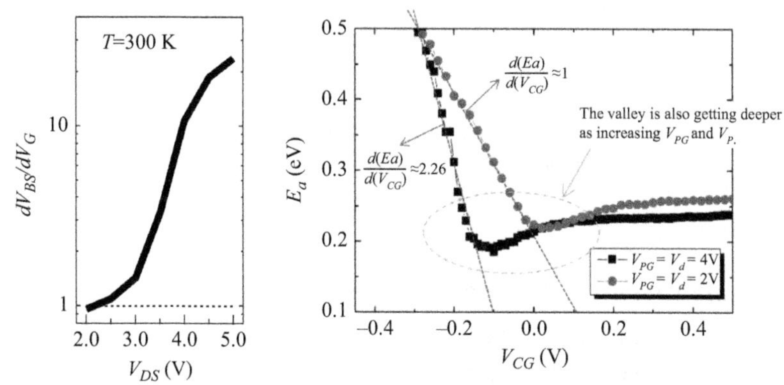

Figure 6.8 Extracted characteristics of the fin-based TIGFET device: (a) dV_{BS}/dV_{GS} versus V_{DS} [27] and (b) E_a versus V_{CG} [28]

a linear change in the barrier height. This relationship means that SS_{min} approaches the thermal limit and, in other words, the carrier transport mechanism is controlled by thermionic emission over the gate barrier. However, whenever a change in E_a exceeds the change in V_g, different carrier transport mechanisms take place such as the weak II followed by positive feedback that allows for sub-thermal SS_{min}. This leads to $dE_a/dV_g > 1$. In Figure 6.8(b), as the potential at V_{DS} and V_{sbb} increases from 2 to 4 V, the dE_a/dV_g factor increases from $\approx$1 to $\approx$2.26. The impact of stronger fields within the channel leads to a larger dE_a/dV_g factor, and thus sub-thermal SS_{min} is possible.

The E_a valley, that occurs for both of the curves in Figure 6.8(b), can be explained by the accumulation of holes that are emitted into the source-side energy barrier. To best explain this property, Figure 6.2(b) shows the mechanism of II with induced positive feedback. In the top bands (n-type configuration), the accumulated holes under the gate bands may fall into the source side and effectively reduce the effective barrier height for electrons and therefore enhance the electron current. The positive feedback effect, in turn, provides enhanced electron current under the right conditions. Next, as the CG voltage continuous to increase, the bands fall, the weak II rate reduces, the holes are depleted from the source side, and the effective barrier height at source increases ($V_g > -0.1$ V). Note that the final E_a is smaller for larger V_{DS} and V_{sbb}. This leads to larger ON-currents and steeper SS for larger voltages.

Next, let us discuss how V_{sbb} affects the SS of the device. As shown in Figure 6.6(c), an increase in V_{sbb} (from 3 to 5 V) for the n-type device leads to a smaller SS_{min}. This device characteristic is explained by how V_{sbb} affects the Schottky barrier at the source side. When V_{sbb} increases, the Schottky barrier becomes thinner and the bands near the SBB at source region fall lower. This theoretically leads to higher ON-currents (since more electrons can tunnel through the Schottky barrier) and lower OFF-currents (due to higher potential difference between the SBB at source and gate regions). These are experimentally demonstrated in Figure 6.6(c) since the ON-currents increase, while the OFF-currents decrease for larger values of V_{sbb}. Now, the increase in electron flow into the source Schottky barrier leads to more electrons for II to occur. When the bands near the SBB at source region lowers, the potential difference between SBB at source and gate increases and thus the hole concentration increases; this leads to a higher hole concentration that will be dragged to the SBB at source region when the device switches (refer to step 3 in the steep SS mechanism of TIGFETS in Section 6.4). These two effects are combined and lead to a beneficial decrease of SS_{min}.

6.6.3 Experimental dependency on temperature

The same device structure as (discussed in Section 6.5.1) is used to experimentally determine the effect of temperature on the SS. Figure 6.9(a) shows the I–V curve of the n-type mode for temperatures ranging between 100 and 380 K when V_{sbb} and V_{DS} are set to 5 V. When the temperature is set to 100 or 200 K, the ON-current and OFF-current are drastically reduced. When the temperature is set from 300 to 380 K, high ON-current values are seen with decent pA OFF-currents. The minimum SS of these readings are all below the thermal limit of MOSFETs as shown in Figure 6.9(b). In this plot, as the temperature decreases and as V_{DS} increases, the minimum SS decreases. At approximately $V_{DS} = 2$ V, the SS of the device approaches the thermal limit for each temperature. The trend lines in Figure 6.9(b) show a reduced sensitivity to temperature for $V_{DS} > 3$ V as compared with the thermal limit line.

6.6.4 Origin of temperature dependency

The n-type measurements, as presented in Figure 6.9(b), show lower SS_{min} for lower temperature and reduced SS_{min} sensitivity to temperature for large V_{DS}. These device

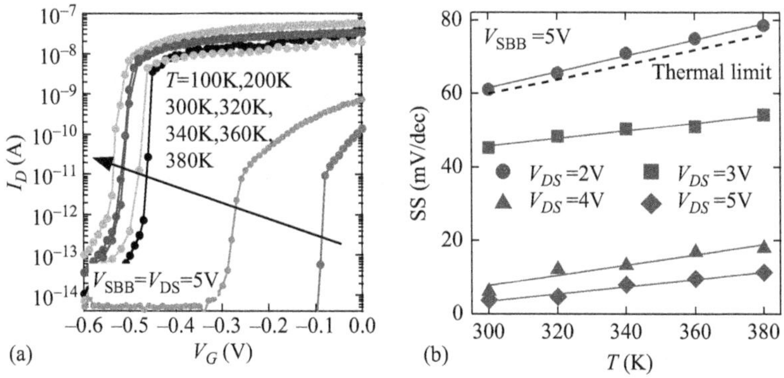

Figure 6.9 Experimental data on the fin-based TIGFET device [27]: (a) n-type experimental I–V curves ranging temperature between 100 and 380 K. (b) n-Type minimum SS versus temperature for different V_{DS}

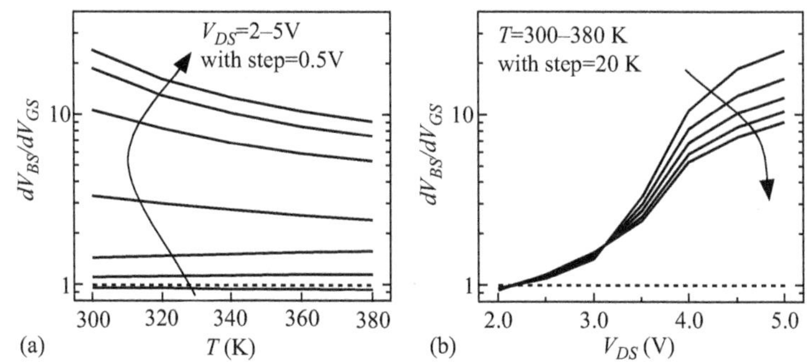

Figure 6.10 Extracted dV_{BS}/dV_{GS} from the n-type device using different (a) temperature and (b) V_{DS} values [27]

characteristics can be explained by (6.6). First, an increase in temperature decreases the term $d(V_{B,CG} - V_{B,PGS})/dV_{CG}$ which causes the SS_{min} to increase. This is also shown in the extracted dV_{BS}/dV_{GS} versus temperature plot in Figure 6.10(a), where $dV_{BS} = d(V_{B,CG} - V_{B,PGS})$. When the temperature increases, the CG electrode has a lower effect on the potential at mid-point of the channel, meaning the II worsens, and the SS_{min} tends to increase. Second, there is a stronger effect on the SS_{min} for higher fields than for lower temperatures. This is shown more clearly in Figure 6.10—the dV_{BS}/dV_{GS} term is more strongly affected by V_{DS} in Figure 6.10(a) than by the temperature in Figure 6.10(b). This suggests SS_{min} has a low sensitivity to temperature at high V_{DS} and is in agreement with the experimental results from Figure 6.9(b).

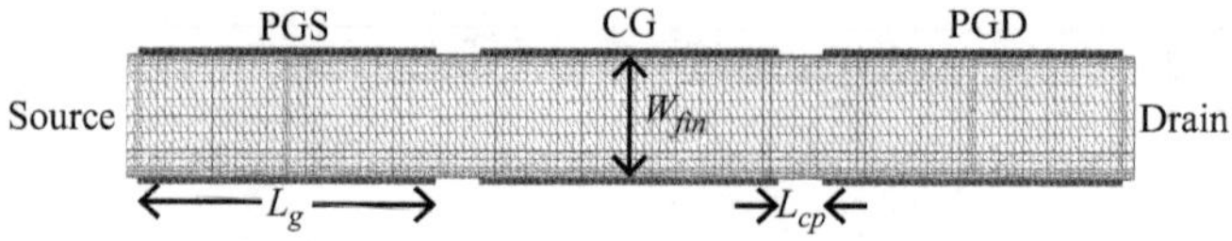

Figure 6.11 TIGFET device-level 2D structure coming from TCAD Sentaurus

6.7 SS behavior from device simulations

The steep SS characteristic of TIGFET devices was studied using the TCAD Sentaurus simulation tool from Synopsys Inc. The channel structure and gate voltages were varied using a 2D slice of a Si-based fin TIGFET. The drain current versus CG voltage was plotted to record the steep slope behavior, while the density of holes and electrostatic potential along the channel were extracted to visualize the switching mechanism of TIGFET devices. This study is an extension of [17] by using TCAD Sentaurus to predict SS_{min} behavior under various TIGFET device parameters. In this section, we refer to the polarity gate at source as PGS, polarity gate at drain as PGD, and control gate as CG.

6.7.1 TCAD Sentaurus design

The physical model used in this work solved Poisson's equations and continuity equations for both electrons and holes. The model used Fermi statistics, the Slotboom bandgap narrowing model, the band-to-band tunneling model in the source and drain electrodes, and the barrier lowering model for Schottky contacts. These settings were extracted from calibrated TCAD models performed in [15] and show well-designed short channel effects below the 45 nm technology node [29]. In order to trigger the steep-slope characteristics of TIGFET devices, the van-Overstraeten–de-Man avalanche model was applied. The associated parameters were calibrated to achieve steep slope characteristics using the dimensions and voltages as presented in [17].

The TIGFET device structure under consideration is a 2D slab of a silicon fin with 100 nm gate length (L_g), 15 nm separation between gates (L_{cp}), 40 nm gate width (W_g), and 2 nm SiO_2 (τ_{ox}) thickness, as shown in Figure 6.11. In this section, the subthreshold characteristics of TIGFET devices are investigated by studying SS dependency on L_g (particularly long- and short-channel effects) and V_{DS} voltage.

6.7.2 SS dependency on long-channel devices

Let us first discuss how large channel dimensions affect the SS of the TIGFET device. As shown in Figure 6.12(a), the length of the gates (L_g) are swept between 60 and 130 nm. The curves for the 60, 70, and 80 nm devices show increasing number of decades of current in the subthreshold region for increasing gate lengths. However, for the devices with L_g greater than 80 nm, the number of decades with steep SS_{min} decreases as L_g increases—leading to a decrease in the average SS_{min}. The left gray

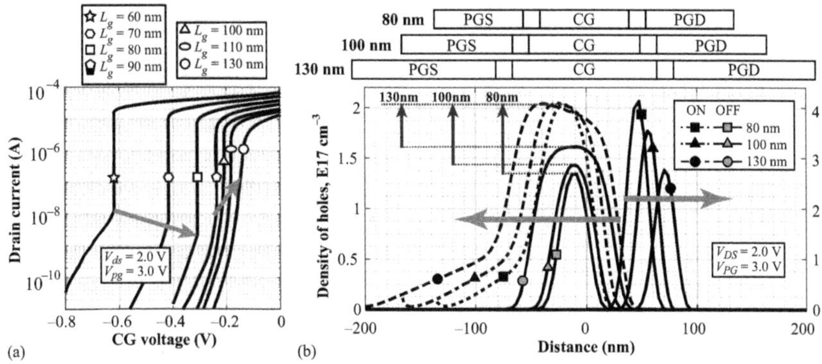

Figure 6.12 *(a) The I–V curve sweeping the length of gates (L_g) between 60 and 130 nm with $L_{cp} = 15$ nm, $V_{DS} = 2$ V, and $V_{PG} = 3$ V. (b) The impact ionization rate (right y-axis) for electrons in the ON-state versus position along the channel for L_g = 80, 100, and 130 nm. The abstract representation for the 80, 100, and 130 nm TIGFET device is shown above the plot to help illustrate the position along the channel ($x = 0$ corresponds to the center of the channel below the CG region). The hole density (left y-axis) versus position along the channel using the same devices. The upper curves (dashed lines) correspond to the ON-state and the lower curves (solid lines) correspond to the OFF-state*

arrow in Figure 6.12(a) shows the increasing number of decades of current for the 60, 70, and 80 nm devices, while the right gray arrow shows the decreasing SS decade count for the 90, 100, 110, and 130 nm devices. To understand the reason behind this effect, the density of holes and the II rate are plotted versus position along the channel for the 80, 100, and 130 nm devices as shown in Figure 6.12(b). The II rate, plotted at the moments before the devices turn OFF, decreases for larger gate lengths. This can be explained by the decrease in the electric fields within the channel when the channel length increases—leading to a decrease in the II rate. On the other hand, plotting the density of holes along the channel shows a secondary effect. First, let us look at the OFF-state curves in the density of holes plot (solid lines). The density of holes decreases linearly for lower gate lengths, while the corresponding ON-currents remain relatively the same. These results show that larger gate lengths can accumulate larger density of holes in the OFF-state. As the devices turn ON, electrons that enter from the source Schottky barrier gather enough energy to overcome the center barrier and create electron/holes pairs in the channel region between CG and PGD. At this time, the hole density in the CG region and between the CG and PGS region increases drastically. This is shown in the ON-state curves (dashed lines) of Figure 6.12(b). Note that the ON-state hole density remains constant while sweeping L_g (approximately 2×10^{17}). Our findings show that a larger change in hole density between the ON-state and the OFF-state allows for steeper slopes. Thus, having larger gate lengths

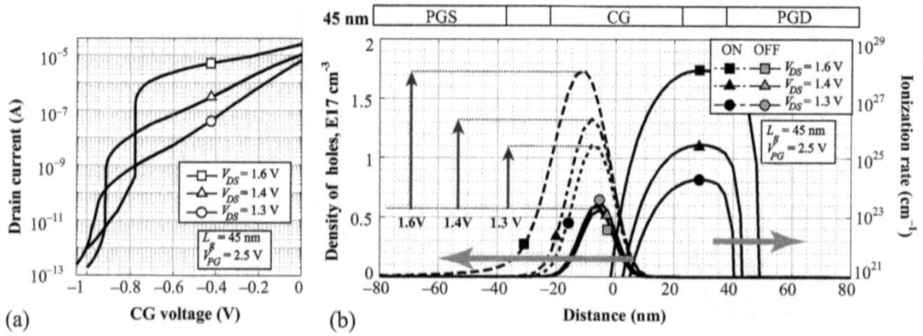

Figure 6.13 *(a) $I_d - V_{CG}$ curves by sweeping the drain-source voltage (V_{DS}) between 1.6 and 1.3 V with $L_g = 45$ nm, $L_{cp} = 15$ nm, $V_{PG} = 2.5$ V. (b) The impact ionization rate (right y-axis) for electrons in the ON-state versus position along the channel with the same voltage sweep. The abstract representation for the 45 nm TIGFET device is shown above the plot to help illustrate the position along the channel. The hole density (left y-axis) versus position along the channel for the ON-state and OFF-state modes (moments before the devices turns OFF or ON, respectively)*

allows for an increase in the hole density in the OFF-state, while the hole density remains constant in the ON-state. This leads to worse transition rates between ON- and OFF-states and gives a larger SS_{min}. This finding is also supported when V_{DS} is swept in Figure 6.13. However, the results there show an increase in the density of holes in the ON-state (rather than the OFF-state) for larger V_{DS}.

6.7.3 SS dependency on voltage

The drain current versus CG voltage is plotted in Figure 6.13(a) for V_{DS} values of 1.6, 1.4, and 1.3 V while L_g is equal to 45 nm and V_{PG} is equal to 2.5 V. The curves show that the average SS_{min} favorably decreases for higher V_{DS}. To explain this result, the density of holes and the II rate are plotted in Figure 6.13(b). As expected, a linear increase in V_{DS} increases the fields within the channel, and this leads to an exponential increase in the II rate in the regions between CG and PGD. This leads to steeper SS characteristics when using larger V_{DS}—as already observed in the TIGFET measurements [17]. On the other hand, the density of holes along the channel can show us a secondary effect as shown in Figure 6.13(b). Here, the curves in the ON-state and in the OFF-state correspond to the moments before the device turns OFF or ON, respectively. First, when the device is in the OFF-state, there is an accumulation of charges near the CG region—similar to Figure 6.12(b). However, now the density of holes in the OFF-state does not change when we sweep V_{DS}. This can be explained by the fact that potential difference between PGS and CG remains relatively the same when V_{DS} increases, and this causes the density of holes to stay constant when the device is in the OFF-state. As V_{CG} increases, and the TIGFET transistor turns ON, the hole density increases. The hole

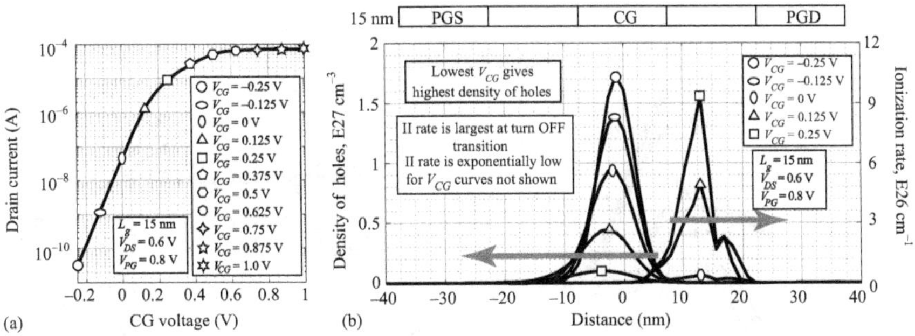

Figure 6.14 (a) I–V curve for the 15 nm device using $V_{DS} = 0.6$ V and $V_{PG} = 0.8$ V. (b) The impact ionization rate (right y-axis) for electrons versus position along the channel for various values of V_{CG}. The maximum II rate occurs at the moment when the device is about to turn OFF (II rate is negligible for V_{CG} values not shown). The hole density (left y-axis) versus position along the channel. Maximum occurs when the device is in the OFF-state. The abstract representation for the 15 nm TIGFET device is shown above the plot to help illustrate the position along the channel

density increases for larger values of V_{DS} as shown in the dark arrows of Figure 6.12(b). These parameters lead to a faster transition rate (steeper SS_{min}) for larger V_{DS} due to larger hole density transition between the ON-state and OFF-state.

6.7.4 SS dependency on TIGFET's short-channel effects

The simulations performed using the 2D model in TCAD Sentaurus show steep SS_{min} for gate lengths as low as 45 nm. For gate lengths below 45 nm, the SS does not go below the thermal limit of MOSFETs. However, showing the plots for these devices is helpful to understand the physics behind the TIGFET steep slope capabilities. The $I_d - V_{CG}$ curve is plotted in Figure 6.14(a), and the II rate for electrons and the density of holes along the channel are plotted in Figure 6.14(b) for the 15 nm device with $V_{DS} = 0.6$ V and $V_{PG} = 0.8$ V. These voltage values were selected to give the best turn ON transition (~80 mV/dec here). The II rate along the channel is shown in Figure 6.14(b) for multiple values of V_{CG}. The II rate magnitude is too small to show for V_{CG} values between 1.0 and 0.375 V and between −0.125 and −0.25 V. This is different than the steep slope devices ($L_g > 45$ nm) since the II rate is largest when the device is ON. For the $L_g = 15$ nm device, the II rate drastically increases when $V_{CG} = 0.25$ V and decreases linearly for decreasing V_{CG}. While the II does increase when the device is about to turn OFF, SS_{min} does not fall below the thermal limit. To understand why this occurs, let us look at the density of holes along the channel as shown in Figure 6.14(b). The hole density is plotted for multiple values of V_{CG}. The density is maximum when the device is in the OFF-state and decreases as the device turns to the ON-state. This is the opposite effect when TIGFETs show steep

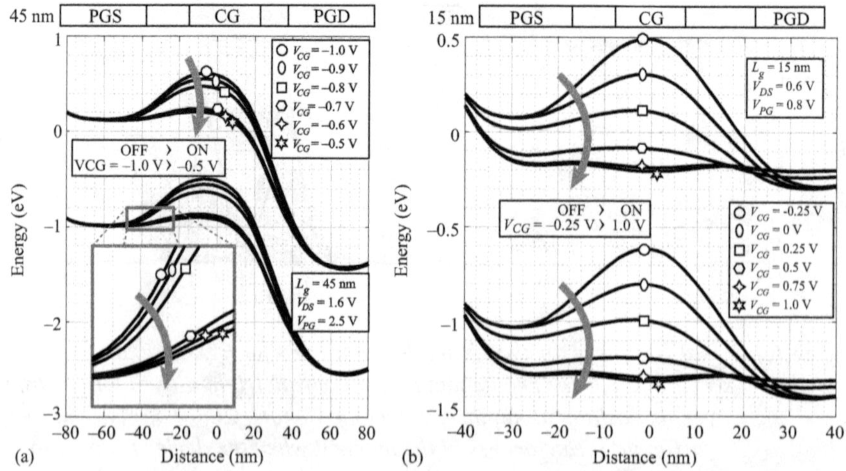

Figure 6.15 Conduction and valence band energy along the channel by varying the CG voltage for the same (a) 45 nm device ($V_{DS} = 2\,V$, $V_{PG} = 3\,V$) and for the same (b) 15 nm device ($V_{DS} = 0.6\,V$, $V_{PG} = 0.8\,V$). The abstract representation for the 45 and 15 nm TIGFET device is shown above the plot to help illustrate the position along the channel

slope behavior (lower than the thermal limit of MOSFETs). While II did occur within the channel, not enough of it was enabled for electron/hole pairs to generate and increase the density of holes within the channel. Thus, the transition between the ON-state and the OFF-state is dependent on how fast the CG region barrier is lowered and is not helped by the density of holes that is generated by the electron–hole pairs between CG and PGS.

Let us look at the band diagrams to understand why smaller gates give worse switching transitions. The conduction and valence band diagrams for the 45 nm device ($V_{DS} = 1.6\,\text{V}$, $V_{PG} = 2.5\,\text{V}$) and for the 15 nm device ($V_{DS} = 0.6\,\text{V}$, $V_{PG} = 0.8\,\text{V}$) are shown in Figure 6.15(a) and (b), respectively. The bands for the 45 nm device come as expected. There is a big change in the bands between the V_{CG} values of −0.7 and −0.8 V since the device switches at this point. Here, we note that the band positioning in the ON-state near the PGS region does not move much between the V_{CG} values of −0.7 and −0.5 V. This comes from the high hole concentration near this region in the ON-state due to the II occurring near the region between CG and PGD. If one can image to remove the II component, the bands would fall down since no holes can enter from the Schottky barrier in the drain side, and the potential at PGS would attempt to bend the bands down. This effect is seen in the bands for the 15 nm device. As shown earlier in Figure 6.14(b), the II rate for electrons in the ON-state of the 15 nm device is low. Since there are not enough electron–hole pairs generated in the ON-state, the bands fall down as CG potential increases as shown in Figure 6.15(b). On a second note, the hole density is highest when the device is completely OFF. The holes that are trapped in the CG region in the OFF-state are dragged into the source Schottky barrier as V_{CG} increases since the bands at PGS become lowered.

6.8 Conclusion

The device characteristics of TIGFET technology, in particular the steep SS, have been studied in detail in this chapter. Notably, we have (1) summarized TIGFET's working principle and fabrication techniques [17], (2) reviewed experimental demonstrations of SS_{min} in TIGFETs with respect to voltage [17,28] and temperature [27], and (3) performed device-level simulations to display and thoroughly explain the SS capabilities of TIGFETs with respect to channel length and voltage. Our results allow us to develop an in-depth explanation into the origin of steep SS and the potential limitations of SS in TIGFETs due to short-channel effects.

Acknowledgment

This material is based upon work supported by the National Science Foundation under Grant No. 1644592 and under CAREER Award No. 1751064.

References

[1] Moore GE. Cramming more components onto integrated circuits. IEEE Solid-State Circuits Society Newsletter. 2006 Sep;11(3):33–35.

[2] International Technology Roadmap for Semiconductors [Online]; 2015. Available from: http://www.itrs2.net/.

[3] Borkar R, Bohr M, Jourdan S. Advancing Moore's Law in 2014!, Intel; Aug 2014. Available from: https://www.intel.com/content/www/us/en/silicon-innovations/advancing-moores-law-in-2014-presentation.html

[4] Esmaeilzadeh H, Blem E, Amant RS, *et al.* Dark silicon and the end of multicore scaling. IEEE Micro. 2012 May;32(3):122–134.

[5] Henkel J, Khdr H, Pagani S, *et al.* New trends in dark silicon. In: 2015 52nd ACM/EDAC/IEEE Design Automation Conference (DAC); Jun 2015. p. 1–6.

[6] Dennard RH, Gaensslen FH, Rideout VL, *et al.* Design of ion-implanted MOSFET's with very small physical dimensions. IEEE Journal of Solid-State Circuits. 1974 Oct;9(5):256–268.

[7] Ionescu AM, Riel H. Tunnel field-effect transistors as energy-efficient electronic switches. Nature. 2011 Nov;479(7373):329–337.

[8] Trommer J, Heinzig A, Mühle U, *et al.* Enabling energy efficiency and polarity control in germanium nanowire transistors by individually gated nanojunctions. ACS Nano. 2017 Jan;11(2):1704–1711.

[9] Salvatore GA, Bouvet D, Ionescu AM. Demonstration of subthreshold swing smaller than 60 mV/decade in Fe-FET with P(VDF-TrFE)/SiO2 gate stack. In: 2008 IEEE International Electron Devices Meeting; Dec 2008. p. 1–4.

[10] Seo J, Lee J, Shin M. Analysis of drain-induced barrier rising in short-channel negative-capacitance FETs and its applications. IEEE Transactions on Electron Devices. 2017 Dec;64(4):1793–1798.

[11] Toh EH, Wang GH, Zhu M, *et al.* Impact ionization nanowire transistor with multiple-gates, silicon-germanium impact ionization region, and sub-5 mV/decade subthreshold swing. In: 2007 IEEE International Electron Devices Meeting; Dec 2007. p. 195–198.

[12] Abele N, Fritschi R, Boucart K, *et al.* Suspended-gate MOSFET: bringing new MEMS functionality into solid-state MOS transistor. In: IEEE International Electron Devices Meeting, 2005. IEDM Technical Digest.; 2005 Dec. p. 479–481.

[13] Kam H, Lee DT, Howe RT, *et al.* A new nano-electro-mechanical field effect transistor (NEMFET) design for low-power electronics. In: IEEE International Electron Devices Meeting, Dec 2005. IEDM Technical Digest.; 2005. p. 463–466.

[14] Heinzig A, Slesazeck S, Kreupl F, *et al.* Reconfigurable silicon nanowire transistors. Nano Letters. 2012 Nov;12(1):119–124.

[15] Marchi MD, Sacchetto D, Frache S, *et al.* Polarity control in double-gate, gate-all-around vertically stacked silicon nanowire FETs. In: 2012 International Electron Devices Meeting; Dec 2012. p. 8.4.1–8.4.4.

[16] Zhang J, Marchi MD, Sacchetto D, *et al.* Polarity-controllable silicon nanowire transistors with dual threshold voltages. IEEE Transactions on Electron Devices. 2014 Nov;61(11):3654–3660.

[17] Zhang J, Marchi MD, Gaillardon PE, *et al.* A Schottky-barrier silicon FinFET with 6.0 mV/dec subthreshold slope over 5 decades of current. In: 2014 IEEE International Electron Devices Meeting; Dec 2014. p. 13.4.1–13.4.4.

[18] Sze SM, Ng KK. Physics of Semiconductor Devices. 3rd ed. New York, John Wiley & Sons; 2007.

[19] Lin YM, Appenzeller J, Knoch J, *et al.* High-performance carbon nanotube field-effect transistor with tunable polarities. IEEE Transactions on Nanotechnology. 2005 Sep;4(5):481–489.

[20] Nakaharai S, Iijima T, Ogawa S, *et al.* Electrostatically reversible polarity of dual-gated graphene transistors. IEEE Transactions on Nanotechnology. 2014 Nov;13(6):1039–1043.

[21] Resta GV, Sutar S, Balaji Y, *et al.* Polarity control in WSe2 double-gate transistors. Scientific Reports. 2016 Jul;6:29448. Article.

[22] Resta GV, Agarwal T, Lin D, *et al.* Scaling trends and performance evaluation of 2-dimensional polarity-controllable FETs. Scientific Reports. 2017 Mar;7:45556.

[23] Lin YM, Appenzeller J, Knoch J, *et al.* High-performance carbon nanotube field-effect transistor with tunable polarities. IEEE Transactions on Nanotechnology. 2005 Sep;4(5):481–489.

[24] Varahramyan K, Verret EJ. A model for specific contact resistance applicable for titanium silicide-silicon contacts. Solid-State Electronics. 1996 Nov;39(11):1601–1607.

[25] Lu Z, Collaert N, Aoulaiche M, *et al.* Realizing super-steep subthreshold slope with conventional FDSOI CMOS at low-bias voltages. In: 2010 International Electron Devices Meeting; Dec 2010. p. 16.6.1–16.6.3.
[26] Fossum JG, Sundaresan R, Matloubian M. Anomalous subthreshold current-voltage characteristics of n-channel SOI MOSFET's. IEEE Electron Device Letters. 1987 Nov;8(11):544–546.
[27] Zhang J, Trommer J, Weber WM, *et al.* On temperature dependency of steep subthreshold slope in dual-independent-gate FinFET. IEEE Journal of the Electron Devices Society. 2015 Nov;3(6):452–456.
[28] Jeon DY, Zhang J, Trommer J, *et al.* Operation regimes and electrical transport of steep slope Schottky Si-FinFETs. Journal of Applied Physics. 2017 Feb;121(6):064504.
[29] Romero-Gonzalez J, Gaillardon PE. BCB evaluation of high-performance and low-leakage three-independent-gate field effect transistors. IEEE Journal on Exploratory Solid-State Computational Devices and Circuits. 2018 Apr;4:1–9.

Chapter 7
Super sensitive terahertz detectors

Mehdi Hasan[1], Ross Walker[1], Pierre Emmanuel Gaillardon[1], and Berardi Sensale-Rodriguez[1]

7.1 Principles of THz detection in FETs

7.1.1 Dyakonov–Shur model

In 1993, Dyakonov and Shur predicted the possibility of exciting electron plasma waves in short channel field effect transistors (FETs) at terahertz (THz) frequencies [1]. As gate lengths in high electron mobility transistors (HEMTs) and silicon metal-oxide-semiconductor field effect transistors (MOSFETs) are made increasingly smaller, charge transport becomes ballistic, and device channels can behave as resonators for electron plasma waves. The THz response in these devices results from plasma waves being generated in the device channel due to modulation of the electron concentration. As an example, when the transistor is biased such that the drain terminal is open circuited in AC and a THz wave is coupled between the source and gate terminals, electron plasma waves arising from perturbations in the electron density in response to the incident THz radiation can be amplified by reflections at the asymmetrically connected terminals. In general, a detector or a mixer operating at THz frequencies must have an antenna structure in order to collect incoming electromagnetic radiation with the wavelength (hundreds of microns) much greater than the longitudinal active region dimensions (typically tens of nanometers as dictated by the gate length). From this perspective, two typical detector configurations are usually studied in the literature. These configurations can be represented by one of the two equivalent circuits depicted in Figure 7.1, namely, gate coupling (Figure 7.1(a)) and drain coupling (Figure 7.1(b)). In short, channel devices (relative to the plasmon decay length), electron plasma wave oscillations result in a gate bias-dependent resonant voltage response at the drain terminal [1,2]. In sufficiently long devices or in devices with short channels but low electron mobility, the electron plasma wave oscillations are damped as they propagate along the channel and vanish before reaching the drain end; this results in a nonresonant DC voltage response, which as mentioned for the previous case results is also dependent on the gate bias [1–3].

[1]Department of Electrical and Computer Engineering, The University of Utah, USA

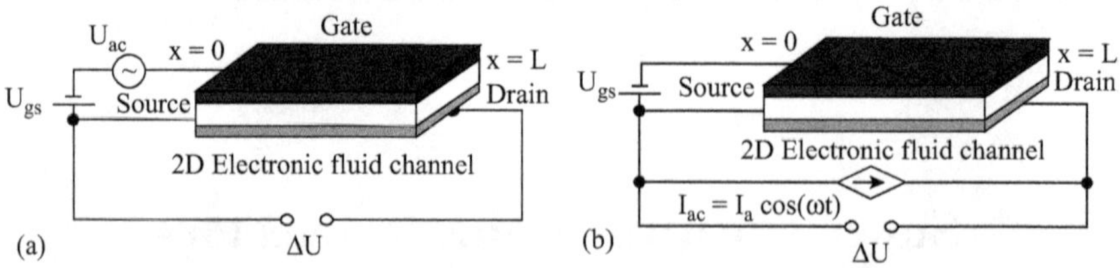

Figure 7.1 Schematic geometry for FETs operating as THz detectors. Two configurations are depicted: (a) THz-induced AC voltage between gate and source terminals and (b) THz-induced AC current at the drain. © 1996. Figure adapted, with permission, from Ref [4]

7.1.2 Theoretical formalism and modes of operation

The basic equations describing the high-frequency response of a transistor channel can be derived from the hydrodynamic equations that govern electron transport as described by Dyakonov and Shur [1,4,5]:

$$\frac{\partial v}{\partial t} + v\frac{\partial v}{\partial x} + \frac{e}{m}\frac{\partial V}{\partial x} + \frac{v}{\tau} = 0, \tag{7.1}$$

$$\frac{\partial n}{\partial t} + \frac{\partial (nv)}{\partial x} = 0, \tag{7.2}$$

where (7.1) is Euler's equation representing the flow of the electron fluid, and (7.2) corresponds to the classical continuity equation. In these equations, $v(x,t)$ is the local electron velocity, $n(x,t)$ is the local electron concentration, e is the electron charge, m is electron effective mass, $\partial V(x)/\partial x$ is the longitudinal electric field in the channel, V(x,t) being the local potential drop from gate to channel, and τ is the electron momentum relaxation time.

Assuming a gradual channel approximation [6], V and n are related by

$$n = \frac{C(V - V_t)}{e}, \tag{7.3}$$

where C is gate capacitance and V_t is the threshold voltage. Equation (7.3) is valid whenever $V \gg V_t$. In general, (7.1)–(7.3) can be solved for any boundary value problem. Boundary conditions need to be set in both AC and DC according to the actual detector configuration.

For a device biased so that, $V_{GS} \gg V_t$ and under a configuration with open boundary conditions at the drain, the detected voltage is small in relation to the gate bias overdrive. In this case, the boundary conditions at the source ($x = 0$) and drain ($x = L$) terminals are given by

$$\begin{cases} V(0,t) - \langle V(0,t)\rangle = V_a\cos(\omega t) & x = 0 \\ v(L,t) = 0 & x = L \end{cases} \tag{7.4}$$

where $\langle \cdot \rangle$ represents time average, $\langle V(0,t) \rangle = V_0$ is the gate bias overdrive, and $V_a \cos(\omega t)$ is the AC voltage induced by the THz radiation.

Solutions of (7.1)–(7.3) under the boundary conditions described by (7.4) can be expressed in the time-harmonic form

$$\begin{cases} v = \bar{v} + v_1 + v_2 + \cdots \\ V = \bar{V} + V_1 + V_2 + \cdots \end{cases} \tag{7.5}$$

where $\bar{v}$ and $\bar{V}$ represent the time-average values for electron velocity and channel potential, respectively, and v_n and V_n are their nth time harmonics. By using (7.1)–(7.5), the following equations are derived for the first harmonic:

$$\frac{\partial v_1}{\partial t} + \frac{\partial u_1}{\partial x} + \frac{v_1}{\tau} = 0, \tag{7.6}$$

$$\frac{\partial u_1}{\partial t} + S^2 \frac{\partial v_1}{\partial x} = 0, \tag{7.7}$$

where $u_1 = eV_1/m$ and $s = (eV_0/m)^{0.5}$ are the electron plasma wave group velocity. By considering solutions to (7.6) and (7.7) of the form $u_1, v_1 \propto e^{i(kx-\omega t)}$, the following dispersion is derived:

$$k = \frac{\omega}{s}\sqrt{1 + \frac{i}{\omega t}}. \tag{7.8}$$

From the above equations, it is possible to completely determine the device dynamics. As discussed in [4,7], the THz responsivity (R) of such FET fed with a matched half-wave dipole antenna can be calculated as

$$R \approx \frac{100}{V_0} F(\omega, S)\,[\mathrm{V/W}], \tag{7.9}$$

where $F(\omega, s)$ is given by

$$F(\omega, S) = 1 + \frac{2\omega\tau}{\sqrt{1+(\omega\tau)^2}} - \frac{1 + (2\omega\tau/\sqrt{1+(\omega\tau)^2})\cos(2k_0^+ L)}{\sin h^2(k_0^- L) + \cos^2(k_0^+ L)}, \tag{7.10}$$

where $k_0^{\pm} = (\omega/s)\,\{[(1+\omega^{-2}\tau^{-2})^{1/2} \pm 1]/2\}^{1/2}$. The above equations formally hold for parabolic band semiconductors; however, they can be modified to be valid for Dirac materials such as graphene if effective mass is replaced by $\hbar(\pi n)^{1/2}/v_F$, where $\hbar$ is reduced Planck's constant and v_F is Fermi velocity.

From (7.9) and (7.10), it is observed that $f(\omega, s)$, thus R, can be alternatively expressed as a function of two variables: $\omega\tau$ and $s\tau/L$. Shown in Figure 7.2 are colormaps of f as a function of $\omega\tau$ and $s\tau/L$. From here, two modes of operation are clearly distinguishable: (i) a resonant mode, in which the detector response is narrowband but exhibits very high responsivity at discrete frequencies (plasmonic resonances) satisfying the condition: $kL = (2n+1)\pi/4$, $n = 1,2,3,\ldots$, and (ii) a non-resonant mode where detection is broadband. The resonance peaks are represented by the dark lines in the upper right quadrant of Figure 7.2. Resonant detection corresponds to the situation where $\omega\tau \gg 1$ and $s\tau/L \gg 1$, while nonresonant detection

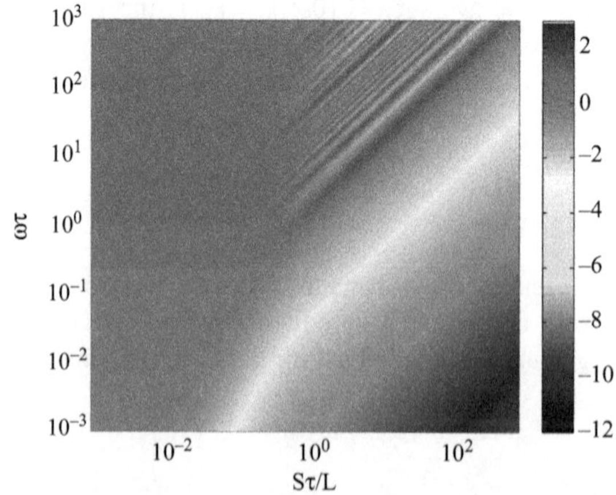

Figure 7.2 Grayscale plot of $\log_{10}(f)$ versus $\omega\tau$ and $s\tau/L$

occurs otherwise. In the region where the response is resonant, the larger $\omega\tau$ and $s\tau/L$, the larger f and thus the larger the responsivity. In practice, several conditions need to be satisfied in order to achieve resonant detection including a relatively high mobility and use of devices with relatively short gate lengths. While modern Si FETs are being designed and fabricated with very short gates, mobility of these devices at room temperature is on the order of 200–500 cm^2/V s, far too low for resonant detection. However, at cryogenic temperatures, nanoscale Si FETs may exhibit resonant response. Resonant detection has been demonstrated in high-mobility materials, including GaAs HEMTs [8], GaAs FETs [9], and GaN HEMTs [10]. Furthermore, graphene [11–20] is a promising material that might lead to resonant responses as the result of a long momentum relaxation time.

7.1.3 Nonresonant detection: principles and advantages of subthreshold biasing

At low frequencies ($\omega\tau \ll 1$) or in long channel devices ($s\tau < L$), the FET THz detector response is nonresonant. In this regime, damping prevents the device from exhibiting sharp resonant peaks and the detector response is broadband. Since overall, the detector response is heavily dependent on the charge density in the channel, having device models that are valid across all biasing space is the key. In this regard, work by Knap *et al.* [21] expanded previous modeling efforts to accurately account for the subthreshold region. In this regime, electron concentration depends exponentially on gate bias [22,23]. The channel resistance exponentially increases below threshold and may become comparable to or even higher than the input resistance of the measurement system. At this point, the response signal decreases due to a simple voltage dividing effect. The effects of load resistance and increased channel resistance in the subthreshold region were taken into account in works by Stillman *et al.* [24,25].

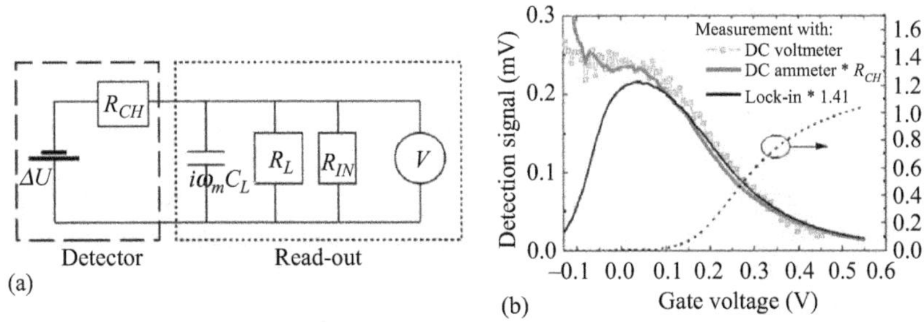

Figure 7.3 *(a) Equivalent electrical circuit used to understand loading effects. Here, V is the measured photovoltage, U is the photo-response of the transistor, R_{CH} is the channel resistance, C_L is a parasitic capacitance, ω_m is the modulation frequency, R_L is the external load resistor (if present), and R_{IN} is the input resistance of the preamplifier. (b) Comparison of the detected signal from a Si MOSFET measured with different instruments: DC voltmeter (squares), DC ammeter (multiplied by the channel resistance) (thick gray line), and lock-in at $\omega_m = 133$ Hz multiplied by $\sqrt{2}$ (thin black line). Right scale (dashed line): DC transfer characteristics for 1 mV of source-to-drain voltage. © 2011. Image extracted, with permission, from Ref [21]*

Furthermore, in [21], it is shown that the photo-response voltage in the device is given by the following phenomenological formula:

$$\Delta U = \frac{V_a^2}{4}\left[\frac{1}{\sigma}\frac{d\sigma}{dV}\right]_{V=V_g} = \frac{V_a^2}{4}\left[\frac{d}{dV_g}Ln[I_{DS}(V_g)]\right], \tag{7.11}$$

where σ represents the channel conductivity. Equation (7.11) can be derived by solving the hydrodynamic equations previously introduced. The current responsivity (R_i) for an intrinsic device is thus given by [26]

$$R_i = \frac{1}{2}\left[\frac{d}{dV_g}Ln[I_{DS}(V_g)]\right]. \tag{7.12}$$

When the device is loaded externally with an impedance Z_L, which can result from the measurement setup, the real photovoltage V, resulting from the rectification of THz radiation, is influenced by the loading effect. To interpret the loading effects, we can use the equivalent circuit of the detector loaded with the measurement setup (cables, connectors, and the preamplifier) as shown in Figure 7.3(a). The registered signal (V) by the readout circuit is thus given by [21]

$$V = \frac{\Delta U}{1 + (R_{CH}/Z_L)} = \frac{A}{1 + (R_{CH}/Z_L)}\left[\frac{d}{dV_g}Ln[I_{DS}(V_g)]\right], \tag{7.13}$$

where R_{CH} is the channel resistance, Z_L is the load impedance of the setup, and A is the coupling efficiency of the THz power to the intrinsic or ideal part of the FET, which becomes maximum for conjugate impedance matching. From these equations, we see that the responsivity is dependent on three factors, (i) responsivity of the intrinsic device, (ii) coupling of the THz power to the intrinsic device, and (iii) ratio of the channel resistance to the external load.

To experimentally establish the relation between the nonresonant THz photo-response of a FET and the channel conductivity, depicted in Figure 7.3(b) are the measured DC transfer characteristics for a Si MOSFET in the absence of THz radiation for a small source-to-drain voltage (1 mV). Photo-response measurements upon THz illumination are also depicted in Figure 7.3(b). The lock-in signal is multiplied by $\sqrt{2}$. Physically, the coupling coefficient, A, is determined by (i) the coupling efficiency of the incoming THz radiation to an antenna feeding the transistor together with (ii) the match of the antenna impedance to the impedance of the transistor. A is proportional to the power of the incoming THz radiation, but a precise determination of this factor, especially at THz frequencies, is still an unsolved problem. When fitting the maximum of the calculated photo-response (using (7.13)) to the experimentally measured DC curves for the Si MOSFET, it was found that $A = 1.25 \times 10^{-4}\ \mathrm{V}^2$.

7.2 Overview of THz detectors and state of the art

Owing to the unprecedented development in THz sources and detectors in the last decade, technologists and researchers have intensified their efforts to develop advanced THz sensing and imaging systems with superior performance and unique functionalities. One of the key elements to realize such systems is the ability to monolithically integrate high-performance semiconductor devices with THz antennas and other passive structures. Existing THz detection systems can be broadly classified according to the detection principle: thermal, optical, and semiconductor or electronic detectors. Current state-of-the-art THz direct detectors employing bolometers, pyroelectric sensors, Golay cells, and Schottky barrier diodes (SBD) have been demonstrated. However, these devices suffer from either large size, low output power, or both. These existing commercially available detectors have specific performance and application limitations arising from their design and principle of operation.

- Bolometers display best detectivity and can reach high modulation bandwidths but require cryogenic temperatures [27–29]. Bolometers are usually of high size, weight, and do not allow monolithic integration.
- Golay cells, which are simple opto-acoustic detectors capable of operating in a wide spectrum range (up to 30 THz), are also bulky and have their modulation frequency limited to a few tens of hertz [30].
- Pyroelectric detectors are characterized by high sensitivity, high bandwidth (up to 30 THz), and room temperature operation [31]. However, the modulation frequency of pyroelectric detectors does not exceed 10 kHz.
- SBDs are the prototypical electronic components for room-temperature detection of high frequency radiation, where the operating exploits the nonlinearity and

asymmetry of the current–voltage characteristic. Schottky diode detectors operate in the bandwidth up to 2–3 THz and in the gigahertz sampling frequency range. Their responsivity and noise-equivalent power (NEP) range from ~100 V/W to 1–2 kV/W and from 2 to 50 pW/Hz$^{1/2}$, respectively [32–36]. The curvature coefficient of SBDs is fundamentally limited to be less than q/kT (38.5 V^{-1} at room temperature), which limits the maximum sensitivity that can be achieved [37].

As mentioned, owing to complementary metal-oxide-semiconductor (CMOS) technology being ubiquitous, interest in using nanometer FETs for THz applications aroused in the early 1990s. However, from an experimental perspective, the real large-scale interest in using FETs as THz detectors started in the early 2000s after the first experimental demonstration of sub-THz detection in Si-CMOS FETs [38–41]. Soon after, in 2006, it was shown that Si-CMOS FETs can attain a NEP comparable to that in the best conventional room temperature THz detectors [39]. Both pioneering works have clearly stated the importance of Si-CMOS FETs, which present the advantages of room temperature operation, very fast response times, easy on-chip integration with read-out electronics and high reproducibility, leading to straightforward array fabrication. In this regard, it is worth mentioning that in solid-state technology, SBDs are the longtime workhorse for THz electronics, but these receivers have a fundamental limit to its responsivity due to the diffusion thermal limit [37,42]. FETs share this limit due to the current turn-on mechanism being thermionic emission, thus from both a responsivity and NEP perspective, these devices can provide similar performances.

7.2.1 Recent progress on nanowire-based THz detectors

Nanowire FETs are promising candidates for THz detectors, due to their one dimensionality, which modifies the spectra of allowed plasma modes. Nanowires can be easily removed from the host substrate and placed on top of a new functional one for individual contacting, even in relatively large numbers, with a simple planar technology very suitable for low-capacitance circuits. Due to their reduced size, these systems offer a very high cutoff frequency together with the possibility to be arranged in ordered arrays for imaging applications.

Both InAs and heterostructure InAs/InSb nanowire lateral FETs have been studied in the literature [43–45]. InAs nanowire FET detectors were first developed for 0.3 THz radiation by the application of both bow-tie and log-periodic antennas between the source and gate electrodes [43]. The THz response was tested in a top-incidence configuration or through the substrate using a Si lens for improved collection. A NEP of a few 10^{-9} W/Hz$^{1/2}$ was measured in these early reports [43]. The concept was then extended to a higher frequency range by appropriate scaling of the antenna geometry. In this case, a NEP of ~6×10^{-11} W/Hz$^{1/2}$ was recorded using a 1.5 THz quantum cascade laser as a source [44].

Hetero-structured nanowires, consisting of two segments (InAs and InSb forming a heterojunction in the middle), exhibited a peculiar behavior [45]. The gate electrode was located on one side of the nanowire, as it was for the InAs devices, and on both sides of the InAs/InSb heterojunction (see inset in Figure 7.4(a)). In general, the devices behaved as n-channel FETs at gate voltages above threshold $Vt = -2$ V and

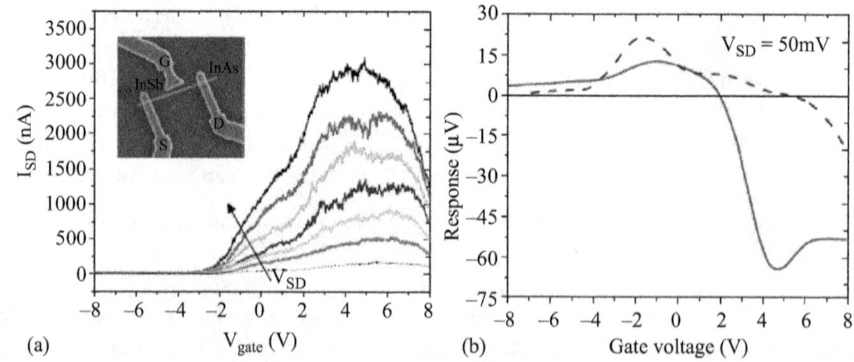

Figure 7.4 *(a) Current voltage characteristics, at different source voltages (from 10 to 60 mV) for an InAs/InSb nanowire FET. In the inset, a SEM image of the device active region. The two arms of the log-periodic antenna realized through e-beam lithography are, respectively, connected to source (S) and gate (G) electrodes, while the current is measured through the drain (D) electrode. (b) Response of the InAs/InSb nanowire to 0.293 THz radiation. Solid line shows the measured response, dashed line is obtained from transfer current–voltage characteristics. © 2012. Image extracted, with permission, from Ref [45]*

up to $V_{GS} = 4$–5 V. The $I - V$ characteristics of the device are shown in Figure 7.4(a). Furthermore, depicted in Figure 7.4(b) are (i) the measured response of the transistors to 0.293 THz radiation (solid line) and (ii) the shape of the gate voltage dependence of the response, which is predicted based on the numerical differentiation (see (7.13)) of the current–voltage characteristic (dashed line). Although the shape obtained from the current–voltage characteristic differs from the experimental one, the dependence can still be qualitatively explained by (7.13).

7.2.2 Recent progress on graphene-based THz detectors

Recently, graphene captured much interest as a material enabling realization of robust and cheap THz detectors operating at room temperature and based on the Dyakonov and Shur theory [4,5]. Graphene has indeed a very high carrier mobility, even at room temperature [12]. Furthermore, it supports plasma waves that are weakly damped in high-quality samples [13,14]. Thus, single-layer (SL) and bilayer (BL) graphene FET (GFET) plasma-based photodetectors could outperform other THz detector technologies. In the last couple of years, various graphene-based structures have been proposed, that is, SL or BL graphene–insulator–graphene heterostructure tunnel FET (GIG TFET) [46–52], vertical graphene-based hot-electron transistors [53], lateral SL or BL GFET [54], vertical cascade multiple graphene layer structures [55], etc.

A promising graphene device consists of a graphene–insulator–GFET as reported in [46]. Such devices have been fabricated and characterized using both SL and BL

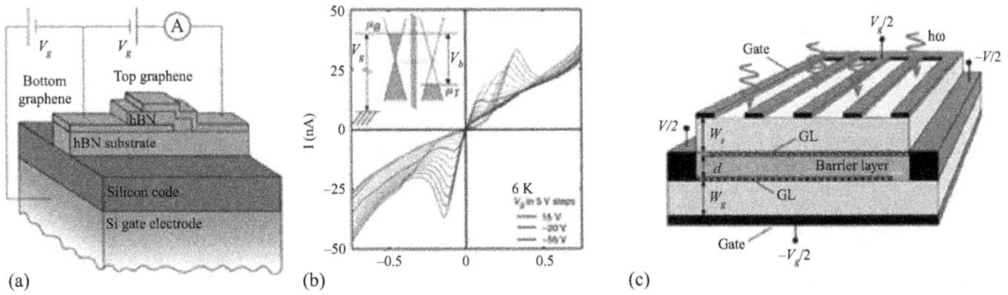

Figure 7.5 (a) Schematic of a graphene-boron nitride-graphene resonant tunneling transistor (from Ref [49], 2013), (b) measured current–voltage characteristics of such a device, the inset shows schematically the relative positions of the Fermi energies of the two graphene layers at the peak of the I (V) curve. (c) Schematic view of a GIG FET structure with "electrical doping" and top gate serving as the grating coupler. © 2014. Image extracted, with permission, from Ref [52]

graphene, see Figure 7.5(a). The DC and RF characteristics for such devices have been extensively studied [46–52,56–58] showing negative differential conductance in both SL and BL GIG FETs as depicted in Figure 7.5(b). Recent works have predicted that such GIG FET structures can show strong resonant THz detection, compared to ungated GIG FET structures, see [47,48,56]. The developed analytical models clearly show that at frequencies close to ω_p (resonant frequency) or its harmonics, the detector responsivity exhibits sharp maxima provided the plasma oscillation quality factor Q $\gg$ 1. Since, the electron plasma wave group velocity (s) depends on bias voltage V_{GS} and ω_p depends on s and L (graphene length), hence the responsivity peak can be tuned by means of either changing V_{GS} or L. Efficient coupling of THz radiation into plasmons in these structures can be done using grating gate topologies [52] as depicted in Figure 7.5(c).

Graphene has also been proposed for broadband THz detection by employing a simple top-gate antenna-coupled configuration for the excitation of over damped plasma waves in the channel of a GFET [16]. Log-periodic circular toothed as well as bow-tie antennas have been used for 0.29–0.4, 2, and 4 THz radiation detection [17–20]. Similarly, a split-bow tie antenna has been used to improve the coupling and hence responsivity at 0.6 THz [17]. Figure 7.6(a) and (b) depicts R_v (voltage responsivity) measured in exfoliated SL and BL GFETs at room temperature, for 0.3 THz, while sweeping V_{GS} from −1 to +3.5 V and modulating the THz source at 500 Hz [16]. Each curve in Figure 7.6(a) corresponds to a different relative orientation between the source electric-field polarization and the antenna axis. In the case of the SL devices, the photo-response drops rapidly with angle until it becomes almost zero when the incoming polarization is orthogonal to the antenna axis, confirming the efficacy of the dipole antenna as coupling element. Figure 7.6(c) and (d) plots R_v and NEP measured in CVD grown SL GFETs at room temperature, for 0.6 THz, while sweeping V_{GS} from −1.5 to +1.5 V [17]. Together with the expected photovoltage

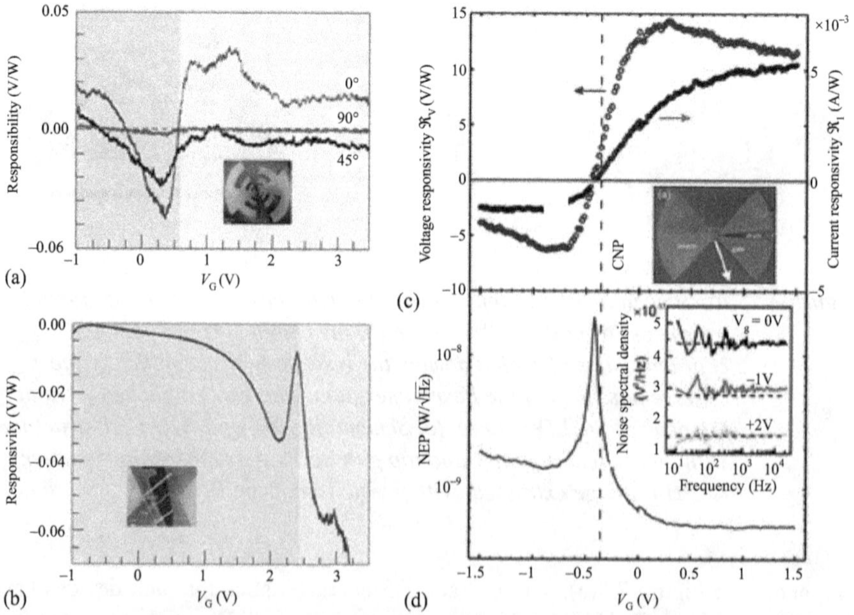

Figure 7.6 Room-temperature responsivity as a function of gate bias for detectorsbased on (a) SL-GFET (with data for different angles between the beam polarization axis and the antenna axis) and (b) BL-GFET; the detectors consist of a log-periodic circular-toothed antenna between source and gate, different background colors identify regions below and above the Dirac point. (c) Responsivity for a SL-GFET with a split bow-tie antenna, (d) NEP for the SL-GFET depicted in (c). © 2012. Figures (a) and (b) reprinted from Ref [16]; © 2014, figures (c) and (d) reprinted from Ref [19]

change in the vicinity of the Dirac point, a further sign switch is observed in all cases, suggesting a possible contribution of photo-thermoelectric origin to detection [59,60]. Similar results were obtained for detection at 1.63 and 3.11 THz using GFETs, where the drain and source contact leads worked as antennas for the incoming THz radiation [60].

7.3 Emerging FET devices for THz detection applications: dual independent gate FinFET with super-steep subthreshold slope

Multiple-independent-gate (MIG) silicon FinFETs were recently shown capable of enabling: (i) device-level polarity control, (ii) dynamic threshold modulation, and (iii) subthreshold slope (SS) tuning down to ultra-steep-slope operation. These operation mechanisms can unlock a myriad of opportunities in digital as well as analog design. In order to reduce the power consumption in digital integrated circuits, increase the trans-conductance generation efficiency (g_m/I_D) of transistors in analog circuits, and for attaining a very sensitive nonlinear response to radio frequency, transistors attaining very steep SS at room temperature are required. The SS in conventional MOSFETs is limited to >60 mV/dec due to the turn-on mechanism for current in such devices being thermionic emission. To overcome this fundamental limit, different types of field-effect transistors, based on alternative current mechanisms, have been proposed during the last decade. In this regard, tunnel FETs (TFETs) [61–65], in which the current turn-on mechanism is band-to-band tunneling rather than thermionic emission, have emerged as an attractive alternative to traditional MOSFETs. Experimental demonstrations of such devices showing *SS* below 30 mV/dec over four decades of current have been recently reported [64]. In addition to TFETs, dual-independent-gate (DIG) FinFETs [66,67] have been also recently demonstrated capable of achieving a very steep *SS* ($SS \ll 60$ mV/dec at room temperature). The reason behind this super-steep slope is a positive feedback induced by carrier multiplication due to weak impact ionization under proper biasing conditions. In this regard, experimental demonstrations of DIG FinFETs have shown *SS* as small as 3.4 mV/dec at room temperature and the possibility of attaining such small slopes over five decades of current swing [66]. Using DIG FinFET, we can overcome the major challenge of 60 mV/dec *SS* limit and thus the current responsivity limit of ~20 A/W in FETs at room temperature.

A sketch of the structure and a scanning electron microscope (SEM) image of a DIG FinFET are depicted in Figure 7.7(a). The device consists of a fin-shaped silicon channel, metallic source and drain contacts, and two independent gate electrodes. In general, in MIG FinFETs, there are two polarity gates (PGs), a source-PG (PG_S) and a drain-PG (PG_D), which modulate the Schottky barriers at the source and drain ends of the device. Moreover, a control gate (CG) controls the potential barrier in the channel and thus can govern whether the device is in ON or OFF state. In DIG FinFETs, the PG_D and PG_S terminals are electrically connected, thus leading to a single PG, as depicted in Figure 7.7(a) [66]. The device shown in Figure 7.7, which will be discussed herein, has a channel length of ~600 nm (CG length 200 nm); the fin height and width are 340 and ~60 nm, respectively [66]. Control of carrier injection at the Schottky barrier by the PG offers the possibility of arbitrarily setting the operation mode of the device as either n-type or p-type [68]. The Schottky barrier height is smaller for holes compared to electrons, so the ON state current is higher for p-type operation of the device.

Depicted in Figure 7.7(b) is the operation mechanism of the device. Application of a positive voltage on the PG creates a potential well under this terminal. When a voltage is applied on the CG, at the onset of conduction, conducting electrons can acquire enough energy so that electron–hole pairs are generated by impact ionization as indicated by step 1 in Figure 7.7(b). The electrons drift toward the drain and the generated holes accumulate in the potential well under the CG as indicted by step 2 in Figure 7.7(b). This lowers the potential barrier and provides for further electrons for impact ionization. In this process, the more electron–hole pairs that are generated, the lower the potential barrier becomes, thus setting a positive feedback. Another

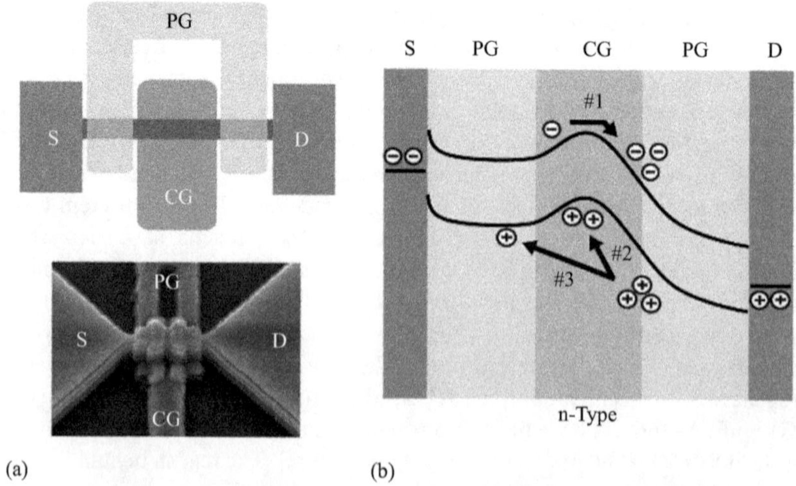

Figure 7.7 (a) Structure sketch and SEM images of fabricated DIG FinFET and (b) conceptual band diagram for an n-type device illustrating the mechanism of impact-ionization during the ON-transition. © 2014. Figure extracted, with permission, from Ref [66]

factor that further enhances this carrier multiplication process is modulation of the Schottky barrier by the holes that are swept toward the source, as depicted in step 3 in Figure 7.7(b). Because of this positive feedback, very steep SS ($SS \ll 60\,\text{mV/dec}$) is possible in DIG FinFETs.

7.3.1 A continuous compact DC model

In order to evaluate the potential of electron devices for analog and RF applications, accurate estimates for the current–voltage characteristics as well as for its derivatives are required. For instance, compact models for TFETs enabled the evaluation of the perspectives of these devices for analog applications [69,70] as well as RF detectors.

The complete current equation for the DIG FinFET can be expressed in the following form:

$$\begin{aligned} I(V_{GS}) &= C_T[b_1(V_{GS}+V_{OFF})^{c_1} H(V_{GS},V_{OFF})+1] \\ &\cdot\left[\frac{1}{2}+\left(\frac{1}{2}-b_0\right)\tan h\left(\frac{V_{GS}+V_{OFF}}{c_0}\right)\right], \end{aligned} \tag{7.14}$$

where $H(V_{GS}V_{OFF})$ is the Heaviside step function evaluated at $V_{GS}-V_{OFF}$. The above-mentioned functions and complete current equation are plotted in Figure 7.8(a). In Figure 7.8(b), results are displayed in semi-logarithmic and linear scale showing the model is in good agreement with the experimental results.

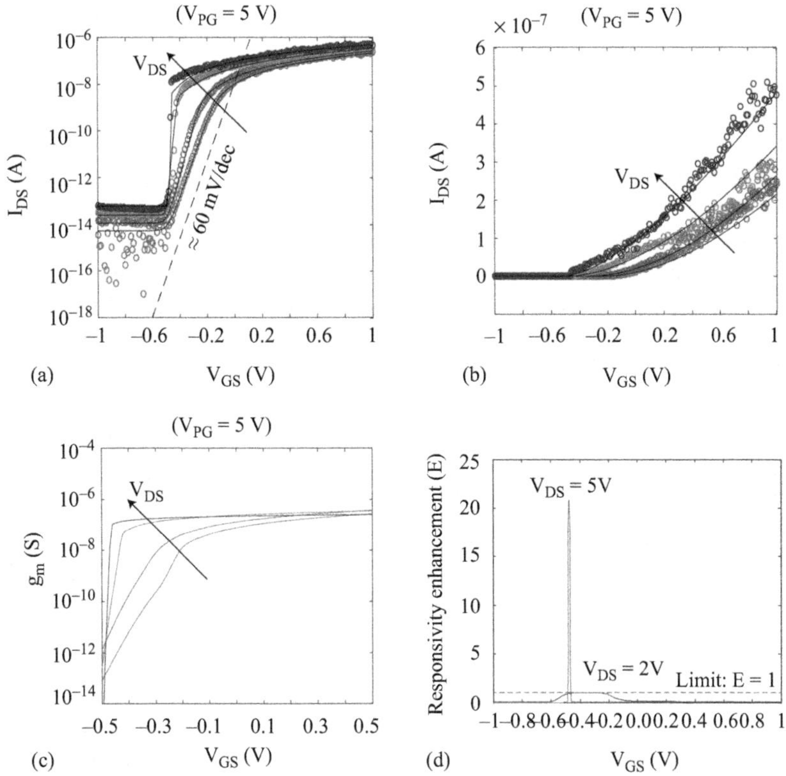

Figure 7.8 (a-b) Experimental (circles) and modeled I -VGS characteristic for a DIG FinFET (a) logarithmic scale, (b) linear scale. (c) Transconductance (gm) evaluated from the discussed compact model (VDS increases from 2V to 5V in 1V steps). (d) Estimated current-responsivity enhancement (E) as a function of VGS bias for two different drain biases for a DIG FinFET configured as RF detector. The RF responsivity is estimated from the developed DC compact model of the device. © 2017. Image extracted, with permission, from Ref [26]

Depicted in Figure 7.8(c) is a typical plot of the transconductance (g_m) as directly extracted from the numerical derivative as evaluated from the model. Assuming current detection, the following upper bound for the detector responsivity can be obtained as in (7.12):

$$R_i = \frac{1}{2}\frac{\partial \ln I_{DS}(V_{GS})}{\partial V_{GS}} = \frac{1}{2}\frac{g_m}{I_{DS}}, \tag{7.15}$$

In traditional FETs, the g_m/I_{DS} ratio is theoretically limited to values below 38.5 V^{-1}, which sets a maximum attainable responsivity. The ratio between the responsivity in

DIG FinFET detectors and the maximum attainable responsivity in traditional FETs is thus given by

$$E = \frac{R_{i,\text{DIG FinFET}}}{R_{i,\max \text{MOSFET}}} = \frac{kT}{q}\frac{\partial \ln I_{DS}(V_{GS})}{\partial V_{GS}} = \frac{kT}{q}\frac{g_m}{I_{DS}}. \tag{7.16}$$

Plotted in Figure 7.8(d) is the responsivity enhancement factor (E), as defined in (7.16), for a DIG FinFET detector configuration where the polarity-gate voltage is set to $V_{PG} = 5$ V. Two cases are analyzed, corresponding to V_{DS} biases of (i) $V_{DS} =$ 5 V, where impact ionization is significant and the device shows maximum $SS \ll$ 60 mV/dec, and (ii) $V_{DS} = 2$ V, where the SS is always >60 mV/dec. We observe that whereas in the first case, in the subthreshold region, much larger responsivity than that in traditional FET detectors is attainable, in the second case, the responsivity is always smaller than the FET limit. This is a result of the capability of attaining a smaller SS under large V_{PG} and V_{DS} biases because of impact ionization processes.

7.3.2 *Noise-equivalent power predictions*

NEP is a very important metric for a THz detector because it determines the signal-to-noise ratio of a detection system [39,71]. In general, NEP(W/$\sqrt{\text{Hz}}$) = ((Noise Spectral Density (V/$\sqrt{\text{Hz}}$))/ (Responsivity(V/W))). In order to estimate the NEP in DIG FinFET detectors, it is important to analyze the different sources of noise in such devices. Overall, the electrical noise can be divided into a thermal contribution and an excess part: $S_V = S_{V,T} + S_{V,ex}$, where the excess part consists of (i) $1/f^{\alpha}$-type flicker noise ($\alpha = 1$) and (ii) generation recombination-type noise, which varies as $1/f^2$. There are essentially two physical mechanisms behind any fluctuations in electrical current: (i) fluctuations in the mobility and (ii) fluctuations in the number of carriers. In general, the excess noise can be attributed to carrier number fluctuations stemming from charge carrier capture, release, and recombination events. In FinFETs as well as DIG FinFETs, the charge fluctuation in the gate dielectric could also induce fluctuations of the carrier mobility, giving rise to so-called correlated mobility fluctuations (CMF). When analyzing all noise contributions, it is found that 1/f noise mostly originates from the CMF due to trapping/de-trapping of channel carriers into slow gate dielectric traps [72,73]. Our estimates for all the noise contributions in the DIG FinFET, under the bias conditions at which maximum responsivity is attained, show that generation recombination and flicker noises dominate over thermal noise, as also observed by other authors [74,75]. The total (current) noise spectral density is estimated to be on the order of 10^{-22} A^2/Hz, which leads to projected NEP levels on the order of 10^{-14} W/$\text{Hz}^{0.5}$, thus two orders of magnitude better performance than current room temperature THz detectors.

7.4 Conclusions

We have discussed the physical mechanisms explaining THz detection in FETs. Such detectors can support two modes of operations: (i) resonant mode and (ii) nonresonant mode. Even though resonant photo-detection is still to be fully demonstrated at room temperature, a lot of work has been done on nonresonant photo-detection using FETs.

In this regard, nanometersize FETs operating as broadband THz detectors currently compete and are suitable alternatives to commercially available Schottky diodes. Recent studies show that FETs can reach responsivity levels, NEP, and speed of the same order of magnitude as Schottky diodes while having advantages arising from CMOS VLSI compatibility. In this regard, we have also presented an overview of recent experimental results on the detection ofTHz radiation by nanowire and GFETs.

In the nonresonant mode, responsivity is limited by the SS of the device (>60 mV/dec in traditional FETs), due to the turn-on mechanism for current in such devices being thermionic emission. From this perspective, TFETs and steep-slope Fin-FETs can be a potential route to overcome the limit of responsivity set by the SS of the device, thus overcome better performances. In addition, further improvements of FET THz detectors should relate to (i) improving the coupling of external radiation and (ii) improving the transistor design as well as its integration with impedance matching amplifiers. With transistor dimensions on the nanometer scale, direct efficient coupling through antennas is difficult due to (i) the unknown transistor input impedance in the THz frequency range and (ii) radiation coupling to the substrate instead of the transistor itself. However, the most spectacular progress can be achieved by improving the channel transport properties and reaching the resonant detection regime. Resonant detection can be more sensitive, spectrally resolved and gate voltage tunable. There are two main ways to reach this goal: (i) increase the carrier mobility and (ii) improve the device geometry. High mobility can be attained by using InSb or graphene-based channels with the carrier mobilities exceeding 10,000 cm^2/V s.

References

[1] M. Dyakonov and M. Shur, "Shallow water analogy for a ballistic field effect transistor: New mechanism of plasma wave generation by dc current," Phys. Rev. Lett., vol. 71, no. 15, pp. 2465–2468, 1993.

[2] W. Knap, M. Dyakonov, D. Coquillat, *et al.*, "Field effect transistors for terahertz detection: Physics and first imaging applications," J. Infrared Millim. Terahertz Waves, vol. 32, no. 5, pp. 618–628, 2011.

[3] W. Knap, V. Kachorovskii, Y. Deng, *et al.*, "Nonresonant detection of terahertz radiation in field effect transistors," J. Appl. Phys., vol. 91, p. 9346, 2002.

[4] M. Dyakonov and M. Shur, "Detection, mixing, and frequency multiplication of terahertz radiation by two-dimensional electronic fluid," IEEE Trans. Electron Devices, vol. 43, no. 3, pp. 380–387, 1996.

[5] M. Dyakonov and M. Shur, "Plasma wave electronics: Novel terahertz devices using two dimensional electron fluid," IEEE Trans. Electron Devices, vol. 43, no. 10, pp. 1640–1645, 1996.

[6] M. S. Shur, *Introduction to Electronic Devices*, New York: Wiley, 1996.

[7] R. E. Colin, *Antenna and Radiowave Propagation*, New York: McGraw-Hill, 1985, Ch. 5.

[8] F. Teppe, W. Knap, D. Veksler, "Room-temperature plasma waves resonant detection of sub-terahertz radiation by nanometer field-effect transistor," Appl. Phys. Lett., vol. 87, p. 052107, 2005.

[9] W. Knap, Y. Deng, S. Rumyantsev, *et al.*, "Resonant detection of subterahertz radiation by plasma waves in a submicron field-effect transistor," Appl. Phys. Lett., vol. 80, no. 18, pp. 3433–3435, 2002.

[10] A. El Fatimy, S. Boubanga Tombet, F. Teppe, *et al.*, "Terahertz detection by GaN/AlGaN transistors," Electron. Lett., vol. 42, no. 23, 2006.

[11] A. Geim and K. Novoselov, "The rise of grapheme," Nat. Mater., vol. 6, pp. 183–191, 2007.

[12] A. H. Castro Neto, F. Guinea, N. M. R. Peres, K. S. Novoselov, and A. K. Geim, "The electronic properties of graphene," Rev. Mod. Phys., vol. 81, pp. 109–162, 2009.

[13] M. Jablan, H. Buljan, and M. Soljaccic, "Plasmonics in graphene at infrared frequencies," Phys. Rev. B, vol. 80, p. 245435, 2009.

[14] A. N. Grigorenko, M. Polini, and K. S. Novoselov, "Graphene plasmonics," Nat. Photonics, vol. 6, pp. 749–758, 2012.

[15] A. Tredicucci and M. S. Vitiello, "Device concepts for graphene-based terahertz photonics," IEEE J. Sel. Quantum Electron., vol. 20, p. 8500109, 2007.

[16] L. Vicarelli, M. S. Vitiello, D. Coquillat, *et al.*, "Graphene field-effect transistors as room-temperature terahertz detectors," Nat. Mater., vol. 11, pp. 865–871, 2012.

[17] D. Spirito, D. Coquillat, S. L. De Bonis, *et al.*, "High performance bilayer-graphene terahertz detectors," Appl. Phys. Lett., vol. 104, p. 061111, 2014.

[18] M. Mittendorff, S. Winnerl, J. Kamann, *et al.*, "Ultrafast graphene-based broadband THz detector," Appl. Phys. Lett., vol. 103, p. 021113, 2013.

[19] A. Zak, M. A. Andersson, M. Bauer, *et al.*, "Antenna-integrated 0.6 THz FET direct detectors based on CVD graphene," Nano Lett., vol. 14, pp. 5834–5838, 2014.

[20] X. Yang, A. Vorobiev, A. Generalov, M. A. Andersson, and J. Stake, "A flexible graphene terahertz detector," Appl. Phys. Lett., vol. 111, pp. 021102, 2017.

[21] M. Sakowicz, M. B. Lifshits, O. A. Klimenko, *et al.*, "Terahertz responsivity of field effect transistors versus their static channel conductivity and loading effects," J. Appl. Phys., vol. 110, no. 5, pp. 054512, 2011.

[22] S. M. Sze and K. K. Ng, *Physics of Semiconductor Devices*, Hoboken, NJ: John Wiley & Sons, 2006.

[23] S. Dimitijev, *Principles of Semiconductor Devices*, 2nd Ed., New York: Oxford Univ. Press, 2012.

[24] W. Stillman, M. S. Shur, D. Veksler, S. Rumyantsev, and F. Guarin, "Device loading effects on nonresonant detection of terahertz radiation by silicon MOSFETs," Electron. Lett., vol. 43, no. 7, pp. 422–423, 2007.

[25] W. Stillman, C. Donais, S. Rumyantsev, *et al.*, "Silicon FinFETs as detectors of terahertz and sub-terahertz radiation," Int. J. High Speed Electron. Syst., vol. 20, p. 27, 2011.

[26] M. Hasan, P. E. Gaillardon, and B. S. Rodriguez, "A continuous compact DC model for dual-independent-gate FinFETs," IEEE Electron Devices Soc., vol. 5, no. 1, pp. 23–31, 2017.

[27] J. Wei, D. Olaya, B. S. Karasik, S. V. Pereverzev, A. V. Sergeev, and M. E. Gershenson, "Ultrasensitive hot-electron nanobolometers for terahertz astrophysics," Nat. Nanotechnol., vol. 3, pp. 496–500, 2008.
[28] L. Liu, H. Xu, R. R. Percy, *et al.*, "Development of integrated terahertz broadband detectors utilizing superconducting hot-electron bolometers," IEEE Trans. Appl. Supercond., vol. 19, no. 3, pp. 282–286, 2009.
[29] J. Yan, M.-H. Kim, J. A. Elle, *et al.*, "Dual-gated bilayer graphene hot-electron bolometer," Nat. Nanotechnol., vol. 7, pp. 472–478, 2012.
[30] L. Fernandes, P. Kaufmann, R. Marcon, *et al.*, "Photometry of THz radiation using Golay cell detector," in General Assembly and Scientific Symposium, 2011 XXXth URSI, 2011.
[31] R. Muller, B. Gutschwager, J. Hollandt, *et al.*, "Characterization of a large-area pyroelectric detector from 300 GHz to 30 THz," J. Infrared Millim. Terahertz Waves, vol. 36, no 7, pp. 654–661, 2015.
[32] L. Liu, J. L. Hesler, H. Xu, A. W. Lichtenberger, and R. M. Weikle, II, "A broadband quasi-optical terahertz detector utilizing a zero bias Schottky diode," IEEE Microw. Wirel. Compon. Lett., vol. 20, no. 9, pp. 504–506, 2010.
[33] J. L. Hesler and T. W. Crowe, "Responsivity and noise measurements of zero-bias Schottky diode detectors," in Proc. 18th Int. Symp. Space THz Technol., Pasadena, CA, USA, 2007.
[34] J. Hesler and T. Crowe, "NEP and responsivity of THz zero-bias Schottky diode detectors," in 15th Int. Conf. on THz Electron., Cardiff, 2007, pp. 844–845.
[35] F. Sizov, V. Petriakov, V. Zabudsky, D. Krasilnikov, M. Smoliy, and S. Dvoretski, "Millimeter-wave hybrid un-cooled narrow-gap hot-carrier and Schottky diodes direct detectors," Appl. Phys. Lett., vol. 101, no. 8, pp. 082 108/1–082 108/5, 2012.
[36] R. Han, Y. Zhang, D. Coquillat, *et al.*, "A 280-GHz Schottky diode detector in 130-nm digital CMOS," IEEE J. Solid-State Circuits, vol. 46, p. 2602, 2011.
[37] L. Liu, S. M. Rahman, Z. Jiang, W. Li and P. Fay, "Advanced terahertz sensing and imaging systems based on integrated III-V interband tunneling devices," Proc. IEEE, vol. 105, no. 6, pp. 1020–1034, 2017.
[38] W. Knap, F. Teppe, Y. Meziani, *et al.*, "Plasma wave detection of subterahertz and terahertz radiation by silicon field-effect transistors," Appl. Phys. Lett., vol. 85, pp. 675–677, 2004.
[39] R. Tauk, F. Teppe, S. Boubanga, *et al.*, "Plasma wave detection of terahertz radiation by silicon field effects transistors: Responsivity and noise equivalent power," Appl. Phys. Lett., vol. 89, pp. 253511, 2006.
[40] F. Schuster, D. Coquillat, H., Videlier, *et al.*, "Broadband terahertz imaging with highly sensitive silicon CMOS detectors," Opt. Express, vol. 19, p. 7827, 2011.
[41] F. Schuster, W. Knap, and V. Nguyen, "Terahertz imaging achieved with low-cost CMOS detectors," Laser Focus World, vol. 47, p. 37, 2011.
[42] S. Rahman, Z. Jiang, P. Fay, and Lei Liu, "Advanced terahertz sensing and imaging systems based on integrated III-V interband tunneling devices," Proc. IEEE, vol. 105, no. 6, pp. 1020–1034, 2017.

[43] M. Vitiello, L. Viti, L. Romeo, *et al.*, "Semiconductor nanowires for highly sensitive, room-temperature detection of terahertz quantum cascade laser emission," Appl. Phys. Lett., vol. 100, p. 241101, 2012.
[44] M. Vitiello, D. Coquillat, L. Viti, *et al.*, "Room-temperature terahertz detectors based on semiconductor nanowire field- effect transistors," Nano Lett., vol. 12, p. 96, 2012.
[45] A. Pitanti, D. Coquillat, D. Ercolani, *et al.*, "Terahertz detection by heterostructured InAs/InSb nanowire based field effect transistors," Appl. Phys. Lett., vol. 101, p. 141103, 2012.
[46] R. M. Feenstra, D. Jena, and G. Gu, "Single-particle tunneling in doped graphene-insulator-graphene junctions," J. Appl. Phys., vol. 111, p. 043711, 2012.
[47] V. Ryzhii, T. Otsuji, M. Ryzhii, and M. S. Shur, "Double graphene-layer plasma resonances terahertz detector," J. Phys. D: Appl. Phys., vol. 45, p. 302001, 2012.
[48] V. Ryzhii, A. Satou, T. Otsuji, M. Ryzhii, V. Mitin, and M. S. Shur, "Dynamic effects in double graphene-layer structures with inter-layer resonant-tunnelling negative conductivity" J. Phys. D: Appl. Phys., vol. 46, pp. 315107, 2013.
[49] L. Britnell, R. V. Gorbachev, A. K. Geim, *et al.*, "Resonant tunneling and negative differential conductance in graphene transistors," Nat. Commun., vol. 4, pp. 1–5, 2013.
[50] A. Mishchenko, J. S. Tu, Y. Cao, *et al.*, "Twist-controlled resonant tunnelling in graphene/boron nitride/graphene heterostructures," Nat. Nanotechnol., vol. 9, pp. 808–813, 2014.
[51] B. Fallahazad, K. Lee, S. Kang, *et al.*, "Gate-tunable resonant tunneling in double bilayer graphene heterostructures," Nano Lett., vol. 15, pp. 428–433, 2014.
[52] V. Ryzhii, T. Otsuji, V. Y. Aleshkin, *et al.*, "Voltage-tunable terahertz and infrared photodetectors based on double-graphene-layer structures," Appl. Phys. Lett., vol. 104, p. 163505, 2014.
[53] V. Ryzhii, T. Otsuji, M. Ryzhii, V. Mitin, and M. S. Shur, "Resonant plasmonic terahertz detection in vertical graphene-base hot-electron transistors," J. Appl. Phys., vol. 118, pp. 204501, 2015.
[54] A. Tomadin, A. Tredicucci, V. Pellegrini, M. S. Vitiello, and M. Polini, "Photocurrent-based detection of terahertz radiation in graphene," Appl. Phys. Lett., vol. 103, p. 211120, 2013.
[55] V. Ryzhii, T. Otsuji, M. Ryzhii, *et al.*, "Graphene vertical cascade interband terahertz and infrared photodetectors," 2D Mater., vol. 2, p. 025002, 2015.
[56] B. Sensale-Rodriguez, "Graphene-insulator-graphene active plasmonic terahertz devices," Appl. Phys. Lett., vol. 103, p. 123109, 2013.
[57] M. Ryzhii, V. Ryzhii, T. Otsuji, *et al.*, "Double injection, resonant-tunneling recombination, and current-voltage characteristics in double-graphene-layer structures," J. Appl. Phys., vol. 115, p. 024506, 2014.
[58] M. Hasan and B. Sensale-Rodriguez, "Effect of the intra-layer potential distributions and spatial currents on the performance of graphene SymFETs," AIP Adv., vol. 5, p. 097104, 2015.

[59] N. M. Gabor, J. C. W. Song, Q. Ma, *et al.*, "Hotcarrier-assisted intrinsic photoresponse in graphene," Science, vol. 334, pp. 648–652, 2011.
[60] J. C. W. Song, M. S. Rudner, C. M. Marcus, and L. S. Levitov, "Hot carrier transport and photocurrent response in graphene," Nano Lett., vol. 11, pp. 4688–4692, 2011.
[61] Q. Zhang, W. Zhao, and A. Seabaugh, "Low-subthreshold-swing tunnel transistors," IEEE Electron Device Lett., vol. 27, no. 4, pp. 297–300, 2006.
[62] A. Seabaugh and Q. Zhang, "Low-voltage tunnel transistors for beyond CMOS logic," Proc. IEEE, vol. 98, no. 12, pp. 2095–2110, 2010.
[63] A. M. Ionescu and H. Riel, "Tunnel field-effect transistors as energy-efficient electronic switches," Nature, vol. 479, pp. 329–337, 2011.
[64] D. Sarkar, X. Xie, W. Liu, *et al.*, "A subthermionic tunnel field-effect transistor with an atomically thin channel," Nature, vol. 526, no. 7571, pp. 91–95, 2015.
[65] L. Liu, D. Mohata, and S. Datta, "Scaling length theory of double-gate interband tunnel field-effect transistors," IEEE Trans. Electron Devices, vol. 59, no. 4, pp. 902–908, 2012.
[66] J. Zhang, M. De Marchi, P.-E. Gaillardon, and G. De Micheli, "A Schottky-barrier silicon FinFET with 6.0 mV/dec subthreshold slope over 5 decades of current," in IEDM Tech. Dig., San Francisco, USA, 2014, pp. 13.4.1–13.4.4.
[67] J. Zhang, J. Trommer W. M. Weber, P. E. Gaillardon, and G. D. Micheli, "On temperature dependency of steep subthreshold slope in dual-independent-gate FinFET," IEEE J. Electron Devices Soc., vol. 3, no. 6, pp. 452–456, 2015.
[68] P.-E. Gaillardon, J. Zhang, M. De Marchi, and G. De Micheli, "Towards functionality-enhanced devices: Controlling the modes of operation in three-independent-gate transistors," in Invited, 10th IEEE Nanotechnology Materials and Devices Conference (NMDC), Anchorage, USA, 2015.
[69] L. Barboni, M. Siniscalchi, and B. Sensale-Rodriguez, "TFET-based circuit design using the transconductance generation efficiency gm/Id Method," IEEE J. Electron Devices Soc., vol. 3, no. 3, pp. 208–216, 2015.
[70] W. Li, T. Yu, J. Hoyt, and P. Fay, "InGaAs/GaAsSb interband tunneling FETs as tunable RF detectors," In Device Research Conference (DRC), Santa Barbara, CA, 2014, pp. 21–22.
[71] S. Kim, D.-W. Park, K. Choi, and S.-G. Lee, "MOSFET characteristics for terahertz detector application from on-wafer measurement," IEEE Trans. Terahertz Technol., vol. 5, no. 6, pp. 1068–1077, 2015.
[72] M. Von Haartman and M. Ostling, *Low-Frequency Noise in Advanced MOS Devices*, Dordrecht: Springer, 2007.
[73] A. Lisauskas, S. Boppel, J. Matukas, *et al.*, "Terahertz responsivity and low-frequency noise in biased silicon field-effect transistors," Appl. Phys. Lett., vol. 102, no. 15, p. 153505, 2013.
[74] E. G. Ioannidis, E. G. Ioannidis, C. A. Dimitriadis, *et al.*, "Improved analysis and modeling of low frequency noise in nanoscale MOSFETs," Solid-State Electron., vol. 76, pp. 54–59, 2012.
[75] R. Talmat, H. Achour, B. Cretu, *et al.*, "Low frequency noise characterization in n-channel FinFETs," Solid-State Electron., vol. 70, pp. 20–26, 2012.

Part II

Applications and design techniques of functionality-enhanced devices

Chapter 8

CNT and SiNW modeling for dual-gate ambipolar logic circuit design

Xuan Hu[1], Wesley H. Brigner[1], and Joseph S. Friedman[1]

As has been discussed in previous chapters, ambipolar carbon nanotube field-effect transistors (CNTFETs) and silicon nanowire field-effect transistors (SiNWFETs) are major contenders for beyond-complementary metal-oxide smiconductor (CMOS) computing. To enable the design and analysis of next-generation large-scale circuits and systems, it is important to develop device models that represent the operating behavior of emerging nanodevices. Though the accuracy of such device models vary, these device models greatly improve the efficiency of the design process by enabling extensive circuit simulation prior to slow, laborious, and expensive circuit fabrication. Additionally, for many beyond-CMOS devices, large-scale circuit fabrication is not only expensive but also difficult due to the relative immaturity of fabrication techniques for these emerging devices as compared to those for conventional CMOS.

It is therefore cost- and time-efficient to simulate circuits with unconventional devices prior to fabrication; fabricating numerous devices and circuits merely to test functionality is not a reasonable prospect. Rather than fabricating a new device following every minor change in device structure and parameters, device and circuit designers can use device models easily and rapidly simulate behavior for a wide range of structures and parameters. A device model is a predictive representation, such as a table or set of equations, of device behavior that can be applied to a range of input and device parameters. In order to ensure the reliability of the simulation results, the models must accurately reflect the device properties by matching the underlying physics and previous fabrication results. The effect of modifications to the device properties can enable large-scale and system-level analysis, providing the opportunity to rapidly explore the device parameters that maximize circuit and system functionality efficiency.

The ambipolar characteristics of functionality-enhanced devices (FEDs) provide unique opportunities and challenges for device modeling. Given the importance of device models for the advancement of FED circuits and systems, much work has been dedicated to the modeling of ambipolar SiNWFETs and CNTFETs. However, the challenge of developing a complete model that enables large-scale circuit design while fully and accurately capturing the device behavior has yet to be overcome.

[1]Department of Electrical and Computer Engineering, The University of Texas at Dallas, USA

Therefore, this chapter surveys available FED models and analyzes the utility of these device models for ambipolar logic circuit design to enable the reader to determine which model is best suited for a particular application. The models have been grouped in Table 8.1 according to their underlying approaches and their ability to perform the desired model characteristics. Where the group names do not specify otherwise, these groups are static with fixed drain and source nodes.

8.1 Desired model characteristics

To reliably enable logic circuit design, the basic requirement for a device model is the ability to represent the behavior of a specified device with minimal error. In addition to matching the characteristics of devices that have already been fabricated, it is preferable that a model be physically derived in order to enable prediction of the behavior of similar, but altered, devices. Specific to the FEDs covered in this book, a model must enable the manipulation of the ambipolar device polarity to permit accurate and complete analysis of device and circuit behavior. In order to be effective, FED models should exhibit the characteristics discussed in the following sections: being physically derived, matching available experimental data, being simulation program with integrated circuit emphasis (SPICE)-compatible, enabling polarity gate (PG) modulation, and enabling polarity switching.

8.1.1 Connection to underlying physics

In order to properly represent device behavior, the ideal FED model fully incorporates the entirety of the underlying device physics from the most basic quantum phenomena to the resulting electrical behavior. However, such a model would be prohibitively complex both in terms of model development and simulation runtime, and simplifications of the underlying device physics are therefore necessary. There is a fundamental trade-off between the connection of a model to the underlying physics and the model's simplicity and ease of use.

Models that are developed to closely reflect the underlying physics are described in this chapter as "physically derived." Such a physically derived model is developed based on the actual physical device parameters such as dimensions, structure, materials, doping density, and carrier mobility, among others. With a physically derived model, the model can reliably predict the device behavior if parameters are modified beyond the range of devices that have been fully measured or analyzed. For example, if a 10-nm-long device functions in a particular manner, a physically derived model is able to predict the functionality of a 20-nm-long device. This ability to extrapolate beyond previously demonstrated device behaviors provides physically derived models with the ability to enable device optimization. A predictive device model can be used by circuit designers to evaluate the effect of device parameters on system functionality, even if few or no devices have previously been fabricated. The simulation-based device optimization can then be used to aid the process of device fabrication. As described in this chapter, *ab initio* physical emulators and models

Table 8.1

Modeling approach group	Section numbers	Experimental matching	Physically derived	SPICE compatible	BG sweeping	Transient polarity	Device type	Reference publication
Single-gate physically derived	8.2.1	X	✓	✓	X	X	CNTFET	[3]
	8.2.2	X	✓	✓	X	X	CNTFET	[4]
	8.2.3	X	✓	✓	X	X	CNTFET	[5]
Behavioral ambipolar	8.3.1	✓	X	✓	✓	X	SiNWFET	[6]
	8.3.2	✓	X	✓	✓	X	CNTFET	[7]
Comprehensive semiconductor compensation simulation	8.4.1	X	✓	X	✓	X	WSe_2 FED	[8]
	8.4.2	✓	✓	X	✓	✓	SiNWFET	[9]
Physical ambipolar	8.5.1	✓	✓	✓	X	X	MIGFET	[10]
	8.5.2	✓	✓	✓	✓	X	CNTFET	[11]
	8.5.3	✓	✓	✓	✓	X	CNTFET	[12]
	8.5.4	✓	✓	✓	✓	X	SiNWFET	[13]
Ambipolar with dynamic polarity	8.6.1	✓	✓	✓	✓	✓	CNTFET	[14]
	8.6.2	✓	X	✓	✓	✓	CNTFET	[15]

based on the underlying device physics can be used for these extrapolations beyond the available measured devices.

Instead of representing the underlying physics, a behavioral model attempts to represent a device's behavior with simplified equations that include fitting parameters disconnected from physical reality. Numerous simulations are generally used to determine the values of these fitting parameters to best match the available experimental or otherwise reliable device data. Behavioral models are simpler and more efficient than either an *ab initio* physical emulation or a complete physically derived model. Behavioral models can therefore provide reasonably accurate simulations while dramatically decreasing the simulation runtime. Behavioral models are not necessarily less accurate, but their ability to extrapolate is limited compared to physically derived models.

The models described in this chapter span the range of the trade-off between simplicity and the connection to underlying physics. Models derived from physical equations and parameters are generally more accurate than a behavioral model due to the thorough representation of the device physics, but simulations with physically derived models also require a considerable amount of resources and/or time to run. Because of this, behavioral models are generally more computationally efficient.

8.1.2 Experimental matching

The fundamental purpose of a model is to mimic the actual behavior of a physical device; a model should therefore be able to accurately represent data that was obtained from experimental device measurements. Ideally, a model should perfectly match these measurements; however, there are always trade-offs between the model accuracy and the computing efficiency and model simplicity. Furthermore, there is always a large degree of complexity related to the experimental setup, and it is often quite difficult to determine the true physical parameters required for physical device modeling. It is therefore often necessary to use fitting parameters to interpret experimental data, suggesting a fundamental trade-off between the connection to the underlying physics and the matching to experimental data.

8.1.3 SPICE compatibility

To advance FED logic circuit design, it is critically important to be able to perform SPICE simulations [1] with FED device models, as many circuit design and simulation tools are based on the SPICE environment. Such models can be used in conjunction with other device models to simulate hybrid systems of devices. If a model is not SPICE-compatible, its utility is severely limited; for example, *ab initio* physical emulations created using technology computer-aided design (TCAD) [2] are not generally SPICE-compatible and therefore cannot be easily applied to logic circuit design. Therefore, though TCAD models generally provide considerably higher accuracy than other models, they are typically used as references for validation rather than for circuit design. Further, it should be noted that to be truly SPICE-compatible, it should be possible to easily modify device parameters and perform complex circuit simulations without convergence issues or other difficulties.

8.1.4 PG modulation

The distinctive advantage of FEDs over conventional devices is that the majority carrier type (i.e., electrons or holes) can be switched through modulation of the PG voltage; it is therefore critical that this capability be included in an FED device model. This requires fundamental changes from conventional transistor models. Early FED models do not enable sweeping of the PG voltage, using this PG voltage merely as a binary switch between electron and hole conduction. More recent models have included modulation of the PG voltage in DC (non-transient, steady-state) simulations, but this V_{PG} has rarely been used to modulate the drain–source current. To enable efficient and accurate FED design and simulation, however, PG modulation should be fully incorporated in a device model.

8.1.5 Dynamic polarity switching (interchangeability)

Given the ambipolar nature of FEDs, a model that enables the full utilization of FEDs should provide dynamic polarity switching in addition to the PG modulation mentioned previously. That is, in addition to being able to switch the majority carrier in a steady-state simulation, it should be possible to switch the majority carrier within a transient simulation. Without this ability, it is impossible to perform simulations to fully evaluate the switching time of even the simple XOR gate that is so efficiently realized with FEDs. It is therefore also impossible to simulate large-scale circuits where timing is critical.

This dynamic polarity switching requires the inclusion of interchangeability of the FED source and drain nodes. In FEDs, as in CMOS transistors, the source node is the node where the majority carrier flows into the device. In n-type devices, the electrons act as the majority carriers and flow form source to drain, resulting in an electric current from drain to source and a positive V_{DS}. On the other hand, holes are the majority carriers in p-type devices; they flow from source to drain, introducing a positive V_{SD} voltage drop across the device. While initialization of static source and drain nodes is sufficient for conventional CMOS circuit, FED circuit design requires inclusion of switching of the majority carrier type, implying that the source and drain nodes must be interchangeable.

As all the nodes in conventional models are defined during initialization of the simulation, FED models require novel solutions for interchanging the source and drain nodes. Without this interchangeability, FED models produce significant errors and physically unrealistic responses to the given stimuli. For instance, some models function properly when the source terminal is connected to a positive voltage and the drain terminal is connected to ground; however, when the source is connected to ground and the drain is connected to a positive voltage, current still flows from source to drain, impossibly implying that current is flowing backwards—from ground to a positive voltage. In contrast, a model capable of dynamic polarity switching always allows the current to flow in the proper direction. Therefore, the dynamic polarity switching and interchangeability of the source and drain nodes is critical in leveraging the full capabilities of FEDs.

8.2 Single-gate physically derived models

This section covers three early models for CNTFETs that have served as a basis for many later FED models. As these works are from the early stages of FED modeling computing applications, no experimental measurements are used to validate the accuracy of these models. The functionality-enhanced behavior of CNTFETs and SiNWFETs is also not incorporated.

8.2.1 Unipolar SPICE model for ballistic CNTFETs

One of the earliest CNTFET models was proposed in 2004 [3]. This model calculates the drain–source current based on the CNTFET surface potential. For single-walled CNTFETs (SW-CNTFETs), the drain–source current is a function of one-dimensional ballistic tunneling [16,17]. The voltage applied to the CNTFET gate can be used to modulate the Schottky barrier (SB) height along the CNTFET channel, modulating the probability of that carriers will tunnel through the barrier [18,19]. I_{DS} can therefore be calculated by first computing the SB height as a function of the source, drain, and gate voltages and then calculating the quantity of charge along the CNT channel according to Fermi–Dirac distribution:

$$n_{CNT} = \int_{E_C}^{+\infty} \frac{D(E)}{2} (f(E - \mu_S) + f(E - \mu_D))\, dE \tag{8.1}$$

where n_{CNT} is the quantity of charge carriers. Given the carrier density through the CNT, the drain–source current I_D can be obtained by

$$I_D = \frac{4ek_BT}{h} \sum_p \left(\ln(1 + e^{-\xi_S}) - \ln(1 + e^{\xi_D})\right) \tag{8.2}$$

where ξ_s and ξ_d are defined as

$$\xi_i = \frac{\psi_S - \Delta_p - \mu_i}{KT}, \quad \text{for} \quad i = s, d. \tag{8.3}$$

ψ_S is the voltage by which the channel potential is decreased resulting from V_G applied at the gate and can be obtained through $\psi_S = V_G - (Q_{CNT}/C_{INS})$, where Q_{CNT} represents the total charge stored at the capacitor C_{INS} formed across the insulator between the gate and CNT. The flat-band voltage V_{FB} represents the required voltage to fully flatten the SB to permit the carriers to travel through the channel with minimum impedance. This V_{FB} is similar to the threshold voltages in conventional CMOS transistors. Combining all the listed variables, the CNTFET drain–source current can be obtained.

As little experimental data was available at the time this work was published, it is difficult to compare the model to experimental measurements. In order to confirm the validity of the model, the authors ran simulations using SPICE and compared the results to numerical data obtained using MATLAB® [20,21] based on the physical equations; the SPICE model matches the numerical data.

This model was applied to single-device and logic gate simulations. Similar to conventional CMOS, I_{DS}–V_{DS} simulations are presented and compared to numerical model results. In addition, for various flat-band voltages V_{FB}, I_{DS}–V_G simulations are included without comparison to other data or measurement. While this model only includes n-type CNTFET behavior, several logic circuits and corresponding simulations based on pseudo-n-type logic styles are also included. By adding a load capacitance, transient simulations can be performed to obtain the transient response of the device and logic circuits, including the propagation delay.

Overall, this work represents the n-type SiNWFET and CNTFET behavior by developing a SPICE-compatible device model. With the possible extension to device variation by tuning physical parameters, this model is determined to be a physically derived model. To confirm the validity of the model, simulation results are compared to numerical results from MATLAB.

8.2.2 *Unipolar Verilog-A model for doped CNTFETs*

Based on the physically derived equations from [3] as discussed above, [4] develops an improved model for doped ballistic SW-CNTFETs. While multiple different structures of CNTFETs have been proposed [4,20] is designed to represent a conventional CNTFET (described in [11] as partially gated CNTFET). As the CNT channel doping is assumed to be uniform, the ballistic electron behavior along the CNT can be derived from the Landauer equations [22].

In this model, a positive voltage applied to the CNTFET gate (V_G) lowers the energy barrier height, thus determining the total charge accumulation. The channel potential is then obtained by

$$V_{CNT} = V_{GS} = \frac{Q_{CNT}}{C} \tag{8.4}$$

where C is the parasitic capacitor across the gate insulator. The channel potential is then used to determine the drain–source current using the Landauer equation as

$$I = \frac{4ek_BT}{h}\sum_{p=1}^{+\infty}\left[\begin{array}{c}\ln\left(1+\exp\dfrac{-\Delta_p+V_{CNT}}{k_BT}\right)\\ -\ln\left(1+\exp\dfrac{-V_{DS}-\Delta_p+V_{CNT}}{k_BT}\right)\end{array}\right]. \tag{8.5}$$

A significant contribution of this work is that the authors considered the CNTFET sub-band minima for higher accuracy.

8.2.3 *Ambipolar VHDL-AMS model for CNTFETs*

Similar to the other models in this group discussed above, [5] describes compact models for both unipolar and ambipolar CNTFETs. This CNTFET model includes conduction mediated by both electrons and holes, as well as the FED ambipolarity.

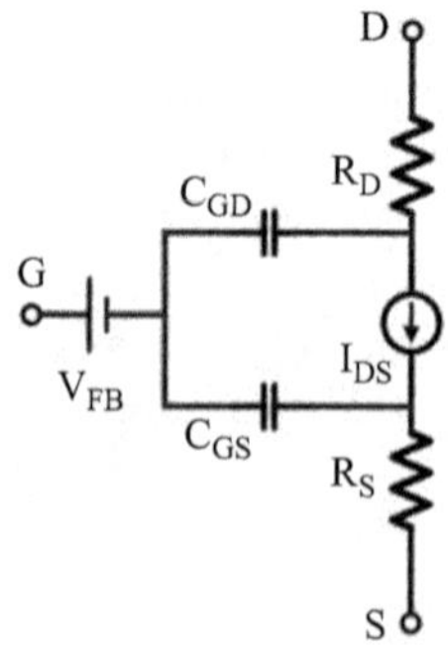

Figure 8.1 Equivalent circuit of the CNTFET model, where V_{FB} is the voltage required to flatten the energy band. Adapted, with permission, from Reference [5]

Though only one gate is included, this work is identified as the first CNTFET model that incorporates ambipolar behavior.

The unipolar model of this work is an improvement of [3]. While the model of [3] was developed using SPICE, some convergence problems limit its application for circuit design and simulation. Reference [5] therefore adapts the model of [3] to VHSIC hardware description language – analog and mixed signal (VHDL-AMS) for improved circuit convergence. As shown in Figure 8.1, some parasitic components such as R_D, R_S, C_{GD}, and C_{GS} are included in the model to resolve circuit convergence issues. For the drain-source current, (8.2) is used. Here, both ξ_s and ξ_d are the same as those presented in [3], though ψ_S is slightly more complex:

$$\Psi_S = V_{GS} - \frac{\left(\alpha \cdot (V_{GS} - \Delta_i) + \sqrt{(\alpha \cdot (V_{GS} - \Delta_i))^2 + 4\xi^2}\right)}{2} \tag{8.6}$$

where α is a precomputed parameter that is treated as an input. With the input voltage V_{GS} and the total charge within the device, the input capacitance can be obtained as

$$C_G = \frac{\delta Q_{CNT}}{\delta V_{GS}} \Rightarrow C_G = \frac{\delta Q_{CNT}}{\delta \psi_S} \cdot \frac{\delta \psi_S}{\delta V_{GS}}. \tag{8.7}$$

As the physical device coefficients are not fully integrated into this model, a set of fitting parameters must be computed before the simulation. These fitting parameters must be updated before simulation of a device with modified parameters. The complexity of this pre-simulation parameter computation procedure limits this model's utility for circuit design.

In addition to the unipolar CNTFET, the functionality-enhanced ambipolar CNTFET is also modeled. The basic flow of this model is presented in Figure 8.2. While in an ideal model the FED PG voltage controls the carrier density and modulates the drain–source current, this work uses two separate unipolar models to represent the n- and p-type current. As shown in Figure 8.2, the input voltage V_{GS} is used to determine

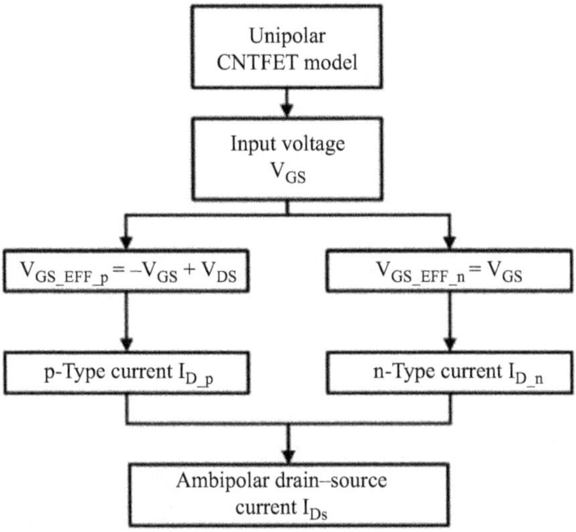

Figure 8.2 Pseudo-ambipolar model flowchart. Input voltage V_{GS} is used in the calculation of both n- and p-type current to represent the ambipolar behavior. Adapted, with permission, from Reference [5]

whether the CNTFET is operating under electron- or hole-drive current; this V_{GS} is also used to determine the n-type current I_{D_n} while $V_{GS\_EFF} = -V_{GS} + V_{DS}$ is used for the p-type current I_{D_p}. This approach provides a simplified method for modeling FEDs by sacrificing the simulation accuracy.

The unipolar CNTFET is simulated and compared to the model from [3]. While the original model causes glitches that may result in convergence issues, the proposed model smooths out the glitch and provides a smooth transition between states. Similarly, the sharp step transition in the SB height of the original model is smoother in this work. This work provides good agreement to the numerical model [21] in terms of I–V characteristics; however, due to the lack of comparison to experimental measurements, it is difficult to determine the ability of this model to represent the behavior of physical devices.

While no actual PG is modeled in this work, the framework proposed in Figure 8.2 provides the CNTFET ambipolar behavior shown in Figure 8.3. As V_{DS} is varied, the cutoff region between the n- and p-type conduction is also changed. The device polarity is used as an input, and no simulation is presented in which polarity is switched during the simulation. Therefore, it cannot support PG modulation and the voltage threshold cannot be modified as a function of changes in a PG voltage. This model therefore does not enable PG modulation. It should also be mentioned that the ambipolar SB CNTFET model is not compared to fabricated device data, as no such device had yet been fabricated.

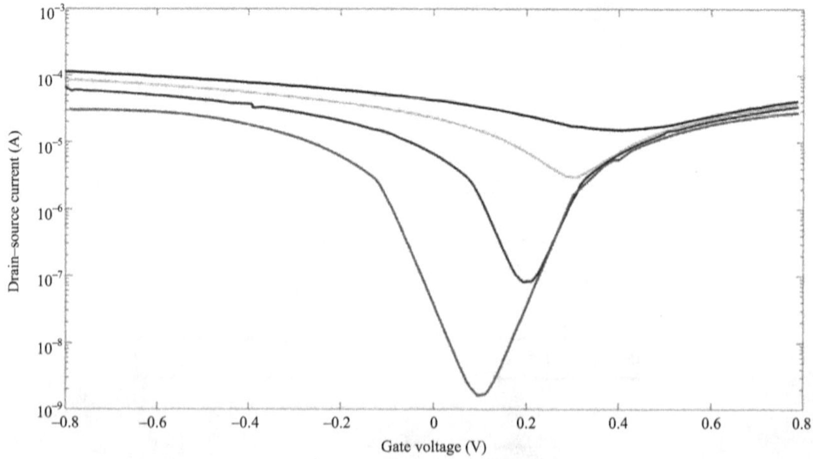

Figure 8.3 I–V simulation results. The device ambipolar behavior is demonstrated with a single gate that modulates the majority carrier type and density. Adapted, with permission, from Reference [5]

Boolean logic circuit schematics based on unipolar CNTFETs are presented in Figure 8.4 and used in transient simulations. Since these device polarities are predetermined during circuit design, no transient polarity switching is demonstrated in simulation of these circuits.

This work proposed SPICE-compatible models for both unipolar and ambipolar CNTFETs. The unipolar CNTFET model reduces the intermediate signal glitch from the original model and improves the convergence problem that limits the application of this model in large-scale circuit simulations. The comparison between proposed work and numerical model exhibits minimal error in the *I–V* simulation. This model is the first modeling work that includes the CNTFET ambipolar behavior of Figure 8.3, which is controlled by a single gate.

8.2.4 Experimental matching

To validate the accuracy of a model, a comparison between simulation results and experimental measurements is generally included. However, device fabrication is expensive and time-consuming, especially if the technology is immature. As a result, the models in this group are compared to numerical data from MATLAB that is based on similar physical equations. Therefore, no matching to experimental data has been provided.

8.2.5 Connection to underlying physics

For the models listed in this section, physically derived equations are used in the drain–source current calculations. For example, parameters such as C_{INS}, V_{FB}, and T can be used to modify the device performance and therefore predict the behavior

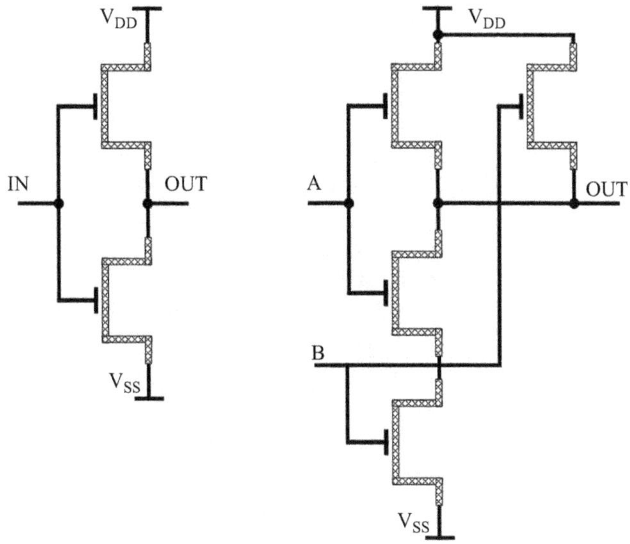

Figure 8.4 Schematic of CNTFET-based inverter (INV) and NAND. Adapted, with permission, from Reference [5]

of other devices with a similar structure. Though fitting parameters are also included within these models, the physical equations provide these models with the ability to make reliable predictions for similar devices.

8.2.6 SPICE compatibility

A SPICE-compatible model provides a drastic increase in the computational efficiency compared to physical emulation tools. The models introduced with this approach are implemented in SPICE-compatible languages such as SPICE and VHDL-AMS, and the fitting parameters are precomputed before being integrated into the model. Therefore, these models are SPICE-compatible and should enable efficient circuit design.

8.2.7 PG modulation

The models of Sections 8.2.1 [3] and 8.2.2 [4] do not directly pertain to the topic of FED modeling, since they were designed to model conventional n-type-only CNTFETs. However, these works were highly influential in the development of SPICE-compatible device models for CNTFET-based circuits. Future researchers adapted this model to FEDs to permit functionality-enhanced performance. The model of Section 8.2.3 [5] presents the first ambipolar current behavior simulation in Figure 8.3, but no independent gate is modeled. Reference [5] only models one gate that covers both metal–semiconductor contacts. As no device in this modeling approach group includes a PG that can be used to modulate the majority carrier type, none of the models in this group support static polarity switching.

8.2.8 Dynamic polarity switching (interchangeability)

No transient simulation with dynamic polarity switching is provided with the models in this group.

8.3 Behavioral ambipolar models

To enable exploration of FED circuit designs that exploit the independent gates and the ambipolarity, these features must be included in the device model. As a first step, behavioral models with limited accuracy have been developed that permit FED circuit simulations that provide a general analysis of circuit behavior. The models in this group permit static, but not transient, polarity switching.

8.3.1 Ambipolar model for double-independent-gate FinFETs

The work in [6] proposed a compact DC model for FinFET FEDs represented by the device diagram presented in Figure 8.5 [23,24]. With expressions derived from [24,25], the equation

$$I_{DS} = (wI_{VGS} + (1-w)I_{V_{DS}\approx 2V})\left(1 - e^{-\left(\frac{V_{DS}^{B}}{A}\right)}\right) \tag{8.8}$$

is used in this model to represent the drain-source current of the FinFET FED, where I_{VGS}, A and B are from (8.9), (8.10), and (8.11).

$$\begin{aligned} I\left(V_{GS}\right) &= C_T[b_1\left(V_{GS}+V_{OFF}\right)^{c_1} H\left(V_{GS}, V_{OFF}\right)+1] \\ &\quad \times\left[\frac{1}{2}+\left(\frac{1}{2}-b_0\right)\tanh\left(\frac{V_{GS}+V_{OFF}}{c_0}\right)\right]. \end{aligned} \tag{8.9}$$

$$\begin{aligned} A &= s_0 + s_1V_{GS} + s_2V_{PG} + s_3V_{PG}^2 + s_4V_{GS}V_{PG} + s_5V_{GS}^2 + s_6V_{GS}^3 \\ &\quad + s_7V_{GS}V_{PG}^2 + s_8V_{GS}^2V_{PG} \end{aligned} \tag{8.10}$$

$$B = t_0 + t_1V_{PG} + t_2V_{GS} + t_3V_{GS}V_{PG} + t_4V_{PG}^2. \tag{8.11}$$

In the above equations, C_{T} is a V_{DS}- and V_{PG}-based factor that modulates the I_{DS}, H(V_{GS}, V_{OFF}) returns the unit step response at V_{GS}-V_{OFF}, and b_0, b_1, c_0, c_1, and s_0-s_8

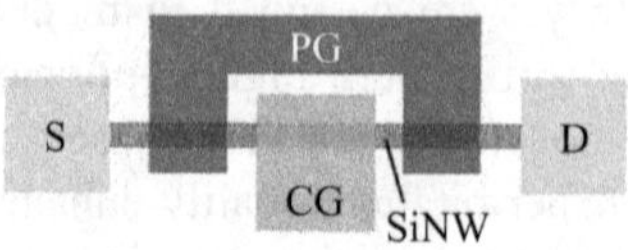

Figure 8.5 Top view of device structure. Both control gate (CG) and PG can be used to modulate the channel carrier density, while PG can also switch the majority carrier type. Adapted, with permission, from Reference [6]

are fitting parameters. Fitting parameters are optimized through iteration according to the technique of the flowchart in Figure 8.6, providing excellent matching to experimental measurements from [23]. While this technique does provide matching to the experimental data, the physical parameters such as device dimensions and material conductivity are not included in the model. The lack of physical parameters prevents the accurate application of this model to devices even slightly different from the previously fabricated devices, as the fitting parameters must be altered in some indeterminate manner to represent the different device behavior.

In the equation for the current, $I = C_T f_{IIF} V_{LF}$, C_T, f_{IIF}, and V_{LF} are the three variables used to compute the FED current. These variables are a function of V_{PG}, V_{CG}, and V_{DS}, which are the actual input voltages used for the current computation. All the device parameters used in this model are fitting parameters gathered from MATLAB [26]. The procedure is presented in Figure 8.6.

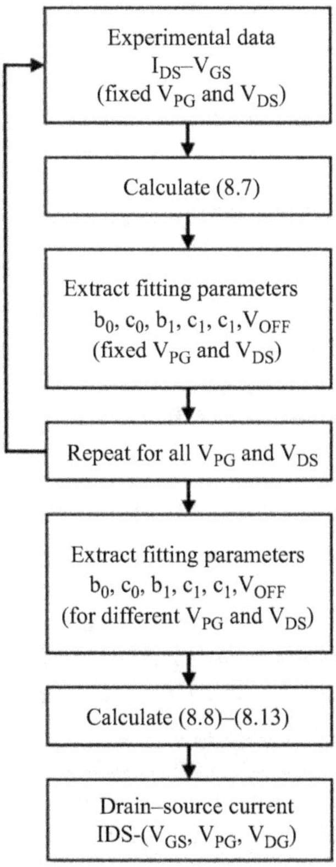

Figure 8.6 Model flowchart. The fitting parameters are calculated through an iteration loop to achieve the desired accuracy. Adapted, with permission, from Reference [6]

This model includes the PG voltage to modulate the drain–source current. In addition, the V_{PG} is used as an input for the fitting parameter C_T according to the equation

$$C_T = \mu_0 + \mu_1 V_{PG} + \mu_2 e^{\sigma_0 V_{PG} + \sigma_1 V_{DS} + \sigma_2 V_{PG} V_{DS}}. \tag{8.12}$$

The PG of this model can modulated in DC simulations but is shown as constant in transient simulations; no dynamic polarity switching is demonstrated. By changing the fitting parameters, this model can be adapted to represent both n- and p-type dual-independent-gate field-effect transistor (DIGFET) CNTFETs. However, as there is no demonstrating of sweeping V_{PG}, it appears that the model may not provide accurate I–V_{PG} simulations.

To validate the accuracy of the proposed model, the simulation results are shown to match the experimental measurements from [23] for both I_{DS}–V_{DS} and I_{DS}–V_{GS} simulations. Different fitting parameter sets are required for different device structure and dimensions; therefore, the flexibility of this model is limited compared to that of physically derived models. This model provides the ability to use the PG voltage as an input, providing an extra degree of freedom for FinFET FEDs.

8.3.2 Ambipolar Verilog-A model for CNTFETs

The CNTFET FED behavioral compact model of [7] includes several fitting parameters to match experimentally fabricated devices. The drain–source current of the CNTFET FEDs can be accurately and efficiently represented with fitting parameters including carrier mobility, the top gate threshold, and the back gate threshold. While each gate modulates I_{DS} according to a metal-oxide-semiconductor field-effect transistor (MOSFET)-like current equation, this model enables independent control of the CG and PG, as in [27]. As the PG is modeled independently, this model enables the first simulation of cascaded logic with FEDs in which the majority carrier type switches.

The authors validated this model by comparing it to experimental measurements from [27]. In one simulation configuration, the two independent gates are connected while the input voltage is varied; in the other configuration, the control gate is used as a voltage input for two constant PG voltages that activate each majority carrier type. Compared to some physically derived models, this behavioral model results in a slightly larger mismatch to experimental data.

Although the model's simulation results are not perfectly matched to the experimental data, this model sufficiently represents the device's ON/OFF behavior for preliminary steady-state logic circuit design and simulation. As the model is developed using Verilog-A, it is fully compatible with the SPICE environment. For the drain–source current computation, the two gate voltages are used as inputs in addition to the drain and source node voltages. In addition, fitting parameters for carrier mobility and threshold voltages are incorporated to improve the accuracy.

To efficiently model the CNTFET FED, it is of crucial importance to include the device polarity switching behavior. Instead of using the PG voltage as a binary switch input, this model permits the drain–source current to be modulated by both the CG and PG, thus allowing simulations of cascaded logic circuits. However, this model does not permit FED polarity switching in transient simulations due to the lack of interchangeability; therefore, the authors have improved the model as presented in Section 8.6.2 [15].

8.3.3 Connection to underlying physics

The models within this group utilize equations that are based on the device behavior rather than the underlying physics. As the fitting parameters are extracted from the experimental results to develop the model, these behavioral models represent the device behavior in a simple manner without requiring significant resources for complex computations with many variables. Therefore, compared to physically derived models, behavioral models can generally provide higher computing efficiency. With properly adapted fitting parameters, these models can be used to represent different devices and structures.

8.3.4 Experimental matching

The simulations with this approach group are validated through comparisons to experimental measurements in various configurations. The matching is generally fairly close, with differences resulting from the use of a model that is far simpler than the underlying physics. These models thus enable modulation of the PG voltage at the cost of a decrease in model accuracy.

8.3.5 SPICE compatibility

Though fitting parameters are used to support the computation, the models in this group are developed using a hardware description language. As the model of Section 8.3.1 [6] does not specify the implementation method, it cannot truly be categorized as a SPICE-compatible model because no related discussion is presented. However, the model of Section 8.3.2 [7] is implemented using Verilog-A, which is fully compatible to SPICE environment.

8.3.6 PG modulation

The model of Section 8.3.1 [6] simulates the fitting parameters with different fixed V_{PG} levels. While no direct I_{DS}–V_{PG} simulation is presented, this V_{PG} can be controlled to result in I_{DS} modulation as a function of both CG and PG. In the model of Section 8.3.2 [7], the PG voltage can be controlled independently from CG, thus providing full PG modulation.

8.3.7 Dynamic polarity switching (interchangeability)

No transient simulation with dynamic polarity switching is provided with the models in this group.

8.4 Comprehensive semiconductor compensation simulation

In contrast to conventional SPICE-level electronic computer-aided design (ECAD) tools that represents device behavior in terms of parameters such as carrier mobility and voltage threshold, TCAD represents the device behavior in terms of the individual electron dynamics. Physical parameters such as materials, structure, dimension, aging, or fabrication variations are all considered in the model for calculation. By incorporating all the physics from the most basic physical level, TCAD approaches can generate extremely precise matches to experimental measurements. Whereas an ECAD model calculates the device current according to the I–V relation equations, a TCAD approach considers every electron within the channel. This produces an accurate output at the cost of a significantly increasing the simulation time.

Since TCAD tools are designed for device design and fabrication, circuit simulations are not the primary goal of TCAD approaches. Unlike ECAD models that are mostly SPICE-compatible, TCAD approaches focus more on the physical behavior than the electrical behavior of semiconductor devices. TCAD simulations generally require greater computational resources and are therefore not suitable for large-scale circuit simulation. The simulation of cascaded logic results in unacceptable simulation runtimes due to the slow convergence and iterative analyses. This section describes TCAD-based SiNWFET and WSe_2FET FED modeling approaches that accurately represent device behavior but are not SPICE compatible.

8.4.1 Ambipolar TCAD simulation for WSe_2FETs

Ambipolar TCAD simulations are described in [8] for undoped two-dimensional FEDs. The two-dimensional material-based FETs [28,29] exhibit higher carrier mobility [30] and scalability [31,32] than conventional FEDs, resulting in improved device performance [33,34]. In addition, the enhanced polarity control capabilities enabled by the PGs permit the carrier mobility to be modulated to perform compact logic within a single device [35,36]. Therefore, the authors developed a TCAD simulation approach to thoroughly analyze and simulate this novel WSe_2FET device.

Two device structures are proposed, as described in Figure 8.7: (a) the top gate structure, in which the gate is only at one side of the two-dimensional WSe_2 channel, and (b) the double-gated structure with CG and PG on both sides of the channel, providing greater control of the drain–source current. Based on the device structure in Figure 8.7, a NanoTCAD ViDES [2,37] approach is proposed for device simulation. This approach utilizes physical parameters related to the channel material and solves the Poisson and nonequilibrium Green's functions iteratively, resulting in a simulation that is highly representative of physical reality.

This approach is used for the simulation of FEDs composed of monolayer and bilayer WSe_2FETs. Various channel lengths are used to simulate the device with both top gate and double gate structure. By switching the PG between ± 1 V, this two-dimensional device behaves as both an n- and p-type transistor, enables the function enhancement. When compared to top gate devices, double gate devices present a

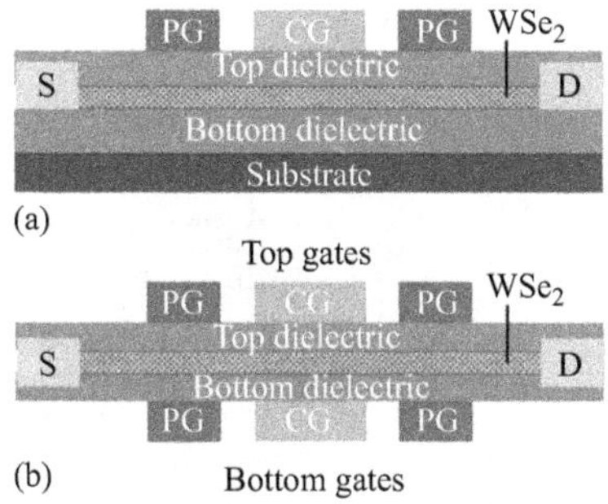

Figure 8.7 Side view of device structures. (a) Top-gate device structure with all gates aligned on one side of the channel. (b) Double-gate structure with gates on both sides of the channel. Adapted, with permission, from Reference [8]

lower I_{OFF} current with a higher I_{ON} current with the same bias conditions, providing superior device functionality. In addition to simulation of the current, this TCAD approach is used to predict device behaviors with different channel materials.

This approach provides the ability to accurately predict device behavior under various conditions such as dimensions, materials, structures, and channel layouts. With parameters relating to the channel material, this approach is able to accurately represent device behaviors via physical emulation tools. Though no direct comparison is shown between the simulations and experimental measurements, TCAD computations are known to provide reliable results; thus, this approach has value in device analysis and design.

8.4.2 Ambipolar TCAD simulation for SiNWFETs

TCAD simulations of ambipolar SiNWFET FEDs are provided in [9] to enable compact circuit design [11,38] based on a fabricated device. Experimentally determined parameters based on device measurements are used to tune the TCAD simulations for better accuracy. The measured device parameters are incorporated into the TCAD approach, such as the SB height. The PG voltage V_{PG} modulates the height of the SB formed at the junctions between the semiconductor and the metal contacts, thus controlling the carrier density within the SiNW channel. While previous works use the PG as a simple switch, this approach permits the PG to be used as a fully functional independent gate, enabling the first demonstrations of transient device polarity switching.

The I_{DS}–V_{CG} simulation from TCAD is compared to the experimental measurements with the PG voltage V_{PG} set to +4 V and −4 V. The device exhibits n-type I–V behavior with V_{PG} = +4 V, and p-type I–V behavior with $V_{PG} = -4$ V. The TCAD results match closely the experimental measurements when the SiNWFET FED is in the highly conductive region of operation for both polarities but matches poorly in the subthreshold region. Though several fixed values were assigned to V_{PG} rather

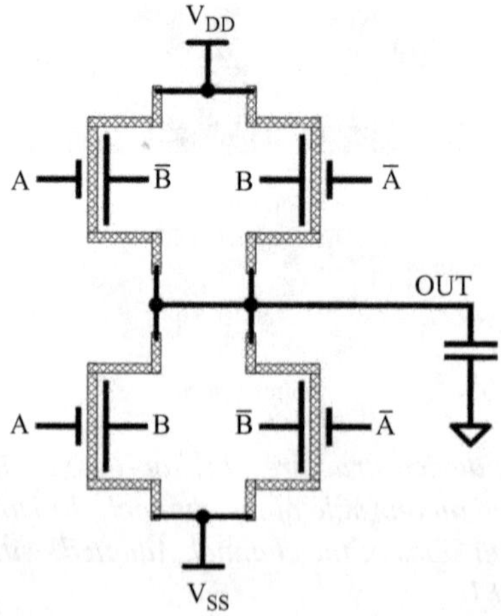

Figure 8.8 Schematic of ambipolar SiNWFET XOR. Adapted, with permission, from Reference [9]

than sweeping the voltage, this dual-gate FED approach can be used to predict static circuit performance. In addition, both V_{CG} and V_{PG} are used as independent inputs to transiently simulate the full-swing four-FED XOR gate, as shown in Figure 8.8.

8.4.3 Connection to underlying physics

As this TCAD approach group represents the ambipolar device behavior at the level of the electron dynamics, there is quite a strong connection to the underlying physics. With physical coefficients determined based on fabricated devices and materials, these approaches accurately represent and predict device behavior. Therefore, TCAD simulations are generally considered as authoritative results that can be used to validate other models when experimental measurements are not available.

8.4.4 Experimental matching

As expected given the physics-based computations of TCAD, this approach group provides exceptional matching to experimental device measurements and can thus be used to reliably predict device performance when different parameters are used. Device and material parameters determined from measurements are used to increase the simulation accuracy. While the approach of Section 8.4.2, [9] validates the TCAD simulation results against experimental measurements, the approach of Section 8.4.1 [8] includes only the simulation results. In general, TCAD provides highly accurate simulation results, as expected.

8.4.5 SPICE compatibility

Though TCAD provides accurate simulation results based on the underlying physics, these approaches are not generally compatible with ECAD environments such as

SPICE. To achieve its high accuracy, TCAD requires a tremendous amount of computational resources, resulting in a long simulation runtime. Consequently, these approaches cannot be easily applied to logic circuit simulations. In other words, the high accuracy of TCAD simulations can be useful for single-device design and simulation, but this high accuracy inhibits the use of TCAD for large-scale circuit design.

8.4.6 PG modulation

While many other FED simulation approaches use the PG only as a binary switch, these TCAD approaches permit the PG to modulate the drain–source current in a fashion similar to that of the control gate. These TCAD approaches thus provide full PG modulation, enabling the analysis of the complex device behaviors involved in circuit design.

8.4.7 Dynamic polarity switching (interchangeability)

Although static PG voltages are used for both approaches, Section 8.4.2 [9] provides the first transient polarity switching simulation of an FED, as in Figure 8.8. Though the transient capabilities of this FED approach cannot be used directly for large-scale circuit simulation due to the lack of SPICE-compatibility, these approaches do provide authoritative data which can enlighten and support further development of SPICE-based transient models.

8.5 Physical ambipolar models

One of the primary advantages of using FEDs is that the dual-gate control of the ambipolarity enables a reduction of circuit complexity and device count compared to conventional CMOS technologies. Therefore, while previous models permit the independent PG to switch the majority carrier of the device, the missing SPICE-compatibility or physical derivation significantly limits the applicability of these models. In addition to supporting static polarity switching, the SPICE-compatible models described in this section are both physically derived and experimentally validated.

8.5.1 MATLAB model for multiple-independent-gate FETs

The MATLAB-based model for multiple-independent-gate (MIG) semiconductor transistors described in [10] represents the behavior of devices that have higher channel controllability than conventional CMOS transistors [39]. In addition to a charge-based model of MOSFETs [40], the proposed model is extended to MIG structures by adapting novel modeling techniques. This model sets a predetermined arbitrary error for iteration steps in order to achieve better evaluation convergence. A technique depicted in Figure 8.9 is introduced to determine and converge the current and voltages of the model for single-gate gate-all-around MOSFETs. Due to the device nonlinearity, converging the model with closed-form solutions is difficult.

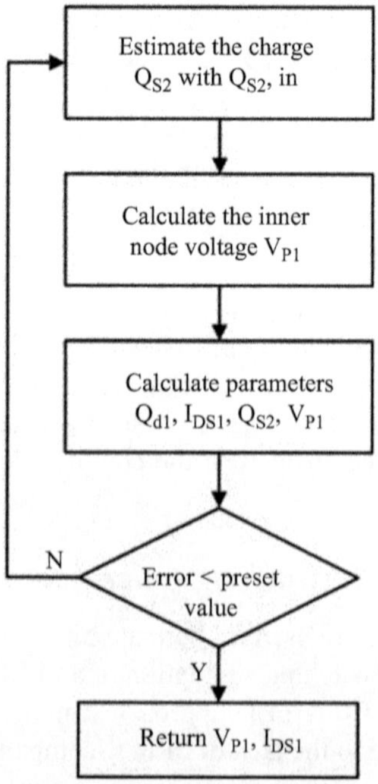

Figure 8.9 Flowchart of modeling technique. Adapted, with permission, from Reference [10]

To solve this problem, charges within the device are first calculated according to

$$Q_{d1} = C_{OX}\left(-\frac{2C_{OX}V_{th}^2}{Q_0} + \sqrt{\left(\frac{2C_{OX}V_{th}^2}{Q_0}\right)^2 + B}\right) \tag{8.13}$$

$$B = 4V_{th}^2 \log\left(1 + e^{(V_{GS1} - V_T + \Delta V_T - V_{p1})/2V_{th}}\right). \tag{8.14}$$

The device inner potential can then be estimated through

$$V_{GSi} - \Delta\varphi - V - \frac{kT}{q}\log\left(\frac{8}{\delta R^2}\right) = \frac{Q}{C_{OX}} + \frac{kT}{q}\log\left(\frac{Q}{Q_0}\right) + \frac{kT}{q}\log\left(\frac{Q+Q_0}{Q_0}\right) \tag{8.15}$$

to derive the drain–source current I_{DS} as

$$I_{DS1} = \mu\frac{2\pi R}{L_1}\left(2\frac{kT}{q}(Q_{s1} - Q_{d1}) + \frac{Q_{s1}^2 - Q_{d1}^2}{2C_{OX}} + \frac{kT}{q}Q_0\log\left(\frac{Q_{d1}+Q_0}{Q_{s1}+Q_0}\right)\right) \text{ [41].} \tag{8.16}$$

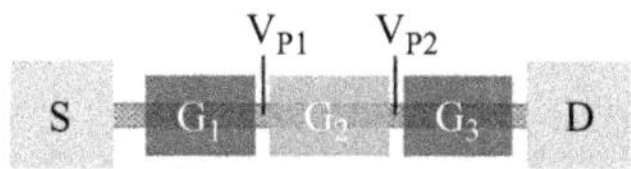

Figure 8.10 Structure diagram of a three-section device. V_{P1} and V_{P2} are the voltages of the gateless regions between the gates. Adapted, with permission, from Reference [10]

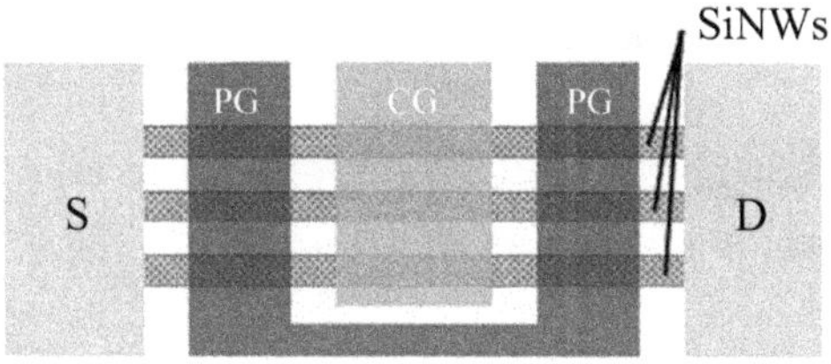

Figure 8.11 Side view of device structure. Adapted, with permission, from Reference [10]

While this unlimited iteration could cause an infinite loop, an arbitrary error threshold ε is used to manually control the trade-off between convergence time and accuracy according to

$$Err(m) := ||V_{p1}(m+1) - V_{p1}(m)|| < \epsilon. \tag{8.17}$$

As the device process technology scales down, nonidealities such as quantum mechanical effects and short-channel effects can result in alternative device behaviors. To improve the model accuracy while extrapolating to sub-10 nm devices, these nonideal effects are incorporated into model.

In addition to the single-gate MOSFET, advanced multi-gate devices are also modeled. As illustrated in Figure 8.10, the FED consists of gateless regions marked as V_{P1} and V_{P2}. The charge and capacitance within these regions differ from the regions that are encircled by the gate metal. Therefore, improved techniques are developed for the three-section structure and the gateless regions between adjacent gates. The model is implemented in MATLAB and then validated via comparison to TCAD Atlas simulation results. The voltages applied to the two additional PGs (V_{G1} and V_{G3}) are only introduced to modulate the carrier density and internal node voltages of the device. As all the models relate only to devices that exhibit n-type behavior, no polarity switching is presented.

For the three-section structure (Figure 8.11) proposed in [42], gates close to the two metal-semiconductor contacts can be used to control the channel carrier density by modulating the local SB height. In the simulations presented in this work [10], gates one and three of the device are always connected to one another to ensure efficient control over the device polarity.

This work described a MATLAB model for gate all-around transistors. The arbitrary error improves the model convergence speed while providing accurate results compared to TCAD simulations. Multiple techniques are presented to enable accurate

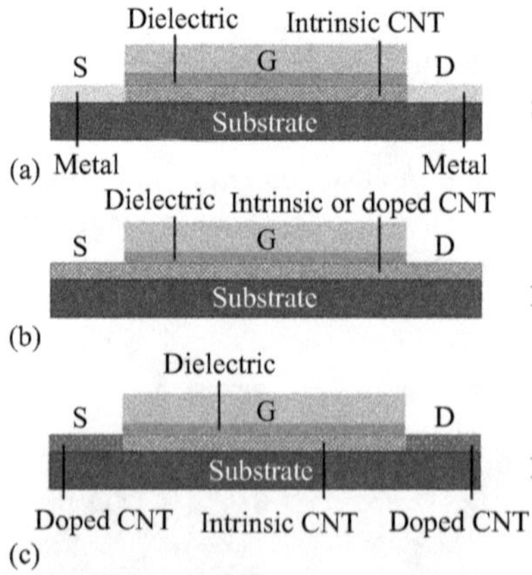

Figure 8.12 Side view of device structures. Three CNTFET structure are depicted, where (a) has direct contact between the metal and semiconductor channel (grid), while (b) and (c) do not include any metal-semiconductor contact that introduces a Schottky barrier. The source and drain terminals of (c) are doped for higher carrier density as compared to an intrinsic CNT (adapted from [11])

representation of the behavior of N-section structures and the intermediate gateless regions. Nonidealities such as quantum mechanical effects and short-channel effects are also integrated for better matching [43,44]. It may be possible to implement the expressions used in this model within a SPICE environment, but this is not clear.

8.5.2 Ambipolar VHDL-AMS model with binary PG for CNTFETs

A SPICE-compatible model for SB CNTFETs is proposed in [11]. This model utilizes the PG as an independent input terminal, enabling various reconfigurable logic circuits that make efficient of use of FEDs. Three possible types of CNTFETs [20] are discussed in this work, as shown in Figure 8.12: the SB, *Partially Gated*, and MOS-like doped CNTFETs. While both the PG and MOS-like devices exhibit unipolar behavior similar to conventional CMOS transistors [17], the majority carrier type of the ambipolar SB-CNTFETs can be modulated by a voltage applied.

While this model is an extension of [3], the convergence problems from that original model are solved by smoothing out discontinuities. This work therefore provides a greater ability to converge in circuit design and simulation. As the original physically derived equations are included in this model, this work also provides the possibility of extrapolating the performance of similar devices. The CNT chirality and diameter are also included in this model, increasing its ability to represent physical behavior. The drain–source current equation used in this model is

$$I_D = \frac{4ek_BT}{h}\sum_{p=1}^{+\infty}\left[\ln\left(1+\exp\frac{-\Delta_P+V_{CNT}}{k_BT}\right)\right. \\ \left. -\ln\left(1+\exp\frac{-V_{DS}-\Delta_P+V_{CNT}}{k_BT}\right)\right], \quad (8.18)$$

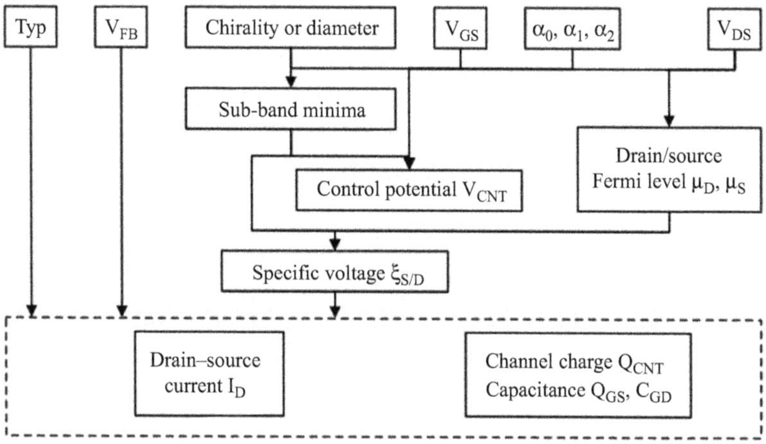

Figure 8.13 Model flowchart. Adapted, with permission, from Reference [11]

where

$$V_{CNT} = V_{GS} - \frac{Q_{CNT}}{C_{OX}} \tag{8.19}$$

In these expressions, e is the electron charge, k_B is the Boltzmann constant, h is the Planck constant, T is the temperature, and Δp is the pth sub-band energy minimum.

The simulation results from this model are compared to both numerical models and experimental data. In comparison to other numerical models [20], this model matches the $I_{DS} - V_{GS}$ relations and provides a smooth transition between discrete states, demonstrating model convergence. In terms of experimental data from [16,45], several $I_{DS} - V_{GS}$ plots with different device parameters are presented to validate this model. This model exhibits moderate matching to these experimental measurements, but does not provide $I_{DS} - V_{GS}$ analyses as presented in [16], as this work does not enable sweeping of V_{GS}.

This work implemented the model using a Verilog-A script to provide better model convergence, following the procedure described in the flowchart (Figure 8.13). Similar to [3], physical parameters such as the flat-band voltage V_{FB}, chirality, and diameter are used in calculation. In addition, fitting parameters α_0, α_1, and α_2 are also included for effective control of the electrical potential V_{CNT} calculation. This model enabled use of the back gate voltage as an independent input for the first time to select the majority carrier type of the device, enabling switching of the device polarity between n- and p-type.

The authors of [11] also propose a double-gate CNTFET structure with multiple gate sections based on the concept of [17] to improve the carrier mobility, with the back gate covering both the CNT source and drain contacts. The SB height at the CNT source and CNT drain contacts can be modulated by the voltage applied to the back gate. If the V_{GS-BG} is lower than a threshold voltage, the holes in the CNT become the majority carriers and the device operates as a p-type CNTFET. On the other hand, a sufficiently high V_{GS-BG} lowers the SB height, causing n-type conductivity with

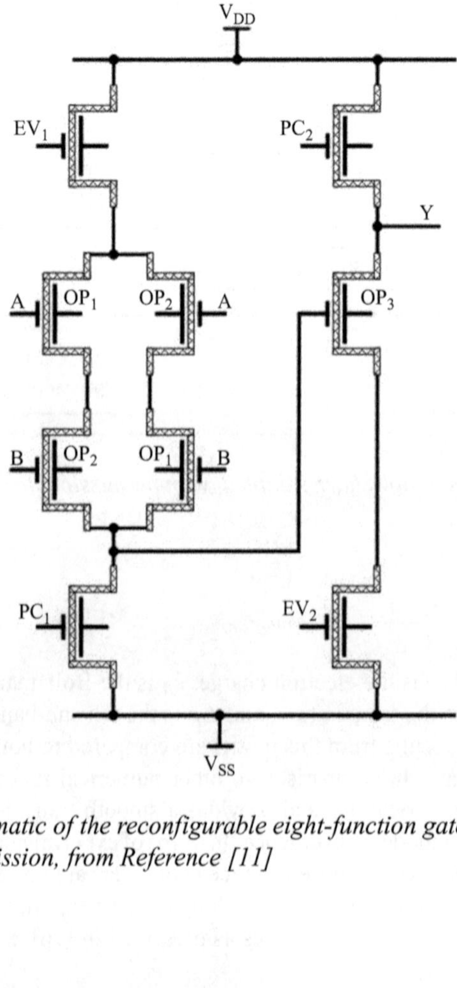

Figure 8.14 Schematic of the reconfigurable eight-function gate. Adapted, with permission, from Reference [11]

electrons as the dominant carriers. When the V_{GS-BG} is floating, the carrier mobility is decreased to a minimum level and the drain–source channel exhibits a relatively high resistance. This double-gate structure enables functionality-enhanced CNTFET behavior.

In this work, the back-gate voltage V_{GS-BG} which controls the polarity is always set to either +1 V or −1 V, switching the device between n- and p-type operation, respectively. Unlike the model of Section 8.2.3 [5], this model provides an independent input terminal for the device which enables comprehensive circuit simulations. An eight-function logic gate based on the CNTFET FED is presented in Figure 8.14. As shown in Figure 8.15, this logical functionality of this circuit can be modified through control of V_{ba}, V_{bb}, and V_{bc}, with transient simulation of this circuit included in [11]. This FED thus enables reconfigurable functionality for secure hardware design, and a variety of logic circuits are presented and simulated thanks to the independent polarity selection achieved with this model.

OP_1	OP_2	OP_3	OUT
+V	+V	+V	$\overline{A\cdot B}$
+V	+V	–V	$A\cdot B$
+V	–V	+V	$\overline{A\oplus B}$
+V	–V	–V	$A\oplus B$
–V	+V	+V	$\overline{A\oplus B}$
–V	+V	–V	$A\oplus B$
–V	–V	+V	$A+B$
–V	–V	–V	$\overline{A+B}$

Figure 8.15 Function table of the eight-function reconfigurable logic circuit. Adapted, with permission, from Reference [11]

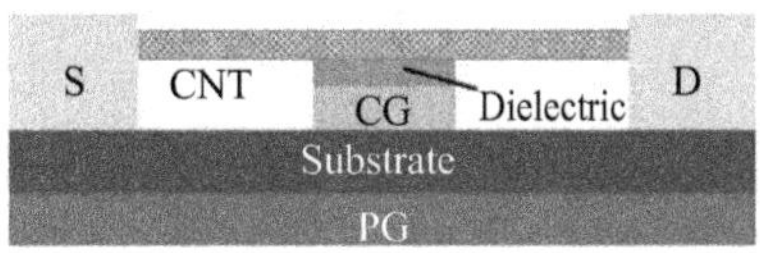

Figure 8.16 Side view of device structure. Adapted, with permission, from Reference [12]

This Verilog-A model enables flexible polarity switching with a back gate voltage V_{BG} that is used as a binary input to control device polarity, enabling transient logic simulations. The device polarities are preset before the simulation, and there is therefore no polarity switching during the transient simulation.

8.5.3 Ambipolar Verilog-A model for CNTFETs

An ambipolar model for dual-gate CNTFETs is developed in [12] based on the unipolar model of Section 8.2.2 [4], with the two independent gates modulate the drain–source current through the CNT channel as shown in Figure 8.16. Along the CNT channel, the energy bands in the region close to the top gate are dominated by the top gate, while the SB near the source and drain contacts are controlled by the back (polarity) gate voltage. The SB height can be obtained by

$$\begin{aligned}\varphi_{\mathrm{Sn}} &= (\varphi_{\mathrm{SB}} - (sbbd[1] - V_{\mathrm{CNTs,Si}})\exp(-d_{\mathrm{tunnel}}/\lambda)) \\ &\quad + (sbbd[1] - V_{\mathrm{CNTs,Si}}). \qquad (8.20)\end{aligned}$$

The drain–source current resulting from the charge-based current can be obtained by

$$
\begin{aligned}
I_{DS-e^-} &= \frac{2Mek_BT}{h}\sum_{p=1}^{nb\_sbbd}\left[\ln\left(1+\exp\frac{-\max\left(\varphi_{\mathrm{Sn}}, sbbd[p]-V_{\mathrm{CNTi,S}}, \varphi_{\mathrm{Dn}}\right)}{k_BT}\right)\right]\\
&\quad -\ln\left(1+\exp\frac{-\max\left(\varphi_{\mathrm{Sn}}, sbbd[p]-V_{\mathrm{CNTi,S}}, \varphi_{\mathrm{Dn}}\right)-V_{\mathrm{Di,Si}}}{k_BT}\right)
\end{aligned}
\tag{8.21}
$$

In addition to this charge-induced drain–source current, the band-to-band tunneling (BTBT) effect can also introduce a carrier flow between the source and drain contacts [46]. Thus, the BTBT current can be determined with

$$
\begin{aligned}
I_{\mathrm{BTBT\_}e} &= \frac{2Mq}{h}T_{\mathrm{BTBT}}k_BT\\
&\quad\times\left[-\ln\left(1+\exp\left(-\left(\frac{-sbbd[1]-qV_{\mathrm{CNTi},s}+qV_{\mathrm{Si},S}}{k_BT}\right)\right)\right)\right.\\
&\quad+\ln\left(1+\exp\left(-\left(\frac{\max\left(\varphi_{\mathrm{Sn}},\varphi_{\mathrm{Dn}}\right)+qV_{\mathrm{Si},S}}{k_BT}\right)\right)\right).\\
&\quad+\ln\left(1+\exp\left(-\left(\frac{-sbbd[1]-qV_{\mathrm{CNTi},s}+qV_{\mathrm{Di},S}}{k_BT}\right)\right)\right)\\
&\quad-\ln\left(1+\exp\left(-\left(\frac{\max\left(\varphi_{\mathrm{Sn}},\varphi_{\mathrm{Dn}}\right)+qV_{\mathrm{Si},S}}{k_BT}\right)\right)\right)
\end{aligned}
\tag{8.22}
$$

A complete equivalent circuit of the CNTFET FED is described in the paper, with node voltages described by

$$
\begin{cases}
V_{\mathrm{CNTs,S}} = \dfrac{C_{\mathrm{SEi}}V_{\mathrm{CNTi,S}}+C_{\mathrm{SEx}}V_{\mathrm{Si},S}+V_{\mathrm{BGiS,S}}C_{\mathrm{BGS}}L_S+Q_S}{C_{\mathrm{BGS}}L_S+C_{\mathrm{SEi}}C_{\mathrm{SEx}}}\\[2ex]
V_{\mathrm{CNTi,S}} = \dfrac{C_{\mathrm{DEi}}V_{\mathrm{CNTd,S}}+C_{\mathrm{SEi}}V_{\mathrm{CNTs,S}}+V_{\mathrm{Gi,S}}C_{\mathrm{FG}}L_I+Flag(0.5_{\mathrm{VBGiS,S}}C_{\mathrm{BGi}}L_I+0.5V_{\mathrm{BGiD,S}}C_{\mathrm{BGi}}L_I)+Q_i}{C_{\mathrm{FG}}L_I+C_{\mathrm{SEi}}+C_{\mathrm{DEi}}+Flag(C_{\mathrm{BGi}}L_I)}.\\[2ex]
V_{\mathrm{CNTd,S}} = \dfrac{C_{\mathrm{DEi}}V_{\mathrm{CNTi,S}}+C_{\mathrm{DEx}}V_{\mathrm{Di},S}+V_{\mathrm{BGiD,S}}C_{\mathrm{BGD}}L_D+Q_D}{C_{\mathrm{BGD}}L_D+C_{\mathrm{DEi}}C_{\mathrm{DEx}}}
\end{cases}
\tag{8.23}
$$

While the single-gate CNTFET model is not compared to any experimental data, the CNTFET FED model is validated against multiple experimental measurements. Sweeping the control gate voltage V_{CG}, the simulation results are compared to measurements from IBM [47]. Sweeping the PG, the model is compared to the measurements obtained in [48] prior to fabrication of the PG. This model is also simulated with different parameters and compared to measurements from [49]. In all of these cases, the matching to experimental data is quite close.

This model was developed using Verilog-A, which is fully compatible with the SPICE environment. The PG is not merely a binary switch, and its voltage is incorporated into the determination of the current through the CNT. V_{PG} is used to adjust the frequency of a ring oscillator formed by three CNTFET FED-based INVs, as

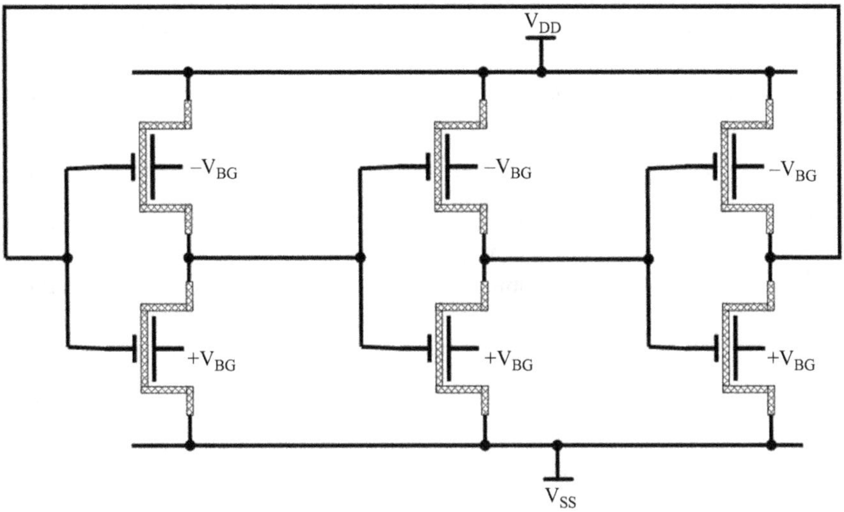

Figure 8.17 Schematic of ambipolar CNTFET ring oscillator. Adapted, with permission, from Reference [12]

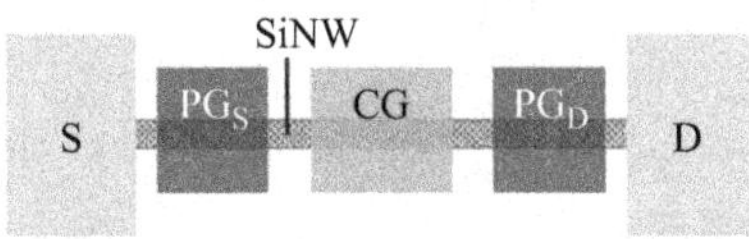

Figure 8.18 Side view of device structure. Adapted, with permission, from Reference [13]

shown in the circuit schematic of Figure 8.17. This circuit is analyzed in a transient simulation, but without dynamic switching of the device polarity.

8.5.4 Ambipolar model for SiNWFETs

To enable SiNWFET FED circuit design, a SPICE-compatible device model is proposed in [13] for the device experimentally demonstrated in [42]. For the triple independent gate structure [50] shown in Figure 8.18, the channel majority carrier type and the corresponding mobility can be modulated by the voltage applied to the PGs. This model thus represents the electrical behavior of this SiNWFET FED in a SPICE-compatible environment.

There are two major components of the drain–source current (I_{DS}) calculation. First, the SiNW surface potential is calculated according to the one-dimensional Poisson's equation; then, the surface potential is used to calculate the SB height for the device. From this derivation, the following two expressions can be obtained:

$$\frac{q(V_{geff} - V_n - \Delta V)}{kT} = \frac{q}{kT} \cdot \frac{Q_n}{C_{ox}} + \ln\left(\frac{Q_n}{Q_0}\right) + \ln\left(\frac{Q_n + Q_0}{Q_0}\right) \tag{8.24}$$

$$\frac{q(V_p - V_{geff} - \Delta V)}{kT} = \frac{q}{kT} \cdot \frac{Q_p}{C_{ox}} + \ln\left(\frac{Q_p}{Q_0}\right) + \ln\left(\frac{Q_p + Q_0}{Q_0}\right) \tag{8.25}$$

In these equations, $Q_0 = 4(kT/q)(\varepsilon_s/R)$, while $\Delta V = (kT/q)\ln(8/\delta R^2)$ and V_{geff} are the effective gate voltage derived from

$$V_{geff} = V_G - V_{FB} \tag{8.26}$$

where V_G is the voltage applied to a gate and V_{FB} is the flat-band voltage for the corresponding charge carrier. When both majority and minority carriers flow through the device simultaneously, the minority carrier density is also considered in order to improve the model accuracy:

$$Q_{\mathrm{n,min}} \approx \frac{qn_i}{R} \int_0^R r e^{(q(\phi_p - v_n))/kT} \mathrm{d}r \tag{8.27}$$

Combining the two carrier densities, the unified carrier density within the SiNW can be derived from

$$Q_{\mathrm{n}} = \sigma_1 \cdot Q_{\mathrm{n,maj}} + \sigma_2 \cdot Q_{\mathrm{n,min}} \tag{8.28}$$

After the carrier density of the channel is computed, the drain–source current can be derived based on the effective SB height determined by the two PG voltages V_{PGs} and V_{PGd} as

$$\begin{aligned} \Phi_{\mathrm{SBeff,n}} &= \Phi_{\mathrm{SB,n}} - (\phi_{\mathrm{PG_S}} - V_{\mathrm{bi}} - V_{\mathrm{S}})(1 - e^{-d_t/\lambda}) \\ \Phi_{\mathrm{SBeff,p}} &= \Phi_{\mathrm{SB,p}} - (V_{\mathrm{bi}} + V_{\mathrm{D}} - \phi_{\mathrm{PG_D}})(1 - e^{-d_t/\lambda}) \end{aligned} \tag{8.29}$$

where $\phi_{SBeff,n}$ and $\phi_{SBeff,p}$ are the SB height for electrons and holes, respectively.

The control gate voltage V_{CG} is used to modulate the SB height in the central portion of the device wrapped by the CG metal as shown in Figure 8.18. The CG-controlling barrier height ϕ_{Cn} and ϕ_{Cp} is determined as

$$\begin{aligned} \Phi_{C,n} &= (V_{bi} + V_s + \phi_{SB,n}) - \phi_{CG} \\ \Phi_{C,p} &= \phi_{CG} - (V_{bi} + V_D - \phi_{SB,p}) \end{aligned} . \tag{8.30}$$

Using the SB height within the regions of the device, the carrier tunneling current [14,48] can be calculated by

$$\begin{aligned} I_{\mathrm{T,n}} &= \pi R^2 A_n^* T^2 \exp\left(-\frac{q\Phi_{\mathrm{SB,n}}}{kT}\right)\left(e^{\frac{q(v_{\mathrm{D}} - v_{\mathrm{n,PG_D}})}{kT}} - 1\right) \\ I_{\mathrm{T,p}} &= \pi R^2 A_p^* T^2 \exp\left(-\frac{q\Phi_{\mathrm{SB,p}}}{kT}\right)\left(e^{\frac{q(v_{\mathrm{p,PG_S}} - v_{\mathrm{S}})}{kT}} - 1\right) \end{aligned} \tag{8.31}$$

In addition to the tunneling current, the drift-diffusion current is also included within this model as

$$\begin{aligned} I_{\mathrm{DD,n}} &= \mu_{\mathrm{n}} \frac{2\pi R}{L_{\mathrm{eff}}}[G(Q_{\mathrm{n,S}}) - G(Q_{\mathrm{n,D}})] \\ I_{\mathrm{DD,p}} &= \mu_{\mathrm{p}} \frac{2\pi R}{L_{\mathrm{eff}}}[G(Q_{\mathrm{p,D}}) - G(Q_{\mathrm{p,S}})] \end{aligned} \tag{8.32}$$

Within these expressions, R and L_{eff} are the radius and length of the silicon nanowire, and $G(Q)$ is defined as

$$G(Q) = 2\frac{kT}{q}Q + \frac{Q^2}{2C_{\text{ox}}} - \frac{kT}{q}Q_0 \ln\frac{Q_0 + Q}{Q_0} \tag{8.33}$$

Combining the two types of current, the following expressions can be obtained:

$$\begin{aligned} V_{\text{n,PG}_\text{D}} &= V_\text{D} - \frac{kT}{q}\ln\left[1 + \exp\left(\frac{q(\Phi_{\text{SB,n}} - \Phi_{\text{eff,n}})}{kT}\right)\right] \\ V_{\text{p,PG}_\text{S}} &= V_\text{S} + \frac{kT}{q}\ln\left[1 + \exp\left(\frac{q(\Phi_{\text{SB,p}} - \Phi_{\text{eff,p}})}{kT}\right)\right] \end{aligned} \tag{8.34}$$

By solving the equations, the device surface potential and drain–source current can be calculated.

To validate this model, the authors compared it to numerical data as well as TCAD. The numerical data is calculated with the Poisson equation, and matches well with the model simulation results. The model is also compared to TCAD, and matches well the TCAD simulation results with both CG and PG control. As the electrical behavior is determined based on physical parameters, this model can be extrapolated to devices that have a similar performance.

With this model, the majority carrier density can be modulated through control of the PG voltage by regulating the SB height within the PG region. Numerous drain–source current simulations validate the accuracy of the proposed model. Rather than using the PG only as a switch, this model integrates the full static modulation behavior of the PG.

8.5.5 Connection to underlying physics

The primary strategy of the models in this group is to calculate the SB height and the associated carrier mobilities. Models in this group incorporate physical parameters such as device diameter, channel length, and carrier mobility; these models can therefore be used to predict the performance of devices with alternative dimensions. However, the strong connection to the underlying physics introduces difficulty in enabling the dynamic polarity switching required for full utilization of the FED characteristics.

8.5.6 Experimental matching

Validation using authoritative data ensures the accuracy of these models. All models in this group are compared to either experimental measurements or TCAD emulation results. In either case, the models are shown to accurately represent the devices.

8.5.7 SPICE compatibility

Comprehensive circuit simulation can be performed with models compatible to SPICE. While the models of Sections 8.5.2–8.5.4 [11–13] include implementation with SPICE-compatible description languages, it is not clear whether the model of Section 8.5.1 [10] is SPICE-compatible; the use of physical expressions to determine the electrical behavior suggests that it should be possible to implement the approach in a SPICE-compatible manner.

8.5.8 PG modulation

The PG of the model of Section 8.5.1 [10] is used to modulate the internal node voltages, thus potentially enabling static PG modulation of the majority carrier concentration; however, this is not demonstrated in the paper. In the model of Section 8.5.2 [11], the V_{PG} values are set to $+/-V_{dd}$ and do not enable modulation of the source–drain current. The models of Sections 8.5.3 [12] and 8.5.4 [13] enable full static polarity switching by applying a sweeping voltage to the PG.

8.5.9 Dynamic polarity switching (interchangeability)

The models of Sections 8.5.2 [11] and 8.5.3 [12] enable transient simulation with static device polarity. In all four models, the device polarity cannot be switched during a transient simulation.

8.6 Ambipolar models with dynamic polarity

Transient simulation is a powerful capability that is absolutely critical in enabling circuit designers to fully explore circuit performance. For transient simulations with conventional unipolar devices, the drain and source terminals can be carefully predetermined without any deleterious impact on circuit performance and simulation accuracy. However, for FEDs, the majority carrier is switched during the course of FED circuit operations, thus requiring that a model include the ability to switch the polarity and interchange the drain and source terminals. If this interchangeability is not fully integrated within an FED model, the simulation results can be inaccurate. It should be noted that FED models that do not enable transient simulation, cannot, by definition provide dynamic polarity or interchangeability. This section introduced models that permit the device polarity to be switched dynamically during the transient simulation; the first does not provide interchangeability, the second does.

8.6.1 Dynamic Verilog-A model with fixed source and drain for CNTFETs

The model of Section 8.5.3 [12] was improved to enable transient simulations with dynamic polarity switching [14]. While dynamic polarity switching had previously been demonstrated through TCAD simulation as described in Section 8.4.2 [9], the model of [14] presents the first transient SPICE simulations in which the FED polarity is dynamically switched. The FED logic gate of [51], shown in Figure 8.19, was

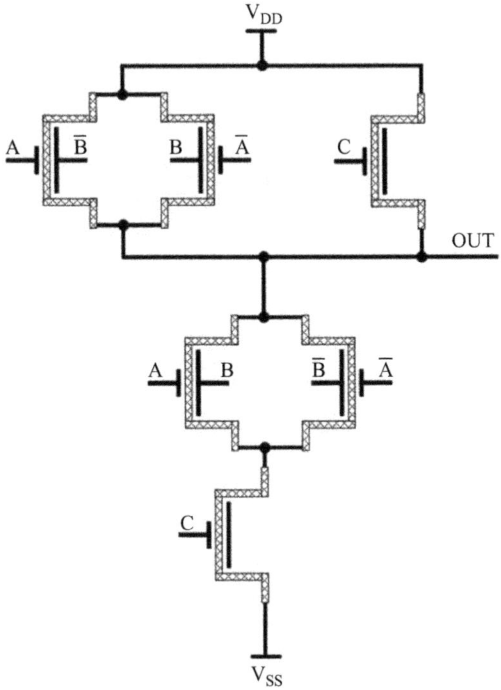

Figure 8.19 Schematic of ambipolar CNTFET logic gate. Adapted, with permission, from Reference [14]

simulated with input signal B connected to the PGs instead of the conventional control gates. As the B signal switches during the simulation, the device polarity switches dynamically between n- and p-type operation. The paper shows simulation results that provide the correct binary behavior, but the shape of the simulation waveforms prevent a conclusive determination of the proper representation of the dynamic polarity switching.

As explained above, source–drain interchangeability is one of the critical challenges encountered when modeling FEDs. When the polarity of a physical FED is switched, the drain and source are dynamically interchanged; however, it appears that the nodes in this model may require initialization before the simulation begins and are therefore fixed during the transient simulation. As it is not clear whether the source and drain can be dynamically interchanged during a transient simulation, this model cannot be relied upon to provide accurate results for dynamic polarity switching.

This work provided the first SPICE-compatible FED simulations that incorporate dynamic polarity switching, which is required to fully evaluate FED-based logic circuit. However, this model does not demonstrate the source–drain interchangeability which is required for accurate FED modeling.

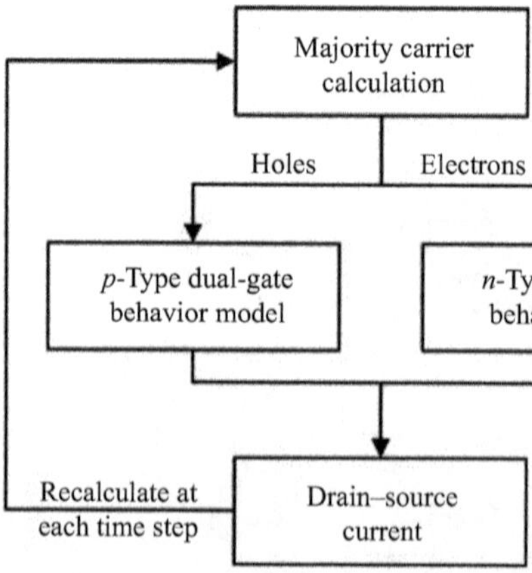

Figure 8.20 Model flowchart. Adapted, with permission, from Reference [15]

8.6.2 Dynamic Verilog-A model with source–drain interchangeability for CNTFETs

This absence of source–drain interchangeability in the model of Section 8.6.1 [14] inspired the development of a model that does provide source–drain interchangeability. Reference [15] presents the first SPICE-compatible FED model that enables both dynamic polarity switching and source–drain interchangeability. This model is based on the behavioral model of Section 8.3.2 [7], with a more complex framework with additional fitting parameters developed in order to model the dynamic polarity switching and enable source–drain interchangeability. The added fitting parameters for device leakage current and output swing range permit better model convergence. This model accurately reproduces the experimental device behavior of [27], enabling its use for large-scale FED circuit design.

Though the model of Section 8.3.2 [7] enables steady-state logic simulations for CNTFET FEDs by integrating the PG sweeping ability, it is ineffective for transient logic simulation because of a lack of source–drain interchangeability. This is resolved by the approach shown in the flowchart of Figure 8.20, with the majority carrier automatically calculated based on the terminal voltages. With this framework, the physical source–drain interchangeability is integrated into the model; therefore, accurate transient simulations with dynamic polarity switching can be realized. This model therefore permits the first transient cascaded FED logic simulations in which the FED polarities are dynamically switched, as applied to a full-swing cascaded FED XOR gate.

As the ability to switch device polarity is the primary feature of FEDs, the development of a model that enables dynamic polarity switching is a major advance for the field. Although the drain and source nodes in fabricated devices are physically interchangeable, no previous SPICE-compatible model had integrated this interchangeability within the model. This model enables the first SPICE-base transient simulation for both single-stage and cascaded logic circuits with dynamic polarity switching, finally enabling a complete behavioral analysis of FED logic circuits.

8.6.3 Connection to underlying physics

The models in this approach group are based on previous work, with added dynamic polarity switching. While the model of Section 8.6.1 [14] is based on the physically derived model of Section 8.5.3 [12], the core computational expressions of the model of Section 8.6.2 [15] are based on the behavioral FED model of Section 8.3.2 [7]. As the primary intent of the model of Section 8.6.2 [15] is to integrate the source–drain interchangeability within the SPICE environment rather than to precisely represent the physical behavior, this model has limitations regarding extrapolation to other devices. However, the interchangeable framework of this model can be adapted to alternative techniques of representing the underlying physics.

8.6.4 Experimental matching

As minimal experimental measurements of the dynamic FED polarity switching are available, the models in this approach group have no opportunity for comparison to transient experimental data. However, the core expressions of these models were previously shown to match static experimental measurements (see Sections 8.5.3 [12] and 8.3.2 [7]). These models can therefore be considered as matching to the available experimental data until future transient experiments enable tuning of the dynamic properties of these models.

8.6.5 SPICE compatibility

Both the models of Sections 8.6.1 [14] and 8.6.2 [15] are SPICE-compatible. While TCAD had previously been the only simulation technique that enabled dynamic polarity switching simulations due to the requirement of source–drain interchangeability, the model of Section 8.6.2 [15] is the first that demonstrates this interchangeability within a SPICE environment. This SPICE-compatibility dramatically increases the applicability of the model while decreasing the simulation time with minimal loss in accuracy.

8.6.6 PG modulation

As these models are dynamic extensions of models (Sections 8.5.3 [12] and 8.3.2 [7]) that provide static polarity switching, the models in this approach group also permit the PG voltage to modulate the carrier density of the CNT channel.

8.6.7 Dynamic polarity switching (interchangeability)

In addition to static polarity switching, these models present dynamic polarity switching of ambipolar CNTFET FEDs. Furthermore, the model of Section 8.6.2 [15] is the first SPICE-compatible model that provides dynamic polarity switching and source–drain interchangeability. This interchangeability framework permits an accurate representation of the FED behavior during dynamic polarity switching.

8.7 Summary

This chapter described several FED models and the capabilities of these models in terms of five crucial model characteristics: connection to underlying physics, experimental matching, SPICE compatibility, PG modulation, and source–drain interchangeability with dynamic polarity switching. This systematic analysis of the history of FED modeling can help to guide the future development of FED device models.

As with conventional device models, a physically derived FED model can provide accurate results while also predicting device behavior for a range of parameters. As device fabrication is a time-consuming and expensive procedure, a model connected to the underlying physics is highly useful for device and circuit optimization.

To validate the correctness and accuracy of a model, comparisons to experimental measurements are critical. The availability of experimental FED data is limited by the fact that only a few devices have been fabricated; there is limited ability to verify that an FED model matches the physics of an experimental device.

In order to design large-scale logic circuits, it is important to use a SPICE-compatible model with parameters that can be easily modified and that can be evaluated rapidly. While TCAD tools can provide better accuracy, SPICE-compatible models provide superior computational efficiency. Therefore, SPICE-compatible ones provide much more utility in large-scale circuit design and analysis.

One of the key features of FEDs is that the device polarity can be modulated to perform complex logic functions using a single device. Beyond the capabilities of conventional CMOS devices, these FEDs with additional PGs can be used to modulate the carrier density within the semiconductor channel. While some models control the device polarity through a binary input, the more advanced models use the PG as a fully functional gate that modulates the drain–source current, thus providing better matching to experimental measurements.

In addition to the aforementioned traits, a model must enable transient simulations in order to design and analyze large-scale circuits. As cascading is required in computing systems [52], static models that do not support the transient simulation of cascaded logic circuits have limited utility. Unlike conventional CMOS transistors where the device polarity is fixed, FED polarity can be switched by modulating the terminal voltages. Given that the determination of source and drain nodes is a function of device polarity relative to the source and drain voltages, it is highly important that an FED model represent this source–drain interchangeability. This is necessary in order to properly represent the dynamic polarity switching of FEDs and provide accurate transient analyses of the dynamic switching process. Though several FED models enable static polarity switching, only one FED model (see Section 8.6.2 [15]) provides the SPICE-compatible interchangeability necessary to evaluate FED logic gates.

This chapter's analysis of the various models and important model features can be used to guide further development of FED model for logic circuit design. Therefore, the authors of this chapter are developing a model that provides all the model features discussed in this chapter. By applying the interchangeability framework [15] to a model that accurately represent the underlying physics, the resulting closed-form functionality-enhanced FED model will be highly effective for large-scale transient circuit simulations.

References

[1] L. W. Nagel and D. O. Pederson, "SPICE: Simulation Program with Integrated Circuit Emphasis," in *Proc. 16th Midwest Symp. Circuit Theory*, 1973.

[2] G. Fiori and G. Iannaccone, "Nano TCAD ViDES," 2016. [Online]. Available: http://vides.nanotcad.com/vides/.

[3] A. Raychowdhury, S. Mukhopadhyay, and K. Roy, "A circuit-compatible model of ballistic carbon nanotube field-effect transistors," *IEEE Trans. Comput. Des. Integr. Circuits Syst.*, vol. 23, no. 10, pp. 1411–1420, 2004.

[4] C. Maneux, J. Goguet, S. Fregonese, T. Zimmer, H. C. D'Honincthun, and S. Galdin-Retailleau, "Analysis of CNTFET physical compact model," in *Int. Conf. Des. Test Integr. Syst. Nanoscale Technol. 2006. DTIS 2006*, 2006.

[5] F. Pregaldiny, C. Lallement, and J.-B. Kammerer, "Design-oriented compact models for CNTFETs," in *Int. Conf. Des. Test Integr. Syst. Nanoscale Technol. 2006. DTIS 2006*, 2006, vol. 1, pp. 34–39.

[6] M. Hasan, P. E. Gaillardon, and B. Sensale-Rodriguez, "A continuous compact DC model for dual-independent-gate FinFETs," *IEEE J. Electron Devices Soc.*, vol. 5, no. 1, pp. 23–31, 2017.

[7] X. Hu and J. S. Friedman, "Closed-form model for dual-gate ambipolar CNTFET circuit design," in *Int. Symp. Circuits Syst.*, 2017.

[8] G. V. Resta, D. Lin, I. P. Radu, F. Catthoor, P.-E. Gaillardon, and G. De Micheli, "Scaling trends and performance evaluation of 2-dimensional polarity-controllable FETs," *Sci. Rep.*, vol. 7, pp. 1–9, 2017.

[9] M. De Marchi, D. Sacchetto, S. Frache, *et al.*, "Polarity control in double-gate, gate-all-around vertically stacked silicon nanowire FETs," in *IEDM*, 2012, pp. 183–186.

[10] A. Antidormi, S. Frache, M. Graziano, P. E. Gaillardon, G. Piccinini, and G. De Micheli, "Computationally efficient multiple-independent-gate device model," *IEEE Trans. Nanotechnol.*, vol. 15, no. 1, pp. 2–14, 2016.

[11] I. O'Connor, J. Liu, F. Gaffiot, *et al.*, "CNTFET modeling and reconfigurable logic-circuit design," *IEEE Trans. Circuits Syst. Regul. Pap.*, vol. 54, no. 11, pp. 2365–2379, 2007.

[12] S. Fregonese, C. Maneux, and T. Zimmer, "A compact model for dual-gate one-dimensional FET: Application to carbon-nanotube FETs," *IEEE Trans. Electron Devices*, vol. 58, no. 1, pp. 206–215, 2011.

[13] J. Zhang, P. E. Gaillardon, and G. De Micheli, "A surface potential and current model for polarity-controllable silicon nanowire FETs," in *Eur. Solid-State Dev. Res. Conf.*, 48–51, 2015.

[14] C. Maneux, S. Fregonese, and T. Zimmer, "Innovative dual-gate CNTFET logic cell: Investigation of technological dispersion impact through compact modeling," *IEEE Trans. Nanotechnol.*, vol. 13, no. 6, pp. 1053–1062, 2014.

[15] X. Hu and J. S. Friedman, "Transient model with interchangeability for dual-gate ambipolar CNTFET logic design," in *IEEE/ACM NANOARCH*, 2017, vol. 5, pp. 61–66.

[16] A. Javey, H. Kim, M. Brink, *et al.*, "High-κ dielectrics for advanced carbon-nanotube transistors and logic gates," *Nat. Mater.*, vol. 1, no. 4, pp. 241–246, 2002.

[17] S. J. Wind, J. Appenzeller, and P. Avouris, "Lateral scaling in carbon-nanotube field-effect transistors," *Phys. Rev. Lett.*, vol. 91, no. 5, p. 58301, 2003.

[18] J. Guo, M. Lundstrom, and S. Datta, "Performance projections for ballistic carbon nanotube field-effect transistors," *Appl. Phys. Lett.*, vol. 80, no. 17, pp. 3192–3194, 2002.

[19] J. Guo, S. Datta, and M. Lundstrom, "A numerical study of scaling issues for Schottky-barrier carbon nanotube transistors," *IEEE Trans. Electron Devices*, vol. 51, no. 2, pp. 172–177, Feb 2004.

[20] J. Guo, S. Datta, M. Lundstrom, *et al.*, "Assessment of silicon MOS and carbon nanotube FET performance limits using a general theory of ballistic transistors," in *Digest. International Electron Devices Meeting,* 2002, vol. 2, no. c, pp. 711–714.

[21] G. Seol, Y. Yoon, J. K. Fodor, *et al.*, "CNT Bands," *NANOHUB Online Simulations and More*, 2003. [Online]. Available: https://nanohub.org/resources/1838/.

[22] R. Saito, G. Dresselhaus, and M. S. Dresselhaus, *Physical Properties of Carbon Nanotubes*, vol. 3. London: Imperial College Press, 1998.

[23] J. Zhang, M. De Marchi, P.-E. Gaillardon, and G. De Micheli, "A Schottky-barrier silicon FinFET with 6.0 mV/dec subthreshold slope over 5 decades of current," in *IEDM*, 13.4.1–13.4.4, 2014.

[24] J. Zhang, J. Trommer, W. M. Weber, P. E. Gaillardon, and G. De Micheli, "On temperature dependency of steep subthreshold slope in dual-independent-gate FinFET," *IEEE J. Electron Devices Soc.*, vol. 3, no. 6, pp. 452–456, 2015.

[25] J. G. Fossum, R. Sundaresan, and M. Matloubian, "Anomalous subthreshold current–voltage characteristics of n-channel SOI MOSFET's," *IEEE Electron Device Lett.*, vol. 8, no. 11, pp. 544–546, 1987.

[26] J. H. Mathews and K. D. Fink, "Curve fitting," in *Numerical Methods Using MATLAB*, 4th ed., Upper Saddle River, NJ, USA: Prentice-Hall, 2004, pp. 263–275.

[27] Y. Lin, J. Appenzeller, P. Avouris, I. B. M. T. J. Watson, and Y. Heights, "Novel carbon nanotube FET design with tunable polarity," in *IEDM*, 2004, pp. 687–690.

[28] L. Yu, A. Zubair, E. J. G. Santos, *et al.*, "High-performance WSe_2 complementary metal oxide semiconductor technology and integrated circuits," *Nano Lett.*, vol. 15, no. 8, pp. 4928–4934, 2015.

[29] M. Tosun, S. Chuang, H. Fang, *et al.*, "High-gain inverters based on WSe_2 complementary field-effect transistors," *ACS Nano*, vol. 8, no. 5, pp. 4948–4953, 2014.

[30] N. R. Pradhan, D. Rhodes, S. Memaran, *et al.*, "Hall and field-effect mobilities in few layered p-WSe 2 field-effect transistors," *Sci. Rep.*, vol. 5, no. 1, p. 8979, 2015.

[31] Y. Yoon, K. Ganapathi, and S. Salahuddin, "How good can monolayer MoS_2 transistors be?," *Nano Lett.*, vol. 11, no. 9, pp. 3768–3773, 2011.

[32] L. Liu, S. B. Kumar, Y. Ouyang, and J. Guo, "Performance limits of monolayer transition metal dichalcogenide transistors," *IEEE Trans. Electron Devices*, vol. 58, no. 9, pp. 3042–3047, 2011.

[33] Q. H. Wang, K. Kalantar-Zadeh, A. Kis, J. N. Coleman, and M. S. Strano, "Electronics and optoelectronics of two-dimensional transition metal dichalcogenides," *Nat. Nanotechnol.*, vol. 7, no. 11. pp. 699–712, 2012.

[34] D. Jariwala, V. K. Sangwan, L. J. Lauhon, T. J. Marks, and M. C. Hersam, "Emerging device applications for semiconducting two-dimensional transition metal dichalcogenides," *ACS Nano*, vol. 8, no. 2. pp. 1102–1120, 2014.

[35] S. Das and J. Appenzeller, "WSe2 field effect transistors with enhanced ambipolar characteristics," *Appl. Phys. Lett.*, vol. 103, no. 10, p. 103501, 2013.

[36] G. V. Resta, S. Sutar, Y. Balaji, *et al.*, "Polarity control in WSe2 double-gate transistors," *Sci. Rep.*, vol. 6, no. 1, p. 29448, 2016.

[37] G. Fiori and G. Iannaccone, "Multiscale modeling for graphene-based nanoscale transistors," *Proc. IEEE*, vol. 101, no. 7, pp. 1653–1669, 2013.

[38] M. De Marchi, M. H. Ben Jamaa, and G. De Micheli, "Regular fabric design with ambipolar CNTFETs for FPGA and structured ASIC applications," in *Proceedings of the 2010 IEEE/ACM International Symposium on Nanoscale Architectures, NANOARCH 2010*, 2010, pp. 65–70.

[39] J. P. Colinge and A. Chandrakasan, *FinFETs and Other Multi-Gate Transistors*. Boston, MA: Springer US, 2008.

[40] B. Iñiguez, A. Lázaro, H. A. El Hamid, *et al.*, "Charge-Based Compact Modeling of Multiple-Gate MOSFET," in *Proceedings of the IEEE 2007 Custom Integrated Circuits Conference, CICC 2007*, 2007, pp. 49–56.

[41] B. Iñiguez, T. A. Fjeldly, A. Lázaro, F. Danneville, and M. J. Deen, "Compact-modeling solutions for nanoscale double-gate and gate-all-around MOSFETs," *IEEE Trans. Electron Devices*, vol. 53, no. 9, pp. 2128–2141, 2006.

[42] M. De Marchi, D. Sacchetto, S. Frache, *et al.*, "Polarity control in double-gate, gate-all-around vertically stacked silicon nanowire FETs," in *Technical Digest – International Electron Devices Meeting, IEDM*, 2012.

[43] N. Chevillon and M. Tang, "FinFET compact modeling and parameter extraction," in *MIXDES*, 2009, pp. 55–60.

[44] M. Tang, F. Pregaldiny, C. Lallement, and J.-M. Sallese, "Quantum compact model for ultra-narrow body FinFET," in *2009 10th International Conference on Ultimate Integration of Silicon*, 2009, pp. 293–296.

[45] A. Javey, R. Tu, D. B. Farmer, J. Guo, R. G. Gordon, and H. Dai, "High performance n-type carbon nanotube field-effect transistors with chemically doped contacts," *Nano Lett.*, vol. 5, no. 2. pp. 345–348, 2005.

[46] S. Fregonese, C. Maneux, and T. Zimmer, "Implementation of tunneling phenomena in a CNTFET compact model," *IEEE Trans. Electron Devices*, vol. 56, no. 10, pp. 2224–2231, Oct 2009.

[47] Y. M. Lin, J. Appenzeller, J. Knoch, and P. Avouris, "High-performance carbon nanotube field-effect transistor with tunable polarities," *IEEE Trans. Nanotechnol.*, vol. 4, no. 5, pp. 481–489, 2005.

[48] J. Knoch and J. Appenzeller, "Tunneling phenomena in carbon nanotube field-effect transistors," *Phys. Status Solidi*, vol. 205, no. 4, pp. 679–694, 2008.

[49] A. Javey, J. Guo, D. B. Farmer, *et al.*, "Carbon nanotube field-effect transistors with integrated ohmic contacts and high-k gate dielectrics," *Nano Lett.*, vol. 4, no. 3, pp. 447–450, 2004.

[50] W. Lu, P. Xie, and C. M. Lieber, "Nanowire transistor performance limits and applications," *IEEE Trans. Electron Devices*, vol. 55, no. 11. pp. 2859–2876, 2008.

[51] M. Haykel Ben-Jamaa, K. Mohanram, and G. De Micheli, "An efficient gate library for ambipolar CNTFET logic," *IEEE Trans. Comput. Des. Integr. Circuits Syst.*, vol. 30, no. 2, pp. 242–255, Feb 2011.

[52] J. Von Neumann, "First draft of a report on the EDVAC," 1945.

Chapter 9

Physical design of polarity controllable transistors

Odysseas Zografos[1], Pierre-Emmanuel Gaillardon[2], and Giovanni De Micheli[3]

This chapter deals with the lack of electronic design automation (EDA) tools that would enable industrial adoption of functionally enhanced devices (FEDs), limiting the abilities we have to explore the true potential of these devices. More specifically, we focus on an embodiment of FEDs, namely, silicon nanowire field effect transistors (SiNWFETs) three independent gate FETs (TIGFETs). TIGFETs offer new properties for logic design, including compact XOR and majority gates. We present a tool-flow that utilizes well-known standard EDA tools for synthesis and placement and routing (P&R) in order to map modern real-world designs onto the SiNWFET technology and compare them with current complementary metal-oxide-semiconductor (CMOS) technology. Also, in this chapter, we give emphasis to the concept of structured-ASIC (application-specific integrated circuit) (sASIC) design and combine it with the fabrication regularity that SiNWFETs demand. We evaluated the performance of the tool-flow using SiNWFET technology by a series of runs, where the SiNWFET always outperformed the reference technology of FinFETs at 22 nm node, in terms of delay being up to ~35% faster and for exclusive OR (XOR)-dominated designs being ~15% smaller.

9.1 Introduction

This section introduces the basic concepts and motivation behind the work needed to enable physical design of a novel technology such as SiNWFETs.

9.1.1 IC design and FPGA–ASIC gap

The broad domain of digital IC design can be divided in the areas shown in Figure 9.1. Many of the hardware solutions available can be either too power-consuming or too costly for a wide range of applications. The particular areas of interest in this work is

[1]IMEC, Belgium
[2]Department of Electrical and Computer Engineering, University of Utah, United States
[3]Laboratory for Integrated Systems, École Polytechnique Fédérale de Lausanne (EPFL), Switzerland

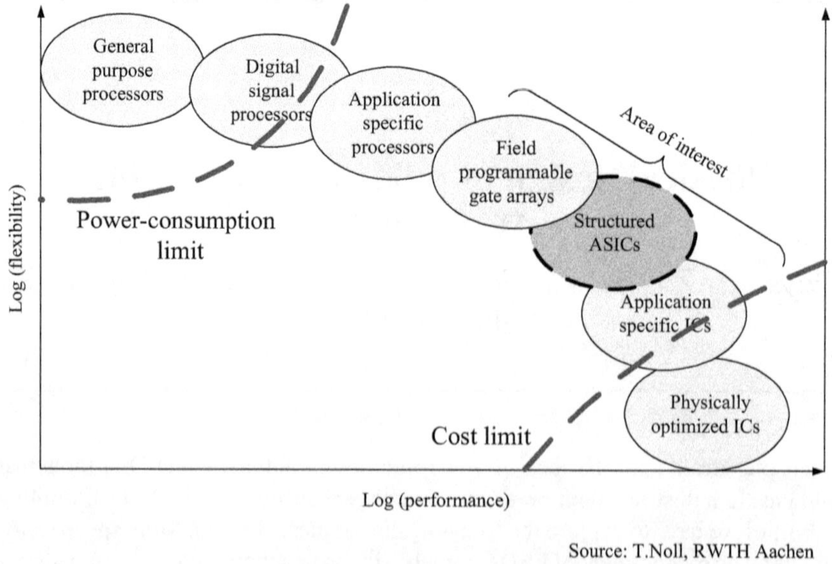

Figure 9.1 Spectrum of implementation solutions for digital IC design [1]

that of the field programmable gate arrays (FPGAs) and ASICs, and this is because the middle area between these two is exactly the kind of design we intend to implement, namely, a sASICs design. In that middle area, we should find an interesting balance between the power-cost and flexibility-performance trade-offs.

One of the most popular hardware solutions, especially in applications of high-volume production, is a circuit that specifically targets the solution of a single problem/application. This circuit has no reconfigurability and can be built by giving as input the gate library and a hardware description language description of the design to EDA tools. This methodology is called ASIC design, and it is the most used in high-volume industry circuit integrations.

The current dominant form of ASICs is that of *standard cell* devices. This approach is enabled by the use of predefined logic and function blocks available to the designer in a cell library. Typical libraries begin with gate level primitives such as AND, OR, NAND, NOR, XOR, inverters, flip-flops, registers, etc. Industrial libraries generally include more complex functions such as adders, multiplexers, decoders, arithmetic-logic units (ALUs), shifters, and memory elements (RAM, ROM, FIFO, etc.). The ASIC solution, as also shown from Figure 9.1, is the one offering the best performance metrics (either through standard cell libraries or full-custom IC design), but it also is the most expensive and time consuming to develop.

In contrast to ASICs that have fixed functionality, it is possible to build chips containing circuitry that can be configured by the user to implement a wide range of different logic applications. These chips have a very general structure and include a collection of programmable switches that allow the internal circuitry in the chip to be configured in many different ways [2]. The dominant form of the programmable logic

Table 9.1 The FPGA–ASIC gap [4]

Property	FPGAs	ASICs
Easy to design	☺	☹
Development time	☺	☹
Design size	☹	☺
Design complexity	☹	☺
Performance	☹	☺
Power consumption	☹	☺
Per-unit cost	☹	☺
Design respins	☺	☹

solution is the FPGA which by eliminating the need to cycle through an integrated circuit production facility, time to market and financial risk can be substantially reduced.

However, FPGAs also have significant disadvantages, in that their designs consume significantly more power (on average consume 14× more dynamic power) and have much lower performance (3.4×–3.6× slower) than equivalent ASIC implementations [3]. Furthermore, FPGAs have a high per-unit cost, which makes them an extremely expensive option for anything other than prototyping applications or relatively small production runs [4].

The aforementioned differences lead to an area, performance and power-consumption gap between ASIC or full-custom designs and FPGAs [3]. More specifically, Table 9.1 presents the aspects in which the two popular methods of digital IC design are different.

The above points serve to illustrate that there is indeed a need of a solution that fills the FPGA–ASIC gap. A solution that has comparable power and performance figures with ASICs but flexibility, cost and ease of designing similar to the FPGA solution. This may well be the intermediate class of circuits, namely, the sASICs class, the concept of which is presented in Section 9.2.1.

9.1.2 Ambipolar devices for Moore's law extension

The main motivation for evolution and device/circuit exploration in the IC field is Moore's law [5]. With the aggressive scaling applied on the electronic devices, their fabrication variability increases and hence manufacturability deteriorates. Along with these factors, power-performance figures become more and more challenging to maintain under viable thresholds. As the limitations of CMOS scaling are reached, there is the need to find and exploit new device-working principles.

Ambipolar conduction is observable in several nanoscale FET devices (45 nm node and below), built using silicon [6], carbon nanotubes [7], graphene [8], as well as transition metal dichalcogenide monolayers including MoS_2 [9,10] and WSe_2 [11,12]. The trend toward devices with intrinsic channels at the 22 nm node and below makes this phenomenon a potential limitation for circuit design, whereas ambipolarity is often suppressed by processing steps; this feature can be exploited to enhance logic functionality. Also, the ambipolar SiNWFETs, with which the current work is completed, present a straightforward approach with great potential in terms of scalability, large-scale integration and compatibility with current CMOS fabrication processes [13]. However, since ambipolar devices in general—as well as SiNWFETs—are not a fully developed and mature technology, there is a significant lack of EDA tools that enable the actual fabrication of designs on such technologies.

9.1.3 Physical design objectives

In order to enable physical design for SiNWFETs and explore how well the sASIC approach performs, the following objectives need to be completed:

- **Library generation**, in order to work with any EDA tool-flow, a library of standard cells is required. However, in order to produce a large number of standard cells, we will show a way to generate SiNWFET standard cells which will constitute our library.
- **Tools configuration**, the tools we are going to use in EDA tool-flow will need some tinkering to accommodate the needs of the SiNWFET library.
- **Benchmark selection**, in order to compare the custom SiNWFET tool-flow and library with CMOS, we need to carefully select the benchmark designs that can exploit the advantages of SiNWFETs and sASICs.

9.2 Background

9.2.1 Structured ASICs

As introduced in Section 9.1, sASICs represent an intermediate digital IC design solution in terms of flexibility and performance. It stands in the middle of the gap between FPGAs and ASICs. The sASIC concept is a very "tunable" one, where the flexibility and performance figures can vary depending on the way the user constructs the design.

9.2.1.1 General concept

The underlying concept behind sASICs is based on a fundamental element called a "tile" by some or a "module" by others (in this chapter, we will use the term "tile" henceforth). The tile contains a small amount of generic logic implemented either as gates and/or multiplexers and/or a lookup table (LUT). Depending on the particular architecture, the tile may contain one or more registers and possibly a very small amount of local RAM.

An array (sea) of these tiles is then prefabricated across the face of the chip. Structured ASICs also typically contain additional prefabricated elements, which may include configurable general-purpose I/O, microprocessor cores, gigabit transceivers, embedded (block) RAM, and so forth (Figure 9.2) [4].

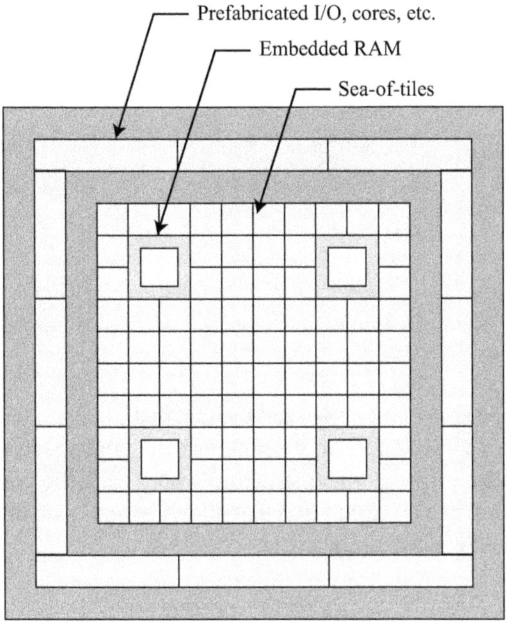

Figure 9.2 The sASIC concept [4]

All components and tiles that may be included in a sASIC (as shown in Figure 9.2) can be configured by the end-user/designer just by designing the final configuration (if the chip contains configurable logic) and interconnection layers of metal.

9.2.1.2 Tile granularity

The most important parameter of a sASIC is the size of the tiles that are going to be used. There can be defined several kinds of "grain sizes" depending on the purpose that the architecture is going to be used. Here are some examples:

- **Extremely fine grain:** This architecture comprises only unconnected components such as transistors and resistors. This approach is very similar to high-end gate array devices.
- **Medium grain:** This may contain some generic logic in the form of multiplexers along with flip-flops. Also, this architecture can be based on LUTs instead of multiplexers.
- **Hierarchical grain:** Here, a kind of a base tile is defined (such as a computational fabric [14]) and is combined with flip-flops and other logic to form a master tile, which is then repeated to create a big set. The master tile can reach the size and computing power of a processor, in which case the design can be characterized as a multiprocessor system-on-chip.

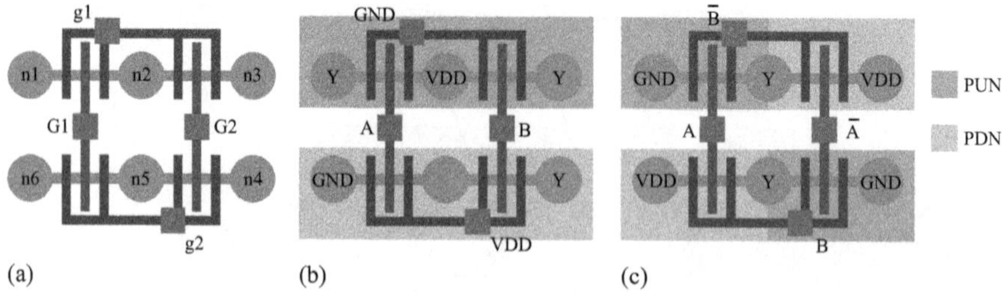

Figure 9.3 Type of tile considered for the sea-of-tiles. (a) TileG2 containing four SiNWFETs. (b) Implementation of NAND gate. (c) Implementation of XOR gate [17]

Considering the above categories and the fact that tile granularity is one the most important factors in sASICs design, we state that the implementation in this chapter considers an extremely fine-grained structure where each tile consists only of four transistors, organized in a configuration that allows several gate implementations per tile (as described in Section 9.2.2).

9.2.2 SiNWFET physical design concepts

Here we present important advances concerning the physical design enablement of SiNWFETs.

9.2.2.1 Sea-of-tiles with SiNWFETs

In the latest technology nodes (45 nm and below), it has become increasingly clearer that layout regularity is a very important factor on maximizing the manufacturability of the layout [15]. It seems that as we continue on the aggressive scaling of devices, just complying with design rules will not be sufficient [15]. The complex interactions among the millions/billions of devices on an IC are poorly understood, hard to model and difficult to abstract [16]. In modern IC technologies, in order to ensure manufacturability, it is important to be aware of neighboring cell interactions, which is a very difficult task. The most efficient way to control the neighborhood interactions is to impose a regular layout with known printing patterns [15]. Additionally, in the case of SiNWFETs, the fabrication process intrinsically produces regular transistor arrays; therefore, it is essential that we build a methodology for designing on the physical level taking into account the regular arrangement of the devices.

In [17,18], the authors present a sea-of-tiles (SoT) approach for layouting the circuit onto SiNWFET arrays. The tile of this configuration would include a small number of transistors (from 2 to 6) with some predetermined interconnections (common control gate (CG), common polarity gate (PG) or common source/drain terminals). The type of tile that was ruled as the optimal in terms of area and delay is the one presented in Figure 9.3.

The tile configuration TileG2 shown in Figure 9.3(a) is presented in a top-view schematic and consists of four SiNWFETs. Using the four transistors of a TileG2, one can combine the node connections and realize six gates (BUF, INV, NAND2, NOR2,

XNOR2, XOR2) [17]. The implementations of a NAND and a XOR gate are shown in Figure 9.3(b) and (c). In the implementation of the NAND gate, we can see that the transistor consisting the pull-up network (PUN) and pull-down network (PDN) is grouped together, instead in the XOR implementation, the transistors are not grouped in terms of PUN and PDN which is possible because the fabrication process does not define any wells or other regions that keep *n*-fets and *p*-fets at distance.

This approach can be considered very similar to a fine-grained SoT sASIC approach, as presented in Section 9.2.1. This actually means that there is now a new possibility to consider sASICs not only because of their flexibility but also because of their manufacturability and closeness to emerging technology layouts.

9.2.2.2 Satisfiable SoT (SATSoT)

After selecting the type of tile that we are going to use in the SoT, the need of figuring out a way to implement all possible logic functions onto a single (or multiple) TileG2 emerges. In other words, how are we going to know which node of the tile to interconnect in order to realize a gate (e.g., XOR2, AOI222, etc.)? The answer is provided by the tool presented in [19] called SATSoT.

Using Boolean satisfiability criterion, SATSoT is able to map netlists of transistors (given logic gates) onto a SoT configuration. The criterion taken into account is the minimization of wiring complexity. This is done through maximizing the sharing of different terminals, which are mostly connected to an actual net of the circuit/gate and not to a power distribution (VDD/GND) net. This tool is able to automatically generate compact mappings with wiring complexities similar to manual layouts [19].

9.2.2.3 Power routing of ambipolar designs

In the previous sections, we have seen what kind of physical design we are going to follow (SoT with TileG2s) and how we are going to map each logic function/cell to this structure (SATSoT). An important matter that needs to be addressed is the power routing of the ambipolar tiles. Reference [18] introduces a novel concept of power routing which deals with its complexity in mapped SiNWFET tiles. In Figure 9.4, we see a XOR2 and a NAND2 gate implemented in two neighboring TileG2 tiles.

We can easily observe that in order to minimize net wiring of non-powering signals, we result with a spread of power supply (VDD) and ground (GND) connected nodes. The solution shown in Figure 9.4 is to alternate the power network and distribute VDD vertically and on a different metal layer. This solution will be implemented, studied and compared with the standard power-routing technique of standard rows in Sections 9.4 and 9.5.

9.3 SiNWFET tile layout and placement and routing

This section describes the technical work needed in order to create a complete SiNWFET library that enables both synthesis and P&R. First, the tool-flow needed is presented, highlighting the points where this work contributes. We present the tool that generates the Library Exchange Format (LEF) files, which are used for P&R.

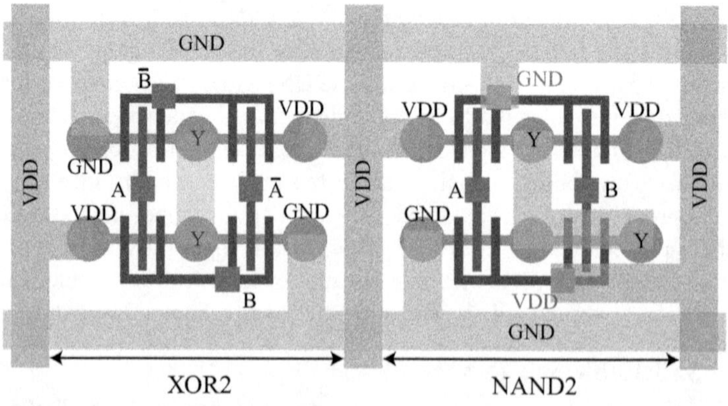

Figure 9.4 Alternative power distribution network layout for SoT with TileG2 [18]

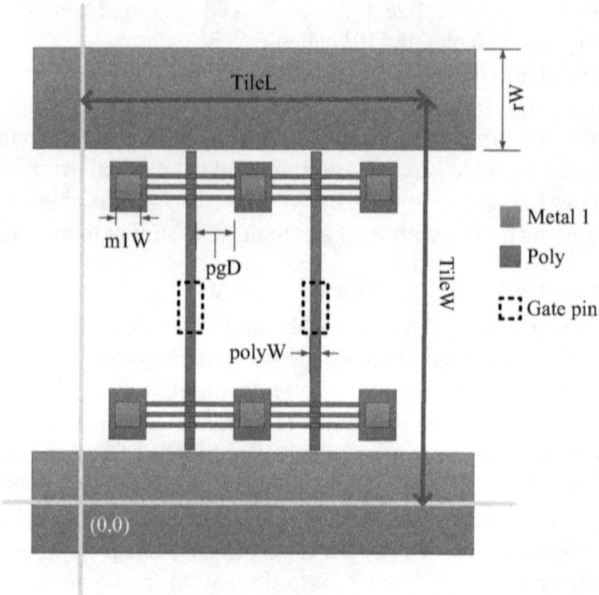

Figure 9.5 TileG2 configuration for 22 nm FinFET technology

9.3.1 Tile configuration for FinFETs

Initially, we configured a TileG2 for 22 nm FinFET technology which is presented in Figure 9.5. Although tile structures are not usually used for FinFET implementation, we defined the following structure in order to be able to have a fair comparison between the FinFET and SiNWFET technologies.

As depicted in Figure 9.5, the TileG2 consists of four FinFETs (number of fingers is symbolic) that are paired by one common terminal and one common gate. We can

Table 9.2 Main dimensions for TileG2 FinFET configuration, as used in this chapter

Name	Value (μm)	Description
m1W	0.060	Metal1 width
m1S	0.060	Metal1 spacing
m2W	0.070	Metal2 width
m2S	0.065	Metal2 spacing
polyW	0.025	Polysilicon width
polyG	0.140	Polysilicon spacing
gpD	0.095	Minimum distance between pillar and gate
rW	0.240	Power/Ground rail width

see that the origin point is inside the GND rail, which is a standard cell approach to enable overlapping of the VDD/GND of neighboring cells. So, for example, if a simple gate consists of four or less transistors, then it would be mapped into a single-tile cell (e.g., INV_X1). If a gate requires more than four transistors, then we produce a multi-tile cell (e.g., XOR2_X1), where the multiple tiles will be interconnected and the set of the tiles will be considered as one cell. The concepts of single and multi-tile cells will become clearer in Sections 9.3.3.1 and 9.3.3.2.

The sizing of the configuration was based on a predictive/estimating 22 nm back-end process, and its main dimensions are presented in Table 9.2 [20].

9.3.2 Tile configuration for SiNWFETs

Using the same configuration (TileG2) and the same back-end process information, we built the tile configuration for SiNWFET devices. But in this case, because of the complex routing that may be involved in designing an ambipolar cell with a tile configuration (see Section 9.2.2.3), we opt to create two configurations. The first one ignores every complication inserted by SiNWFETs except the extra pins of the PGs. The second one implement a more complex VDD/GND pin scheme that enables for the more sophisticated power routing introduced in Section 9.2.2.3.

The first configuration, from hereon referred to as "Type1," is depicted in Figure 9.6(a). There we can actually see that the only real difference between this tile and the FinFET tile configuration is the routing of the PG pins (which has made the tile 25% taller than the FinFET one).

The next more complex tile configuration, "Type2," is presented in Figure 9.6(b). We can clearly see that this tile configuration is larger and invokes the use of more than one metal layer.

In Figure 9.6(b), we observe the reconfiguration of the tile (from Type1 to Type2) in order to be compatible with the power networking that has the VDD distributed vertically on the metal2 layer. Here, the VDD pin of the tile is located on the left-hand side in metal2 layer. The length of the tile ("TileL") is increased in order to leave enough space between the tiles so that the global VDD lines can pass through. Also,

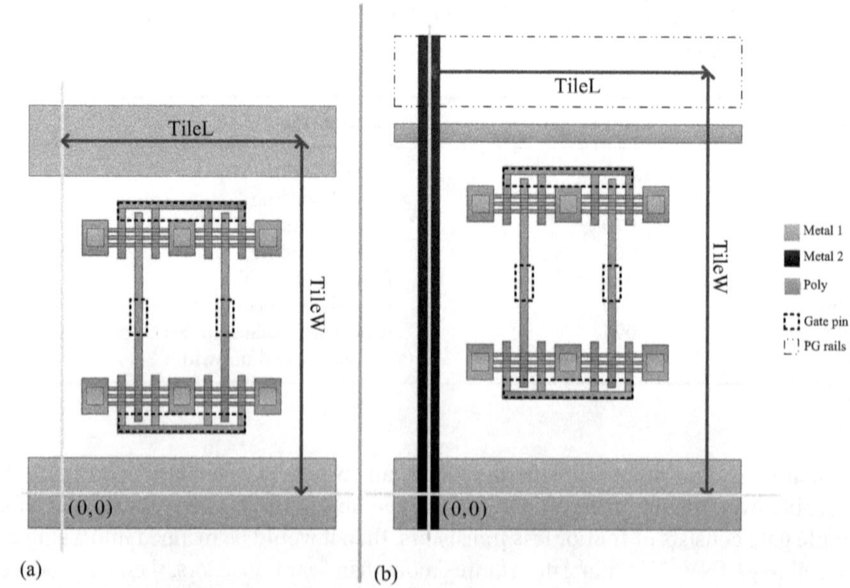

Figure 9.6 TileG2 configuration for 22 nm SiNWFET technology, (a) Type1 and (b) Type2

Table 9.3 Dimension comparison among the different TileG2 configurations

	FinFET	Type1	Type2
TileL	0.840	0.840	0.970
TileW	0.960	1.200	1.440

we can observe a thinner metal1 line on the upper side of the tile which is placed as an auxiliary GND line and will be connected with adjacent auxiliary lines. There is no auxiliary VDD line because each tile is going to have access to two metal2 lines, one connecting its own pin and one connecting the adjacent tile's pin. With this configuration, we can generate cells that comply with technology rules, can be successfully placed and routed (using industry standard EDA tool, System-On-Chip Encounter (SOCE)), and also emulate the power routing structure shown in Figure 9.4.

Finally, Table 9.3 compares the dimensions of all three tile configurations. Clearly, the SiNWFETs libraries have larger dimensions, but we will see later on (Section 9.5) that they can compensate this drawback by the use of their ambipolar functionality.

After defining the tile structures for each technology, we implemented the most essential tasks of a standard cell library generation. This is to read in the node configuration/connectivity of the desired gate (given from SATSoT), apply it on the appropriate structure (one of the aforementioned) and then interconnect the nodes with common net. This interconnection is divided in two kinds, the intra-tile connections (Section 9.3.3.1) and the inter-tile connections (Section 9.3.3.2).

9.3.3 Pin and layout generation

In this section, we present the work done concerning the LEF generation tool we created called LEFGen. This tool was created in MATLAB®, and its purpose is to create LEF cell descriptions for a given cell. LEFGen can produce three possible cell configurations, all of them using tiles—namely, TileG2 (see Section 9.2.2.1). We describe the way we represent the routing wires in each cell to get a realistic estimation of the what is the actual cell's layout.

9.3.3.1 Intra-tile connection generation

We define an intra-tile connection simply as the wiring needed to connect two nodes of the same tile. This wiring is represented in the LEF file in the form of a polygon, which can be defined as a pin, or as an obstruction zone. In order to complete the task of intra-tile connections, we have employed two methodologies:

1. An algorithm that defines a detailed set of polygons (pins or obstructions) which resemble the actual layout.
2. An arbitrary big obstruction zone is placed over the whole routable area of the tile, ignoring the actual connections, but ensuring the correctness of the above-routed layers (added by SOCE in later steps).

The first method is applied in the generation of the FinFET LEF file, while the second method is applied in the generation of both types of SiNWFET LEF files. This distinction was made to avoid dealing with complex wiring that an ambipolar SiNWFET presents.

A rough view of the first detailed methodology's algorithm is given in Algorithm 1. This algorithm is repeated for as many tiles a cell might contain. Initially, the intra-connection predefined structure-dependent wires are loaded into the tile entity flagged as unused. Then all six nodes of the tile are grouped into an array according to their connectivity and eventually store one set of nodes for each net. Afterward, we choose to ignore power nets and then select the appropriate wires. The wire selection is a trivial task since all wires available have an identity which declares the two nodes they are connecting. Next the selected wires are merged and as a result all nets have finished polygons to represent their connectivity. Finally, running through the pin list of the cell, we define if each polygon is an actual pin or an obstruction zone.

The second simplistic methodology of intra-tile connection is quite trivial to implement. We define a large obstruction zone (on metal1 layer) over the tile and then define all the actual pins on a metal higher (metal2). A single-tile NOR2_X1 gate is shown as an example of both methodologies in Figure 9.7(b).

Algorithm 1: Detailed intra-tile connections

```
load wires;
group nodes by net;
keep only one recurrence of each net;
clear out VDD/GND nets;
select wires;
merge connected wires;
for i:all nets found do
  for j:all pins of tile do
    if net(i) == pin(j) then
      label net(i) as pin(j);
      label pin(j) as done;
    else
      label net(i) as obstruction;
    end if
  end for
end for
```

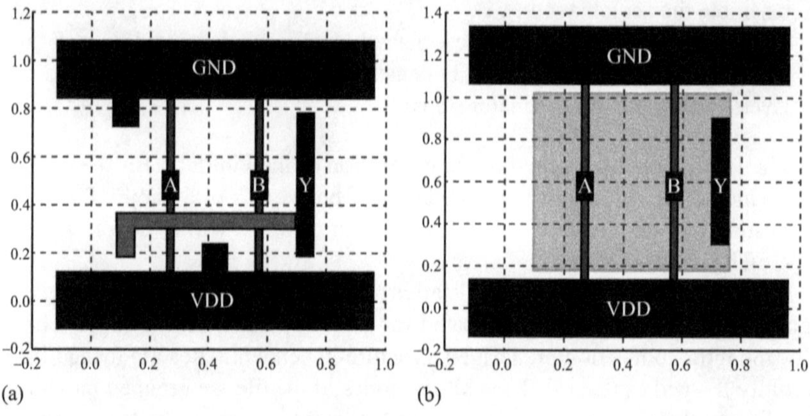

Figure 9.7 Intra-tile connection examples, (a) detailed intra-connection on a FinFET TileG2, (b) large obstruction zone intra-connection on a SiNWFET TileG2. Note: The lines running through the "A", "B" pins are symbolic indications on the polysilicon common CG connections

The simplistic approach for the intra-tile connections is applied to simplify the complexity of creating the LEF files. It is an effort of generating a LEF file that has approximately the same amount of obstruction area. In ambipolar SiNWFET TileG2 cells, the VDD/GND-connected nodes are sparse and their interconnection with the

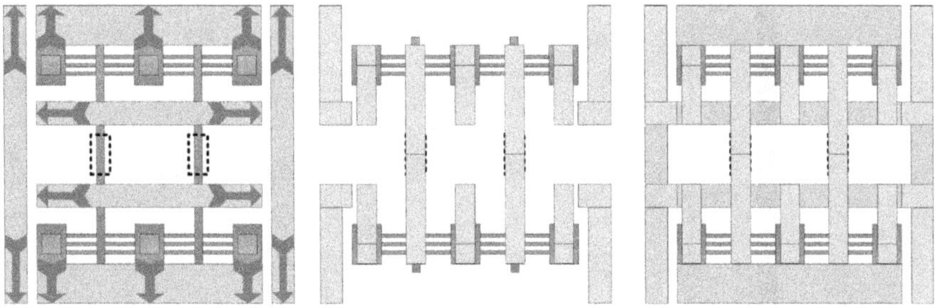

(a) (b) (c)

Figure 9.8 Inter-tile connection routes, (a) all possible paths and directions, (b) all contact/node extensions to enable full connectivity toward every direction and (c) complete view of both (a) and (b)

power network is complicated. In fact, it is too complicated to try and abstract with the LEFGen tool, that is why we applied the large obstruction zone to emulate it.

9.3.3.2 Inter-tile connection generation

With these two methodologies for the intra-tile connections in place, we implemented a unified inter-tile connection method, applied in all the type of TileG2 configurations. It is clear that if a gate consists of more than one tile, then two nodes from different tiles would need to be interconnected. To address this need, we built the following methodology.

Similar to the detailed intra-tile connection methodology, we defined tile structure-dependent wires that are shown in Figure 9.8. There we can see that the predefined wires have the ability to route a signal outside of the tile, from all possible nodes toward all possible directions. In order to handle the interconnections' complexity, the following simplifications were made:

1. A single interconnection may represent more than one signal.
2. All interconnections are kept on the same layer, even if they cross each other, in order to avoid layer checks and add complexity.
3. Gate-to-gate interconnection is considered to be done with polysilicon, so no routing metal is used. This means that if gates of the same tile belong to the same net, they are ignored as interconnected through polysilicon.

These assumptions assisted the tool implementation in terms of complexity but without compromising too much of the realistic abstraction of what a real layout would be like.

Figure 9.8(a) presents the way the initial positions and directions of the interconnecting paths were laid out. Taking advantage of all the spacing left in between the gate pins and node contacts, we defined two horizontal routes (one for each triad of nodes) toward the sides of the tile. Also, two horizontal routes were defined above

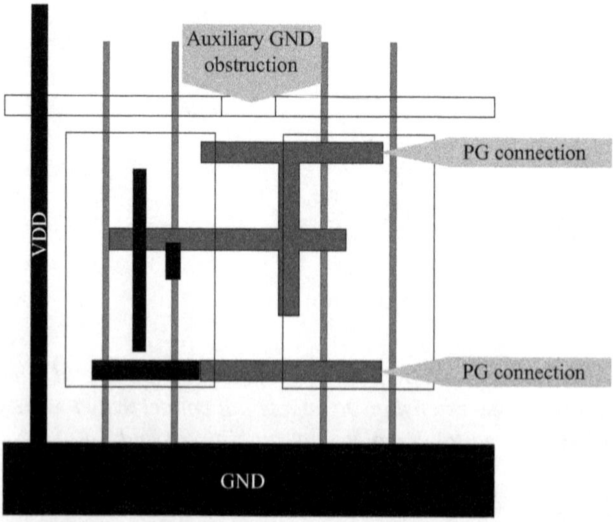

Figure 9.9 Example of multi-tile SiNWFET XOR2 gate. Note: The magenta lines are symbolic indications on the polysilicon common CG connections

and below the tile that would enable interconnection with adjacent tiles on the upper of lower standard row. Finally, we considered two vertical routes that would enable connection with nonadjacent tiles (different row and column). Figure 9.8(b) shows all the possible extensions that may be implemented on each node in order to be connected with one of the aforementioned routes, and Figure 9.8(c) demonstrates the complete network of interconnections. Note that the different shading of interconnections in Figure 9.8(a) and (b) is done only for visualization purposes and that these diagrams do not contain any information about the metal layer the interconnections are implemented on.

The actual algorithm that creates the final interconnections is almost the same as the one described in Algorithm 1. The common net nodes are examined by pairs, and for each connection, appropriate predefined wires are selected and "activated."

Figure 9.8 presents the possible interconnection paths created and adjusted only to the TileG2 structure of FinFETs. For SiNWFET structures, we have defined very similar paths including the connectivity of the PG (shown in Figure 9.9).

An example of a multi-tile cell with the appropriate interconnections is shown in Figure 9.9. The structure of the tile is TileG2 Type2 and is based on SiNWFET, so we can demonstrate the PG connectivity.

Since in SiNWFET tiles we employ the brute large obstruction zone method for intra-tile connections, the PG is done by simply a piece of extra wire into the obstruction area; the rest of its connectivity is ignored.

The limitations of generating LEF files using abstract generated layouts are shown by the fact that it is too complex to produce detailed results when the tile intra-/interconnections are nontrivial. Immediately the problem becomes too complex and we pass from the detailed version of LEF files to the completely approximate one. This lack of flexibility is basically the drawback of the methodology of LEFGen, which however does not restrict the extent of our P&R exploration, rather it limits the depth of the tile cell structure and its impact on the complete design.

9.4 SOCE configuration

In this section, we describe the approaches taken toward implementing the final stage of our tool-flow, namely, the P&R. The main issues addressed and discussed hereafter concern the placement and power routing approaches, along with the general routing complexity and optimization. The EDA tool used for P&R is SOCE, and all further considerations are centered around its functionality.

9.4.1 Placement schemes

As presented in Section 9.3, we have produced a new kind of cell library which contains cells in the form of tiles instead of standard cells. This kind of library is structured with tiles so that it enables hardware implementation onto a SoT of polarity-controllable SiNWFET devices. Two approaches (tile and standard cell) were configured in SOCE and are presented in Sections 9.4.1.1 and 9.4.1.2.

9.4.1.1 Standard cell approach

The standard cell approach is most commonly used in industrial cell libraries. Standard cell design common rules are as follows [21]:

1. All cells must have the same height, to enable easy placement in standard rows.
2. All n-type FETs are placed at the same height in the upper part of the cell and all p-type devices on the lower one, so that we have a continuous *nwell* layer for all p-type devices.
3. All cells have a power (VDD, on top) and a ground (GND, at the bottom).
4. All cell layouts have a fixed origin in the lower left corner of the layout.

The placement of standard cells is what SOCE is made and optimized for, so the P&R procedure of our "CMOS22_std" library did not have any issues to be addressed, and the placement results of a benchmark are shown in Figure 9.10.

As depicted in Figure 9.10, the cells have different width and are placed irregularly inside each row. This placement approach is not suitable for the implementation of a SoT since it ignores the utmost important feature of a SoT, regularity. This regularity can be achieved only if all cells have width multiple of a certain value and are placed in positions with coordinates multiple of this value. In other words, the placement grid has to be as coarse as the minimum width cell, and all other cells must have an integer multiple of that width.

9.4.1.2 Tile cell approach

Contrary to standard cells, placement of tile cells is not trivial for SOCE since they do not have the same height. An example of tile cell placement is shown in Figure 9.11,

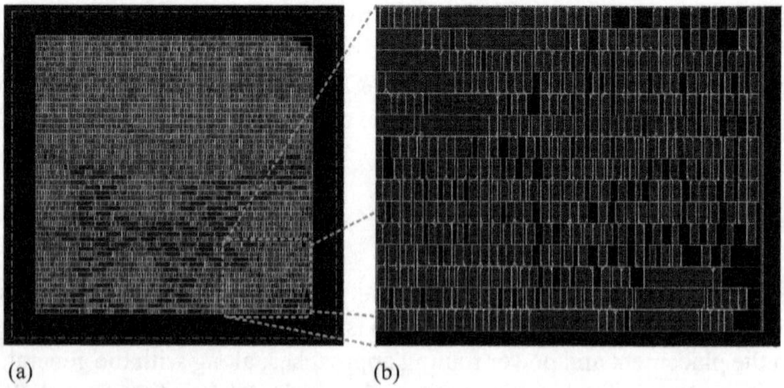

Figure 9.10 Placement for standard cell library: (a) complete view of design and (b) zoomed region where we clearly see the same height/different width cells that all match in standard rows

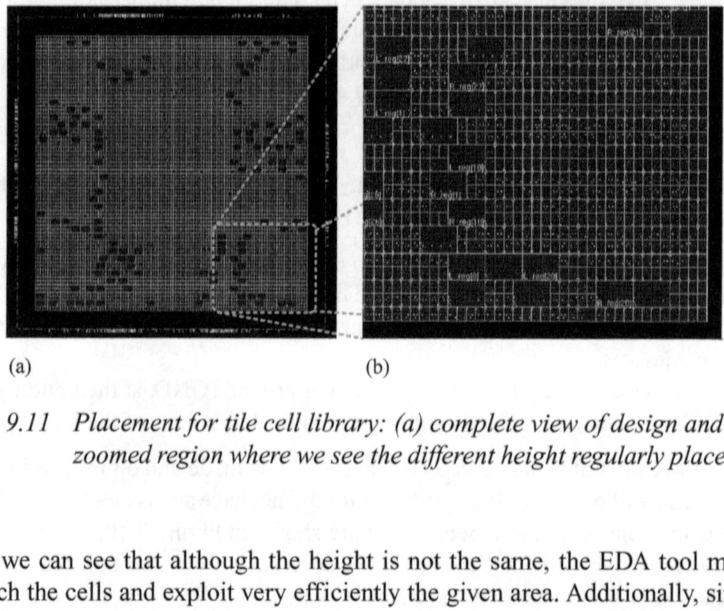

Figure 9.11 Placement for tile cell library: (a) complete view of design and (b) zoomed region where we see the different height regularly placed cells

where we can see that although the height is not the same, the EDA tool manages to match the cells and exploit very efficiently the given area. Additionally, since the tile cells are supposed to be configured on a SoT, the placement of all cells must be regular and every tile of each should be on a placement grid that allows only positions multiple of the width of one tile. This regularity is also shown in Figure 9.11 and can be summed up in the following rules:

1. One minimum-sized element (a tile).
2. Placement grid allow positions only according to the dimensions of the minimum-sized element
3. Multi-tile cells can extend to upper/lower rows of placement.

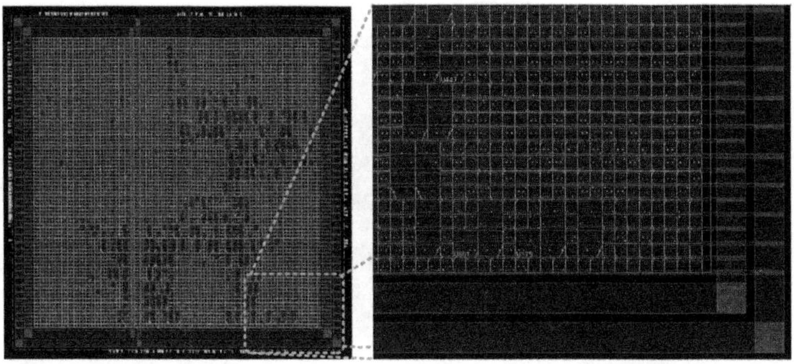

(a) (b)

Figure 9.12 *Standard cell power routing scheme applied onto tile cell library: (a) complete view of design and (b) zoomed region where we can see the horizontal metal1 buses running through each cell/tile*

4. All single-row cells have the VDD pin on the upper part of the cell and the GND on the lower.
5. Multi-tile/multi-row cells do not have obstruction zones of metal1 in between the rows to allow the power rail routing.

9.4.2 Power routing schemes

As we presented in Section 9.3, the two of the three tile configured libraries use a standard horizontal bus power routing scheme, as does the standard cell library "CMOS22_std." The last of the tile configured libraries has a power routing as the one proposed in [18] and presented in Section 9.2.2.3. These two approaches are presented hereafter.

9.4.2.1 Standard cell power routing scheme

In Figure 9.12, we present the standard power routing scheme as it is usually applied in standard cell methodology. Here, we present how this standard cell approach is applied to one of the tile configured libraries. We can see that all cells are placed regularly and that indeed everyone is connected with two horizontal metal1 buses.

The power ring around the design is created using the SOCE command `addRing`, and the horizontal metal1 buses of VDD/GND are created by the special router command (`sroute`) which is employed to create the network of all (or several) power nets.

9.4.2.2 Tile cell power routing scheme

The novel power routing scheme presented in Section 9.2.2.3 is shown in Figure 9.13, applied on the "Type2" SiNWFET library. We can see the very dense grid of vertical metal2 lines that route the VDD signal to each column of tiles.

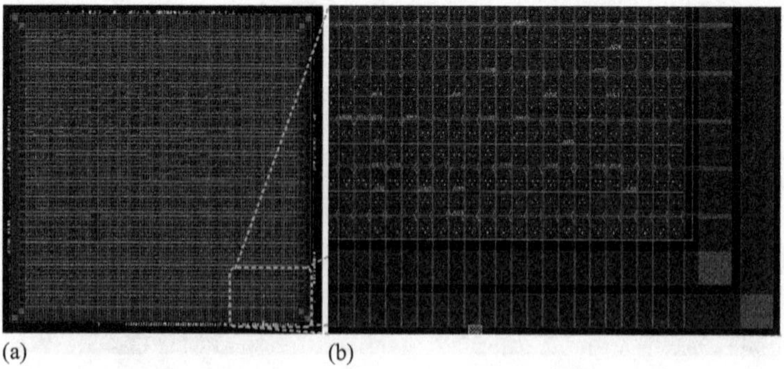

Figure 9.13 Alternative power routing for tile SiNWFET cells: (a) complete view of design and (b) zoomed region

Here, contrary to the previous standard power routing commands, we use the `addStripe` SOCE command to create vertical metal2 VDD lines spaced with the a tile's width. We finally used the special route command (`sroute`) only to create the network of the GND net, since the VDD one is consisted only by the metal2 stripes.

9.5 Results and comparisons

In this section, we present the benchmarking methodology we followed, the performance metrics we applied and the results produced from our tool-flow, while implementing the different benchmarks on the available technologies. Finally, we compare and discuss the results presented, evaluating the performance of the tool-flow and the SiNWFET technology used.

9.5.1 Benchmarking methodology

In the following parts, we describe the benchmark categories and the benchmarks themselves thoroughly in order to introduce a complete view of the benchmarking procedure. We also review the metrics we applied in order to evaluate the results of the tool-flow.

9.5.1.1 Benchmark categories

In order to study the performance of our technology in a variety of design implementations, we devised the following categories of design benchmarks that we will use to characterize the SiNWFET technologies performance when used in a complete design flow.

Control and sequential designs

This category contains three benchmarks that do not have many things in common in terms of operation as circuits. However, their common ground is that all of them are sequential, all them are NAND/NOR-dominated designs and they belong to the family of *OpenCores* [22] benchmarks. The purpose of this category is to gauge the performance of our technology in real-life complex designs.

Triple-DES core: This design [23] implements a common core for Data Encryption using the data encryption standard (DES) and applying it three times on the input data. This benchmark is the smallest in terms of gate count used from the *OpenCores* family. This specific core was designed to be optimized in terms of area. The code-name of this benchmark hereafter is **des3_area**.

Memory controller IP core: This design [24] implements an advanced memory controller intended for embedded applications featuring many memory devices support (SDRAM, SSRAM, FLASH, etc.), 8 Chip selects—each uniquely programmable, flexible timing and up to 8×128 MB. The code-name of this benchmark hereafter is **mem_ctrl**.

Wishbone interconnection matrix IP core: In this benchmark, we use the implementation of an industry standard, Wishbone Interconnection Matrix [25]. This IP core contains multiple master–slave interfaces and enables the parallel and interconnected interaction among many master/slave systems (e.g., DMAs, VGA controllers, ethernet ports, etc.). This matrix enables and controls the operation of a SoC. The code-name of this benchmark hereafter is **wb_conmax**.

Combinational designs

This category of benchmarks covers the class of circuits that SiNWFET technology is most immediate to improve. Arithmetic circuits use strongly binate functions which, as aforementioned, in ambipolar SiNWFET are implemented much more compactly than in standard CMOS. These benchmarks, that are XOR-dominated, give us the opportunity to study how much this compactness of binate functions impacts the total figures of the design.

All benchmarks presented below where generated using the ARITH Project generator from Tohoku University [26].

Parallel multiplier: A 64×64-bit parallel multiplier utilizing modified booth encoding for partial product generation, (4:2) compressor tree for partial product accumulator and a Brent–Kung adder for the final adding stage. This is a large arithmetic fully combinational stage that can limit the performance of a common CPU and so it is interesting to study. The code-name of this benchmark hereafter is **mult64**.

Multiplier-accumulator: This design implements a three 16-bit operand multiplier-accumulator that utilizes similar structures to the mult64, namely, modified Booth encoding, Wallace tree and a Ripple-block Carry Look-ahead adder. The code-name of this benchmark hereafter is **mac316**.

Irreducible polynomial multiplier: A Mastrovito multiplier is a class of parallel multipliers over Galois fields based on polynomial basis representations; this design implements a multiplier over the polynomial basis $x^{17}+x^8+x^3+x+1$. This kind of

Table 9.4 Benchmark summary

	Type	Gate count	Binate ratio
des3_area	SEQ	6,185	0.0254
mem_ctrl	SEQ	15,414	0.0338
wb_conmax	SEQ	68,618	0.0000
mult64	COMB	38,399	0.4306
mac316	COMB	6,352	0.4637
mastrovito	COMB	995	1.1073

multiplier is proven to have equal amount of binate (XOR) and unate (NAND) gates. Hence, it gives as an opportunity to measure the best case scenario for our ambipolar technology. Irreducible polynomial multipliers are often used in cryptography. The code-name of this benchmark hereafter is **mastrovito**.

9.5.1.2 Benchmark summary

In this section, we present a summary of the benchmarks introduced above and we also introduce some useful metrics that help us sort the benchmarks and quantify the results later on.

In Table 9.4, the field "Binate ratio" is a metric of the arithmetic part of the design compared to the control one, and it is defined as the ratio between the XOR (binate) gates and the NAND/NOR (unate) gates included in the design. This metric is defined as $\#XOR/(\#NAND + \#NOR)$. Both "Gate Count" and "Binate ratio" fields were extracted as results after P&R-ing the benchmarks using the Type2 SiNWFET library but are only dependent of the benchmark itself, so they are given as benchmarks characteristics.

9.5.1.3 Performance metrics

Here we review the performance metrics chosen to evaluate and compare the performance of our libraries and tool-flow based on the benchmarks.

Area

One of the most important and critical parameters of the resulted designs that we have to consider is the total area they occupy. This is one of the metrics of evaluation and helps us understand the impact of our libraries (and the different technologies and implementations) on the resulted designs. All area results presented are in μm^2.

Worst negative slack (WNS)

The Worst negative slack (WNS) is a parameter that gives the delay of the critical path of every design. This metric characterizes the actual performance/speed of the resulted design and will be used only in comparisons of results from the same benchmark. All WNS results presented are in *ns*. The measurements for delay/period were done taking into account the nominal VDD values of each type of device (FinFET and SiNWFET) which are $VDD_{FinFET} = 0.9$ V and $VDD_{SiNWFET} = 1.2$ V. This is a fair comparison because the two supply voltage values are the nominal one for each technology.

Power consumption
The predictive technology model (PTM) used to create the CMOS libraries is far more accurate in modeling the dynamic power consumption in the CMOS device, than the model used to create the SiNWFET libraries. However, the leakage power modeling is accurate in both models, and since leakage power is a very important parameter of hardware designs, we compare only that. This way we have fair comparisons between the two technologies. All leakage power results presented are in nW.

Speed-up and area increase
We also define two comparative metrics, speed-up and area+. These metrics measure in percentage (%) the delay decrease and the area increase, respectively, between two implementations of the same benchmark. Both metrics are normalized with the figures produced from the CMOS22 library which uses TileG2 cells with FinFET transistors.

Speed-up is not used as the standard metric of parallelization,[1] but it is defined as

$$\textbf{speed-up} = \left(1 - \frac{P_{lib}}{P_{cmos22}}\right) \cdot 100. \tag{9.1}$$

where P_{lib}, is the period or delay of the benchmark. Negative speed-up percentage represents performance diminishing compared to CMOS22, and conversely positive speed-up percentage represents performance improvement.

Area increase is another percentage metric that denotes the increase (or decrease) of the resulted area of a design compared to the CMOS22 library resulted design. This metric is defined as

$$\textbf{area+} = \left(\frac{Area_{lib}}{Area_{cmos22}} - 1\right) \cdot 100. \tag{9.2}$$

Negative area+ percentage represents area reduction compared to CMOS22, and conversely positive area+ percentage represents area increase. It is expected that P_{lib} will be smaller than P_{cmos22} and that the area $Area_{lib}$ will be larger than $Area_{cmos22}$, when we take into account one of the SiNWFET libraries.

Routing distribution
The final metric defined is actually a pie chart that denotes the distribution of the metal layers used by SOCE to route the design. The information for the metal layers is extracted from SOCE reports. This distribution is critical because it shows the overhead that SiNWFET and tile cell libraries impose in terms of metal layers. Extra layers imply additional yield loss and added processing steps and hence extra wafer cost [27].

[1] In computing, speed-up is defined as the gain in throughput and/or latency due to parallel execution.

Table 9.5 Results for des3_area benchmark

des3_area	**CMOS22_std**	**CMOS22**	**SiNFET1**	**SiNFET2**
Area (μm²)	4,754.295	6,186.586	8,061.264	11,170.577
WNS (ns)	1.670	1.876	2.137	2.217
Leakage power (nW)	0.164	0.160	0.069	0.069
Speed-up (%)	−6.594	–	8.355	10.915
Area+ (%)	−23.152	–	30.302	80.561

Table 9.6 Results for mem_ctrl benchmark

mem_ctrl	**CMOS22_std**	**CMOS22**	**SiNFET1**	**SiNFET2**
Area (μm²)	16,318.550	21,225.768	27,070.926	37,512.389
WNS (ns)	1.775	−0.077	0.717	−0.045
Leakage power (nW)	0.862	0.886	0.253	0.254
Speed-up (%)	36.478	–	15.639	0.630
Area+ (%)	−23.119	–	27.538	76.730

Table 9.7 Results for wb_conmax benchmark

wb_conmax	**CMOS22_std**	**CMOS22**	**SiNFET1**	**SiNFET2**
Area (μm²)	50,977.679	63,857.242	84,889.728	117,633.499
WNS (ns)	−0.502	−0.750	0.742	−0.175
Leakage power (nW)	1.138	1.337	0.765	0.757
Speed-up (%)	4.313	–	25.948	10.000
Area+ (%)	−20.169	–	32.937	84.213

9.5.2 *Benchmark results*

Here we present and compare the results of the benchmarks aforementioned. Through these results, we can evaluate the performance of our libraries and draw conclusions about the advantages and drawbacks of the technology, devices and cell configurations used.

9.5.2.1 Results for control and sequential designs

Tables 9.5–9.7 present all the metrics for the des3_area, mem_ctrl and wb_conmax, respectively. Every benchmark was implemented with all four available libraries.

It is clear from all three tables that the leakage power figures are consistently better for the SiNWFET libraries, ~56% on average less leakage. This is to be expected

Table 9.8 Results for mult64 benchmark

mult64	**CMOS22_std**	**CMOS22**	**SiNFET1**	**SiNFET2**
Area (μm^2)	32,068.726	57,406.651	53,910.690	74,705.198
WNS (ns)	0.840	−0.260	1.279	0.710
Leakage power (nW)	0.759	0.793	0.495	0.486
Speed-up (%)	20.913	–	29.259	18.441
Area+ (%)	−44.138	–	−6.090	30.133

Table 9.9 Results for mac316 benchmark

mac316	**CMOS22_std**	**CMOS22**	**SiNFET1**	**SiNFET2**
Area (μm^2)	6,106.250	11,250.624	10,211.040	14,149.526
WNS (ns)	2.949	2.448	3.238	3.162
Leakage power (nW)	0.101	0.108	0.083	0.086
Speed-up (%)	19.632	–	30.956	27.978
Area+ (%)	−45.725	–	−9.240	25.767

since the SiNWFET are supposed to have very low leakage power, as they are GAA devices. It is also clear that, as again expected, the trade-off for implementing tile cell libraries instead of standard cell ones is the increase in area of the design. The area increases also from SiNWFET1 to SiNWFET2 library since the second one implements the alternative power routing that requires larger tiles. Finally, we can observe that the speed-up for SiNWFET libraries is always positive but not that large, compared to the area increase that all designs present, e.g., ∽10% speed-up to ∽80% area+.

9.5.2.2 Results for arithmetic combinational designs

Tables 9.8–9.10 present all the metrics for the mult64, mac316 and mastrovito, respectively. Every benchmark was implemented with all four available libraries.

The results for this category of benchmarks seem more intriguing than the ones presented in the previously. Here we can observe that along with the reduce leakage power, the SiNWFET libraries produce much faster designs (from ∽18% to ∽35%) with much smaller area increase/overhead. In fact, we can see that for all three benchmarks, SiNWFET1 library produces faster design than the standard cell FinFET library and smaller than the tile cell library (CMOS22). Furthermore, the SiNWFET2 library produces a positive speed-up to area+ trade-off for the mac316 and mastrovito benchmarks. This improvement in terms of delay, in SiNWFET libraries, is due to the excellent electrostatic control over the nanowire channels of the SiNWFET devices.

Table 9.10 Results for mastrovito benchmark

mastrovito	CMOS22_std	CMOS22	SiNFET1	SiNFET2
Area (μm^2)	1,042.335	2,148.480	1,815.870	2,516.299
WNS (ns)	4.287	4.222	4.493	4.426
Leakage power (nW)	0.023	0.025	0.013	0.014
Speed-up (%)	8.355	–	34.833	26.221
Area+ (%)	−51.485	–	−15.481	17.120

Table 9.11 Comparative summary of SiNWFET1 and SiNWFET2 libraries

		SiNWFET1		SiNWFET2	
	Binate ratio	Speed-up (%)	Area+ (%)	Speed-up (%)	Area+ (%)
wb_conmax	0.000	25.948	32.937	10.000	84.213
des3_area	0.025	8.355	30.302	10.915	80.561
mem_ctrl	0.034	15.639	27.538	0.630	76.730
mult64	0.431	29.259	−6.090	18.441	30.133
mac316	0.464	30.956	−9.240	27.978	25.767
mastrovito	1.107	34.833	−15.481	26.221	17.120

9.5.3 Comparisons and conclusions

Finally, we present some summarizing and comparative results. Initially, Table 9.11 summarizes the speed-up and area+ values of the two SiNWFET libraries, which are the ones of interest. The benchmarks in Table 9.11 are given in ascending order of their binate ratio. This parameter indicates the possible gain we have using SiNWFET libraries, because of their compactness of binate gate implementations.

9.5.3.1 Area increase analysis

Figure 9.14 depicts area+ metric relation to the binate ratio of each design. There, as expected, we can see that the increase of the benchmark's binate ratio leads to a decrease of the area overhead (area+) that SiNWFET technology, tile configuration and power routing imposes. We can even observe improvement in terms of area compared to the CMOS22 for the SiNWFET1 library.

From Figure 9.14, we can extract the area overhead that the alternative power routing entails. This overhead is shown in Figure 9.15, and it seems to be decreasing as binate ratio of benchmarks increases.

Furthermore, from both Figures 9.14 and 9.15, we can observe a sudden decrease from the mem_ctrl benchmark to the mult64 benchmark. The cause of this decrease is the sudden increase of the binate ratio between the two benchmarks. However, we can consider that some of the improvement in area overhead is also due to the fact that mult64 is a combinational benchmark without flip-flop cells. This could be a

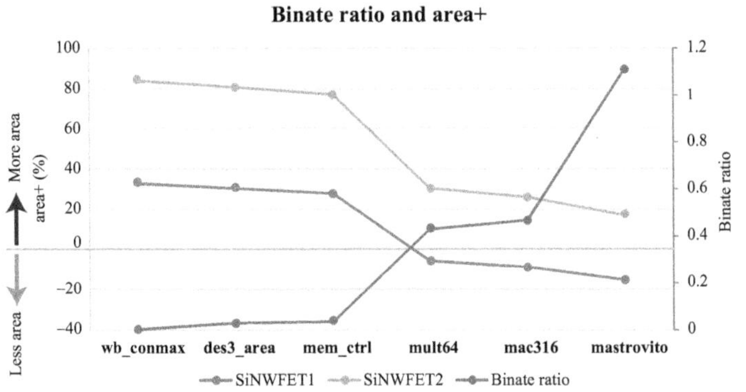

Figure 9.14 Area+ comparison between SiNWFET1 and SiNWFET2 libraries, relation with binate ratio

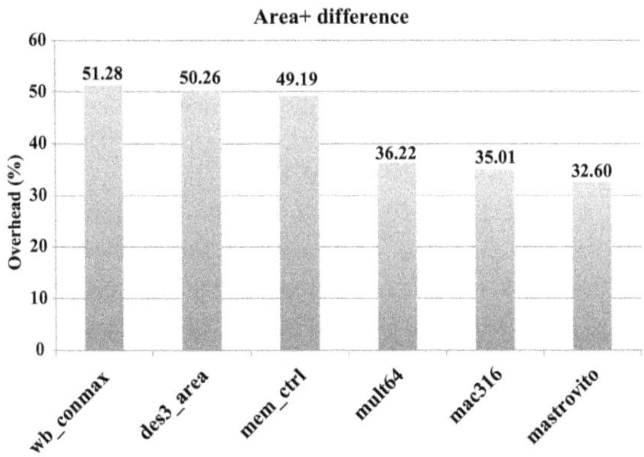

Figure 9.15 Area+ difference between SiNWFET1 and SiNWFET2 libraries, overhead relation

factor of performance of SiNWFET libraries since the flip-flop implementation is not compact. It has not yet been studied extensively which leads us to implement them in a CMOS way without being able to compensate at all for the larger tile size.

9.5.3.2 Speed-up analysis

Figure 9.16 presents the evolution of speed-up extracted from Table 9.11. Here, unfortunately, there are no clear conclusions to be drawn, except the fact that the power routing imposes almost in every case a 15% loss in speed-up and that probably for combinational benchmarks, the delay performance follows the binate ratio of the benchmarks.

Considering the three OpenCores benchmarks we see that the delay performance results are sparse and do not follow any trend in terms of either gate count (wb_conmax is largest of the three) or binate ratio. This is probably due to the critical

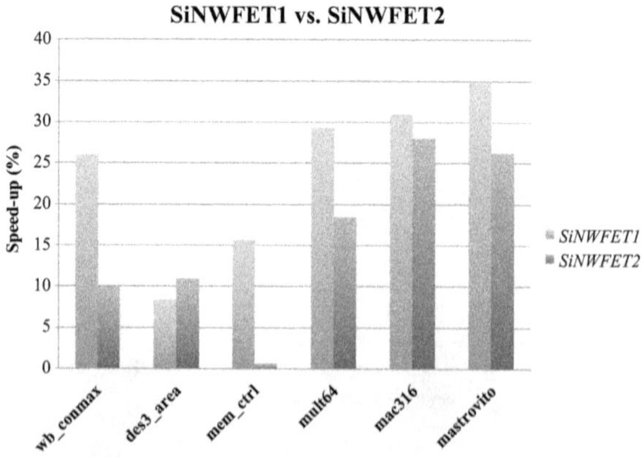

Figure 9.16 Speed-up comparison between SiNWFET1 and SiNWFET2 libraries

path optimization complexities of each circuit. Since SOCE is trying to optimize the design in terms of period and measures critical paths from register to register, it is not obvious that the smallest design will have the shortest/fastest critical path.

9.5.3.3 Design interconnection analysis

In order to further explore the not so clear results of speed-up for the non-XOR-dominated OpenCores benchmarks, we proceed to the further analysis.

The delay of a design is normally dominated by one of the two reasons, the intrinsic delay of the gates included in the critical path or the delay due to high driving load on the outputs of the gates because of long interconnections. Fitting the OpenCores benchmarks into these two categories of delay will probably help us to better understand their performance in terms of speed-up. The first metric considered to evaluate the design's interconnection dominance in delay is *INVratio*, which represents the amount of high-drive strength inverters the design contains. It is defined as

$$INVratio = \frac{High\text{-}drive\ Inverters}{Inverters}. \tag{9.3}$$

High-drive strength inverters are considered the "X4" and "X8." The more high-drive inverters a design contains, the more dominant the intrinsic delays of gates should be considered. Figure 9.17a, presents the INVratio of the OpenCores benchmarks designed with both SiNWFET libraries. As we can see, there is no useful information to be extracted from this graph, since all designs have approximately the same INVratio.

The second metric we explored is iRatio and represents an evaluation of the interconnection lengths in a design normalized by the diagonal of the design's area. It is defined as

$$iRatio = \frac{L_{avg}}{\sqrt{2} \cdot area} \cdot 100. \tag{9.4}$$

where L_{avg} is the average wire length provided by SOCE.

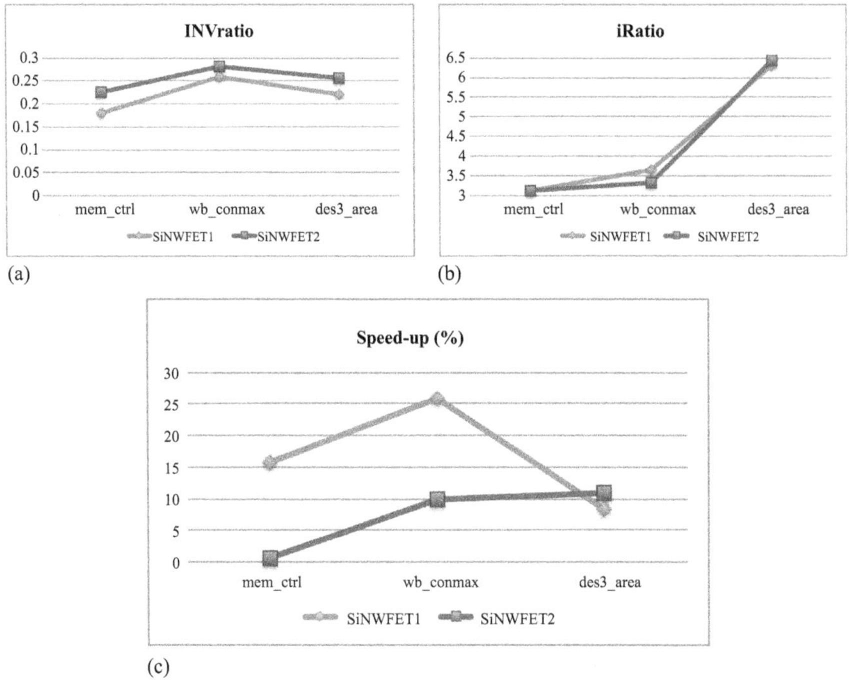

Figure 9.17 Big figure: (a) INVratio, (b) iRatio, and (c) speed-upin terms of iRatio

Figure 9.17b presents the iRatio of the benchmarks implemented in the two SiNWFET libraries. We can observe that both implementations yield a very similar interconnection for each design. We also observe that the des3_area benchmark has longer average interconnections, meaning that its routing is more complex.

Plotting the speed-up of these three benchmarks according to iRatio, we have Figure 9.17c. With Figure 9.17c, we can safely say that the speed-up of a non-XOR-dominated depends on the iRatio of the design and how complex its routing is. The dip in speed-up of the SiNWFET1 des3_area design is because, as shown in Section 9.3, the SiNWFET2 cells are in average much more routable than the SiNWFET1 ones.

9.5.3.4 Metal distribution analysis

The final comparative result that we show in Figure 9.18 is the average metal routing distribution among all the benchmarks. Each pie chart represents the average distribution of metal layers in all designs, independently of categories and size. This averaging helps us determine the general routing (and so the back-end fabrication cost) overhead that the technologies and architectures impose.

Comparing the pie charts in Figure 9.18(a) and (b), there are two differences that should be highlighted. First, we see an increase in the use of metal1 layer in the CMOS22 library. This is due to the fact that in FinFET tile configuration, we use both metal1 and metal2 layers for interconnections, contrary to the standard cell CMOS22_std library that uses only metal1. Hence, more metal1 space is available for

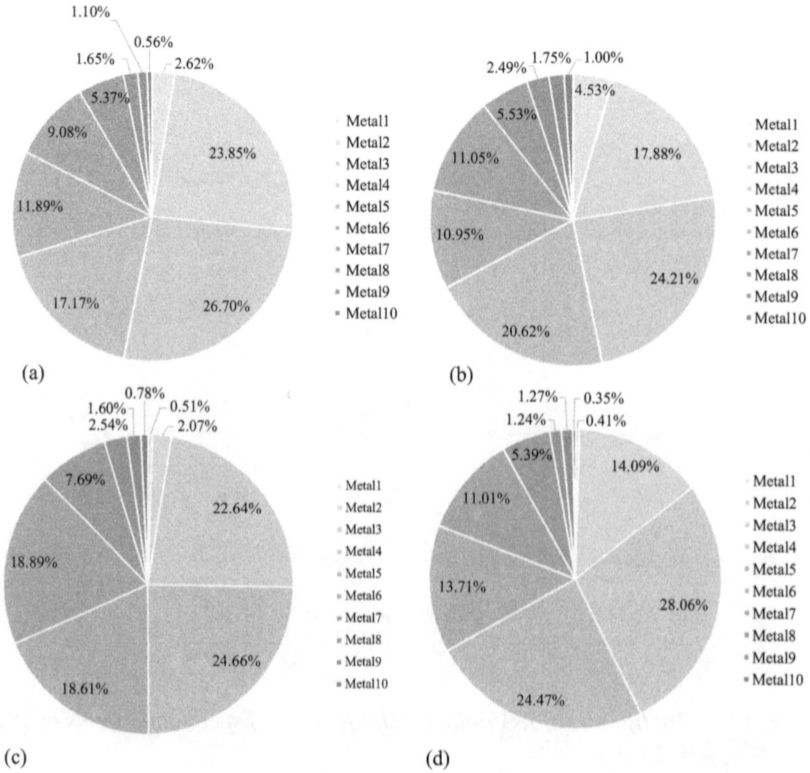

Figure 9.18 Average metal routing distributions for all libraries, (a) standard cell library CMOS22_std, (b) tile cell library CMOS22, (c) tile cell library SiNWFET1, (d) tile cell library SiNWFET2

routing. Second, we observe a decrease of metal2 and metal3 layers and an increase of metal4 layer in library CMOS22. This is due to the fact that some multi-tile cells use metal3 for interconnections, which forces SOCE to push some of the routing one metal layer higher.

Now passing from the FinFET technology and library CMOS22 to the SiNWFET technology and the SiNWFET1 library (Figure 9.18(b) and (c)), we observe the almost complete lack of metal1 routing and the decrease of metal2 routing. Both changes are due to the large brute obstruction zones (metal1) covering a big percentage of the tile's area and the tile interconnections being always on metal2 layer. This has as a consequence the increase in use of each of the metal3, metal4 and metal5 by almost 4%.

Comparing the pie charts of SiNWFET1 and SiNWFET2 libraries, we observe a similar difference where due to the large obstruction zones being on both metal1 and metal2 layers, the interconnections being on metal3 and the VDD routing on metal2, we have a small increase in metal5 and metal6 (∽4% and ∽8%, respectively). If we compare the pie chart in Figure 9.18(a) and (d), we can see that in order to migrate from the standard cell technology to the tile cell ambipolar SiNWFET1, we should sacrifice ∽8% of increase in each of the metal layers, metal4, metal5, metal6.

The main conclusions drawn from this section are

- Following the sASIC approach and moving from standard cell to tile cell library makes almost no difference in terms of metal routing and distribution, so it can be easily considered as a viable choice when layout regularity is needed.
- Implementing an alternative power routing network that enables larger degrees of freedom on the tile routing levels has an impact on the design routing level and sacrifices one metal layer (in this case metal2) without any improvement of any kind. Considering this, we can conclude that it would be preferable to put effort on constraining the tile routing levels and produce a less congested design.

9.5.3.5 Result summary

Finally, to sum up, the main conclusions drawn from all the results presented are as follows:

- All benchmarks have better delay/period performance using SiNWFET libraries, compared to the CMOS22 tile cell library.
- All benchmarks have much less leakage power consumption when implemented on SiNWFET libraries.
- Area performance increases as the binate ratio of the benchmark increases.
- Flip-flops and memory element will reduce the gain in XOR-dominated circuits because of the lack of compacted implementation.
- XOR-dominated combinational benchmarks show the largest gain in terms of area and delay when implemented with SiNWFET libraries, regardless the fact that the tile size is larger (e.g., mastrovito-SiNWFET2: 26.2% speed-up and only 17% area+ with a tile 73% larger compared to CMOS22 tile cell library).
- The area overhead of the alternative power routing network decreases in implementations with high binate ratio but still remains over 30% compared to the standard power routed library SiNWFET1.
- Every transition from CMOS22 to SiNWFET1 to SiNWFET2 library imposes an approximate 4%–8% use of the next/higher available metal layer.

9.6 Conclusions

In this chapter, we present the efforts on building a tool-flow to enable hardware design using novel polarity-controllable devices. The devices used in this work are SiNWFETs and consist of a vertical stack of Silicon nanowires that are control by two gates (double gate (DG)) placed in a GAA configuration. These devices exhibit low leakage power and great electrostatic control of the channel. However, the fabrication of SiNWFETs is done in regular arrays which introduces the need for a regular physical placement pattern. This pattern, named SoT, is identical with an extra-fine grain sASIC approach. Therefore, the goal of the tool-flow is to enable P&R of designs on a SoT of ambipolar SiNWFET tiles.

In order to incorporate the concept of the SoT and the regular fabrication of SiNWFET, we defined tile configurations for both FinFET (used as reference) and SiNWFET 22 nm technologies. Using these configurations LEFGen generated a tile cell library based on abstract approximate layouts of each cell/gate. Since each gate may comprise more than one tile connected together it is evident that the tile cell approach does not have, the same height rule as the standard cell approach.

In order to evaluate and quantify the performance of the polarity-controllable SiNWFET technology and the efficiency of its implementation, we conducted series of runs of the tool-flow using different design benchmarks. These runs proved that in terms of delay performance, the SiNWFET always outperformed the CMOS implemented designs (up to ∽35%) and that if the benchmark was XOR-rich (or XOR-dominated), then the gain in delay would increase and the overhead loss in area would decrease (also eventually become negative, meaning that SiNWFET designs would be both faster and smaller than CMOS ones—e.g., ∽35% faster and ∽15% smaller). It is shown that this technology leveraging ambipolarity has the best results when used for combinational (no bulky flip-flops) arithmetic (XOR-rich) designs. Also, we showed that the migration from the standard cell CMOS technology to the tile cell SiNWFET would cost a ∽8% increase of usage in each of the metal layers metal4, metal5 and metal6.

Although the aforementioned positive results were presented, during the configuration and the evaluation of the tool-flow and technology, some limitations arose that should be addressed. The limitations of the tool-flow presented in this chapter can be focused on the following three issues:

- The **abstraction flexibility of LEFGen** is nonexistent, meaning that we can produce detailed/accurate results for only simple gates/configurations, but then the complexity becomes too high and the only appropriate method is to place a large obstruction zone and basically ignore many interconnections. This issue is a key limitation, since it restricts the depth of the structural exploration that can be done in terms of tile configurations.
- The **dynamic power model** of the SiNWFET devices is not complete and does not produce reliable results comparable in performance to the one produced by the equivalent FinFET model. This limits our possibilities for extensive power-consumption analyses.
- **Sequential benchmarks' performance** is limited by their binate ratio and by the lack a compact and efficient implementation a flip-flop in SiNWFET ambipolar technology.

References

[1] Gaillardon PE. Reconfigurable Logic Architectures based on Disruptive Technologies [dissertation]. Ecole Doctorale Electronique, Electrotechnique, Automatique – Universite de Lyon; 2011.

[2] Brown SD, Vranesic ZG, Brown S, *et al.* Fundamentals of Digital Logic with VHDL Design. McGraw-Hill Series in Electrical and Computer Engineering Series. McGraw-Hill Higher Education; 2004. Available from: http://books.google.ca/books?id=Oygh3IDUTgsC.

[3] Kuon I, Rose J. Measuring the gap between FPGAs and ASICs. IEEE Transactions on Computer-Aided Design of Integrated Circuits and Systems. 2007;26(2):203–215.

[4] Zahiri B. Structured ASICs: opportunities and challenges. In: Computer Design, 2003. Proceedings. 21st International Conference on; 2003. p. 404–409.

[5] Moore GE. Cramming more components onto integrated circuits, Reprinted from Electronics, volume 38, number 8, April 19, 1965, pp.114 ff. IEEE Solid-State Circuits Society Newsletter. 2006;11(5):33–35.

[6] Appenzeller J, Knoch J, Tutuc E, *et al.* Dual-gate silicon nanowire transistors with nickel silicide contacts. In: Electron Devices Meeting, 2006. IEDM '06. International; 2006. p. 1–4.

[7] Heinze S, Radosavljević M, Tersoff J, *et al.* Unexpected scaling of the performance of carbon nanotube Schottky-barrier transistors. Physical Review B. 2003;68:235418. Available from: http://link.aps.org/doi/10.1103/PhysRevB.68.235418.

[8] Novoselov KS, Geim AK, Morozov SV, *et al.* Electric field effect in atomically thin carbon films. Science. 2004;306(5696):666–669. Available from: http://www.sciencemag.org/content/306/5696/666.abstract.

[9] Zhang Y, Ye J, Matsuhashi Y, *et al.* Ambipolar MoS2 thin flake transistors. Nano Letters. 2012;12(3):1136–1140.

[10] Li XX, Fan ZQ, Liu PZ, *et al.* Gate-controlled reversible rectifying behaviour in tunnel contacted atomically-thin MoS2 transistor. Nature Communications. 2017;8(1):970. Available from: https://doi.org/10.1038/s41467-017-01128-9.

[11] Resta GV, Sutar S, Balaji Y, *et al.* Polarity control in WSe2 double-gate transistors. Scientific Reports. 2016;6:29448.

[12] Resta GV, Agarwal T, Lin D, *et al.* Scaling trends and performance evaluation of 2-dimensional polarity-controllable FETs. Scientific Reports. 2017; 7:45556.

[13] De Marchi M, Sacchetto D, Frache S, *et al.* Polarity control in double-gate, gate-all-around vertically stacked silicon nanowire FETs. In: Electron Devices Meeting (IEDM), 2012 IEEE International; 2012. p. 8.4.1–8.4.4.

[14] De Marchi M, Haykel Ben Jamaa M, De Micheli G. Regular fabric design with ambipolar CNTFETs for FPGA and structured ASIC applications. In: Nanoscale Architectures (NANOARCH), 2010 IEEE/ACM International Symposium on; 2010. p. 65–70.

[15] Jhaveri T, Motiani D, Tong KY, *et al.* Maximization of layout printability/manufacturability by extreme layout regularity. Journal of Micro/Nanolithography, MEMS, and MOEMS. 2007;6(3):031011-15. Available from: http://dx.doi.org/10.1117/1.2781583.

[16] Lin YW, Marek-Sadowska M, Maly W. Transistor-level layout of high-density regular circuits. In: Proceedings of the 2009 International Symposium on Physical Design. ISPD '09. New York, NY, USA: ACM; 2009. p. 83–90. Available from: http://doi.acm.org/10.1145/1514932.1514954.

[17] Bobba S, De Marchi M, Leblebici Y, *et al.* Physical synthesis onto a Sea-of-Tiles with double-gate silicon nanowire transistors. In: Proceedings of the 49th Annual Design Automation Conference. DAC '12. New York, NY, USA: ACM; 2012. p. 42–47. Available from: http://doi.acm.org/10.1145/2228360.2228369.

[18] Bobba S, Gaillardon PE, Zhang J, *et al.* Process/design co-optimization of regular logic tiles for double-gate silicon nanowire transistors. In: Nanoscale Architectures (NANOARCH), 2012 IEEE/ACM International Symposium on; 2012. p. 55–60.

[19] Gasnier C, Gaillardon PE, De Micheli G. SATSoT: a methodology to map controllable-polarity devices on a regular fabric using SAT. In: Nanoscale Architectures (NANOARCH), 2013 IEEE/ACM International Symposium on; 2011.

[20] NIMO. Predictive Technology Model; 2013. Available from: http://ptm.asu.edu/.

[21] Brunvand E. Digital VLSI Chip Design With Cadence and Synopsys CAD Tools. ADDISON WESLEY Publishing Company Incorporated; 2010. Available from: http://books.google.ch/books?id=YippPgAACAAJ.

[22] OpenCores. OpenCores Benchmarks; 2005. Available from: http://opencores.org/.

[23] OpenCores. Triple DES IP Core; 2009. Available from: http://opencores.org/project,des.

[24] OpenCores. Memory Controller IP Core; 2010. Available from: http://opencores.org/project,mem_ctrl.

[25] OpenCores. Wishbone Interconnect Matrix IP Core; 2005. Available from: http://opencores.org/project,wb_conmax.

[26] Computer Structures Laboratory TU Graduate School of Information Sciences. ARITH Project; 2013. Available from: http://www.aoki.ecei.tohoku.ac.jp/arith/.

[27] Maly W, Heineken H, Khare J, *et al.* Design for manufacturability in submicron domain. In: Proceedings of the 1996 IEEE/ACM International Conference on Computer-aided Design. ICCAD '96. Washington, DC, USA: IEEE Computer Society; 1996. p. 690–697. Available from: http://dl.acm.org/citation.cfm?id=244522.244973.

Chapter 10
BCB benchmarking for three-independent-gate field effect transistors

Jorge Romero-González[1] and Pierre-Emmanuel Gaillardon[1]

10.1 Introduction

In 1965, Gordon Moore predicted the exponential growth in the number of transistors that can fit in a single chip. This trend, best known as Moore's law [1], was used for *research and development* targets over the following decades. Pushing the limits of traditional *complementary metal–oxide–semiconductor* (CMOS) *field-effect transistors* (FETs) by utilizing high-mobility channel materials, high-κ dielectrics, and three-dimensional architectures has allowed for remarkable energy-efficient devices at the nanometer technologies [2,3]. The *International Technology Roadmap for Semiconductors* (ITRS) [3] was developed to map technology generations and predicted that downscaling transistors below the 5-nm node will be challenging [4]. Below this node, computing performance is expected to decrease—due in particular to scaling issues in the supply voltage and short-channel effects. While the future of Moore's law is uncertain [5], scaling limits of MOSFETs have motivated the need for beyond-CMOS technologies. Higher benchmarking performances at no additional cost to yield and reliability will ensure full integration into leading chip manufacturers.

Beyond-CMOS devices use alternative switching mechanisms not seen in standard MOSFETs. *Tunnel FETs* (TFETs) [6] are a promising class of transistors that utilize quantum-mechanical band-to-band tunneling instead of thermionic injections of electrons over a potential barrier. These characteristics allow TFETs to be promising candidates for low-power electronics due to their very low OFF-currents and steep *subthreshold slopes* (SSs) [7]. However, these devices suffer from low ON-current due to their large band-to-band tunneling barrier—leading to large switching delays. Lower bandgap channel materials are expected to improve their performance such as those used in the *graphene nanoribbon TFET* (gnrTFET) [8] and in the *van der Walls TFET* (vdWTFET) [9] devices. Rather than using the charge of an electron, spin-based devices adopt the spin of an electron to carry information. The interaction between the aligned spin of a carrier and the magnetic properties of the material

[1]Department of Electrical and Computer Engineering, University of Utah, USA

allows for multiple types of devices to be formed. In particular, the *spin torque domain wall* (STT/DW) [10] uses current-driven motion of a single domain wall to store and retrieve information. These devices are promising for memory applications due to their very low voltage operation and high-density integration. On the other hand, passive spintronic devices, such as the spinFET [11], use ferromagnetic materials at the source and drain to inject and detect the spin polarized current flowing through a two-dimensional electron gas. While these spintronic devices are known to consume less power than conventional CMOS and are nonvolatile in nature, implementation into semiconductor chips and performance reliability are challenges to be overcome. Replacing the insulator of a MOSFET with ferroelectric material can also lead to low-voltage/low-power devices. In particular, the *negative capacitance FET* (NCFET) [12] uses the negative slope of polarization versus electric field (in stable configuration) of ferroelectric materials to give a body factor of less than 1—enabling steep SS capability and reducing the lower limit on the operating voltage. While this device is capable of high ON-current/OFF-current (I_{ON}/I_{OFF}) ratios at low V_{DD} [13,14], issues in the hysteresis behavior in circuit implementations should be further investigated [15].

In parallel to transistor scaling, *functionality-enhanced devices* (FEDs) extend the functional diversity of a single transistor to receive a boost in circuit performance. In particular, the *three-independent-gate FET* (TIGFET) technology [16] introduces three MOS gate terminals on a semiconducting channel and Schottky-barriers at source and drain. This design allows for dynamic reconfiguration of the polarity (n- or p-type) [17], dynamic control of the threshold voltage (V_T) [16], and dynamic control of the SS [18]. These devices have no need for highly doped source and drain and thus have less reliance on fabrication steps [19].

The popularity increase in promising beyond-CMOS devices and the urgency escalation ofCMOS limitations created the need for a uniform benchmarking methodology that would compare multiple alternative technologies to standard CMOS. The *Nanoelectric Research Initiative* (NRI) center started this investigation—leading to what is now called the *beyond-CMOS benchmarking* (BCB) [20–23]. Obtaining data from the device community, the NRI group has continually extended the methodology to cover a larger range of transistors and extract more benchmarking circuits. The first benchmarking version (BCB 1.0) [20] evaluated a set of devices with partial common guidelines to calculate the area, switching delay, and energy, while the second release (BCB 2.0) [21] developed a more uniform methodology. The third and most recent release (BCB 3.0) [22] improved the methodology by including standby power, adding sequential logic circuits, and increasing the number of devices. The final BCB methodology (BCB 3.0) covers devices from various genres such as electronic, ferroelectric, straintronic, orbitronic, and spintronic.

The BCB methodology contains a vast amount of benchmarking circuits with evaluations in area, delay, energy, and power. The basic logic gates include inverter (fan-out 1, 2, and 4), NAND (2, 3, and 4 inputs), two-input XOR, 4-to-1 MUX, 6-T SRAM register bit cell, and the gated D-latch state element. The authors of BCB 3.0 then use these elementary circuits to build higher level arithmetic and sequential architecture. These include a 32-bit adder that utilizes a standard ripple-carry architecture and a 32-bit ALU that computes the addition, subtraction, NAND, and NOR of two 32-bit numbers. Once the methodology was written, device data from the community were gathered—including transistor area estimates and device parameters such as the ON-current, OFF-current, nominal V_{DD}, and intrinsic capacitance. These are the basis for electronic-based switching (CMOS and all TFET devices), while other technologies use additional information such as in the case for NCFET and the PiezoFET [24] that include the ferroelectric material response time. Some of the spin-based devices such as the *spin wave device* (SWD) [25] and the *spintronic majority gate* (SMG) [26] have their output dependent on the majority signal of the inputs. These type of devices are referred to *majority gate circuits* and use different set of circuit equations as they are limited to majority-based circuits.

Alternative technologies have three fundamental parametrics as captured by the BCB methodology. They are based on the intrinsic capabilities of the device that make it uniquely different from CMOS. The (1) computational variable, (2) switching mechanism, and (3) circuit architecture must be identified to accurately compare and contrast the multiple technologies. A full integration into the BCB methodology can be made once these fundamental principles are thoroughly understood. In the case of TIGFETs, where the intrinsic device supports enhanced logic functionality and offers distinct transport phenomenon, unique circuit architectures and improved device parameters make it a promising class of transistors. The BCB methodology, with the added TIGFET architecture, is a MATLAB® code open to the public, available online at [27].

In this chapter, a detailed explanation is given for the introduction of TIGFET devices into the BCB methodology as first proposed in [28]. In Section 10.2, the fundamental principles of TIGFET technology are presented. In Section 10.3, the intrinsic device model under consideration is described. In Section 10.4, the basic circuit-level opportunities are investigated. In Section 10.5, the equations for area, delay, energy, and power for the various circuits are listed. The results are outlined and thoroughly examined in Section 10.6, and the chapter is concluded in Section 10.7.

10.2 TIGFET principles

In this section, we briefly review the device structure, fabrication techniques, working principle, and logic behavior of TIGFET transistors.

10.2.1 Generalities

TIGFET devices control their electrostatic properties by modulating two additional independent gate terminals [16]. As illustrated in Figure 10.1, a typical TIGFET device contains three MOS gated contacts on a semiconductor channel structure between metallic (Schottky junction) source and drain regions. The *control gate* (CG) in the inner region controls the carrier conduction by modulating a potential barrier, capable of turning the device OFF and ON-similarly to the gate of a MOSFET. The *polarity gate at source* (PGS) and the *polarity gate at drain* (PGD) modulate the Schottky

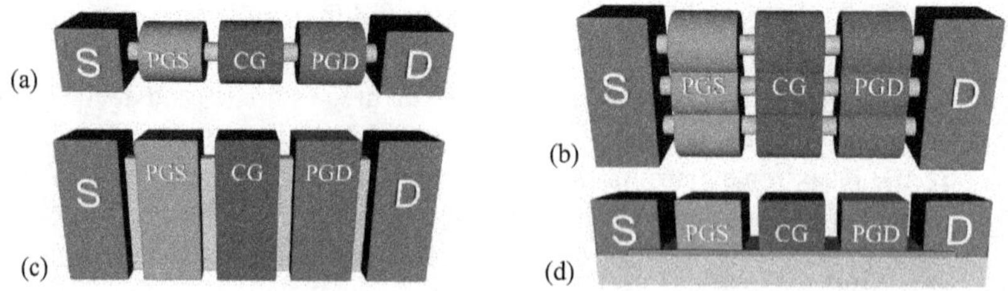

Figure 10.1 Sketches of different TIGFET topologies using (a) single nanowire, (b) multiple nanowires, (c) FinFET, and (d) 2D horizontal layer

barrier thickness to respectively allow electrons or holes to flow into the channel—allowing for reconfiguration of the polarity not possible with conventional MOSFETs.

10.2.2 Fabrication techniques

Multiterminal reconfigurable transistors have been implemented in standard Si technologies using multiple geometric designs, as illustrated in Figure 10.1. In particular, single Si nanowire [29], multiple Si nanowire [16], and Si fin [18] structures were fabricated to help achieve better electrostatic control over the channel. The *gate-all-around* (GAA) channel geometry of the multiple nanowire structure is best suited for suppressing short-channel effects [16,17,29]. Notably, TIGFET architecture was fabricated using four vertically stacked nanowires in [17]. The nanowires were defined using electron-beam lithography on a lightly p-type doped silicon-on-insulator substrate. A deep reactive ion etching step follows to form the stacked nanowires. After thermally grown SiO_2 as the gate dielectric, the gate structures are formed. Polycrystalline silicon PGS and PGD are patterned with a self-aligned CG gate. Silicon nitride spacers are added to isolate the gated structures. Lastly, the Schottky barriers are created by annealing a nickel layer in the source and drain pillars to form nickel silicide. The fabrication procedures are similar for most multi-gated devices. In addition to standard Si, a variety of other channel materials have been used such as two-dimensional (2D) materials [30–32] and carbon nanotubes [33]—offering enhanced control over the channel with increase ON-current by growing nanotubes or mechanically exfoliating WSe_2 (from the transition-metal-dichalcogenide family).

As a major obstacle in modern CMOS technology, source and drain doping is absent in reconfigurable technology. By exploiting dopant-free processes, TIGFET transistors are sought to be more robust to several fabrication-related variabilities [19].

10.2.3 Working principle

The FED family of devices operates within the same set of principles. The MOS gates at PGS and PGD modulate the barrier thickness of the Schottky contacts at source and drain, respectively, while the MOS gate at CG controls the potential barrier in the center of the device.

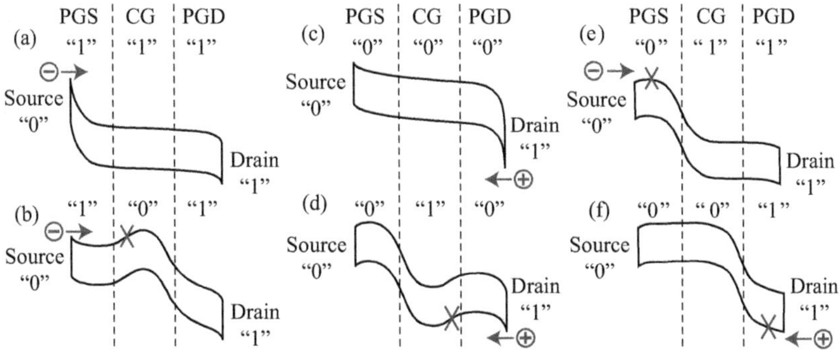

Figure 10.2 Band diagrams under different bias conditions. LVT n-FET in the (a) ON-state and (b) OFF-state. LVT p-FET in the (c) ON-state and (d) OFF-state. HVT low-leakage OFF-state in an (e) n-FET and (f) p-FET. Notation: "0" = G_{ND} and "1" = V_{DD}.

To understand how these principles affect the characteristics of the device, let us first consider the case in which CG modulates the channel conduction. Here, both polarity gates (PGS and PGD) are set to a constant bias, while CG is set to a dynamic bias. If the polarity gates are set high, as shown in Figure 10.2(a) and (b), electrons pass through the Schottky barrier at the source side and become the majority carriers. The transistor is in ON-state if CG is set high (Figure 10.2(a)) and in the OFF-state if CG is set low (Figure 10.2(b)). On the other hand, if the polarity gates are set low, as shown in Figure 10.2(c) and (d), then holes are the majority carriers that are passing through the Schottky barrier at the drain side. The transistor turns ON when CG is set low (Figure 10.2(c)) and OFF if CG is set high (Figure 10.2(d)).

The current transport mechanism in this case is dominated by thermionic emission of electrons over a potential barrier—coming from the modulation in the channel conduction by the CG bias. The current flowing through the TIGFET device has the following expression [16,34]:

$$I_D = AA^*T^2\exp\left(-\frac{q\phi_B}{\kappa_B T}\right), \tag{10.1}$$

where ϕ_B is the effective barrier height, A is the junction area, T is the temperature, A^* is the effective Richardson constant, q is the electron charge, and κ_B is the Boltzmann constant. In the subthreshold region, a direct change in CG will cause a change in the barrier height ($\Delta\phi_B = -\Delta V_{CG}$) and the current will exponentially increase.

Let us now consider the TIGFET transistor behavior when the polarity gates (PGS and PGD) can be controlled independently. While this device is capable of a dynamic polarity control, it can also be switched between two different threshold voltages (V_T): low-V_T (LVT) and high-V_T (HVT). In the first case, when PGS=PGD, the majority carriers only need to overcome the potential barrier at CG to create a current, as previously discussed. This leads to a low-V_T configuration since there is a direct change between V_{CG} and ϕ_B. In the second case, when PGS≠PGD, both electrons and holes are blocked by the Schottky barriers at source and drain, respectively. This leads to a lower OFF-current since majority carriers need to gather enough energy to tunnel through the Schottky barrier to induce a current. Considering the n-FET in Figure 10.2(e), in order to switch the device ON, the potential at PGS must increase to allow electrons to tunnel through the Schottky barrier at source side. The current mechanism may be approximated by (10.1) except that now tunneling of thermally excited carriers dominates. Since the probability of tunneling is always smaller than 1, the potential at PGS will have a lower efficiency on the effective barrier height of the source Schottky barrier [16]. This relationship can be approximated as $\Delta\phi_B = -\lambda\Delta V_{CG}$, where λ is less than 1. This relationship means that a higher potential at PGS is needed to turn on the device—resulting in high-V_T configuration. The same mechanism works for the p-FET configuration. However, PGS and CG are both set low, and PGD is set high to turn the device OFF as shown in Figure 10.2(f). Here, electrons are blocked by the source Schottky barrier, while holes are blocked by the drain Schottky barrier. PGD must be decreased in order to reduce the effective Schottky barrier height and thus start thermally assisted tunneling at the drain side to create a channel conduction.

While TIGFETs are capable of dual polarity and threshold control, the ON-current (and the OFF-current) of both n-FET and p-FET configurations can be optimized to be equal as shown in Section 10.3.1 "Electrical Properties." The n- and p-type current symmetry is possible by designing Schottky barrier heights with mid-gap workfunctions as reported in [17].

The final key characteristic of TIGFET devices is their capability of breaking the thermal SS that impedes further performance development past 5 nm technology nodes in standard MOSFET designs. The ~60 mV/dec at room temperature is the fundamental limitation of SS on MOSFET transistors due to the temperature-dependent thermionic injections of carriers into the channel. TIGFET devices, along with other alternative technologies, such as TFETs [35] and NCFETs [36], offer switching mechanisms that are capable of SS reduction. In particular, the TIGFET device has been simulated to obtain an n-type SS of ~5 mV/dec and a p-type SS of ~35 mV/dec and measured to achieve down to 6.0 mV/dec for five decades of current [18]. The principal switching mechanism for TIGFET devices that allow for steep SS characteristics is the positive feedback induced by weak impact ionization. We refer the reader to Chapter 7 "Triple-Independent Gate FET's Super Steep Slope" for a detailed origin of TIGFET's steep SS capabilities. This device characteristic will not be exploited in this study.

10.2.4 Logic behavior

For a design purpose, an abstract representation of a TIGFET device can be extracted from its electrical behavior. For instance, if a bias potential is applied at CG and the polarity gates are tied to a constant bias, then the channel conduction is controlled by CG—entering a low-V_T configuration and behaving as either a unipolar n-type MOSFET or a unipolar p-type MOSFET, as shown in Figure 10.3(b). If a bias

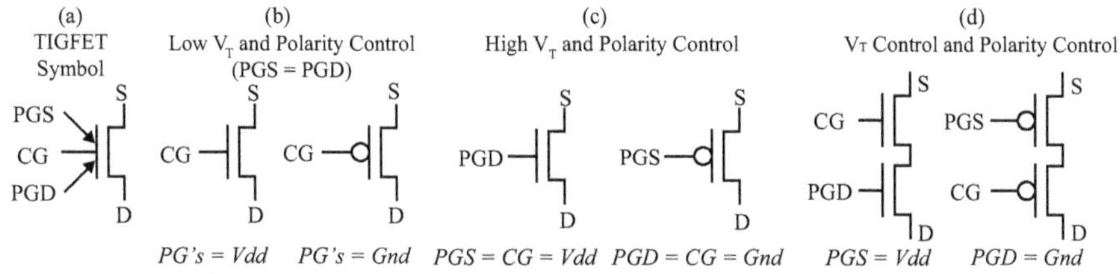

Figure 10.3 (a) Symbol representation of a TIGFET device with the corresponding MOSFET representations: (b) channel conduction by the CG bias, (c) channel conduction by the PG bias, and (d) channel conduction by the series CG and PG biases

potential is applied at one polarity gate and the two other gates are tied to a constant bias, then the channel conduction is controlled by the PG bias and the device's high-V_T mode is enabled—similarly behaving as unipolar MOSFETs, as shown in Figure 10.3(c). Lastly, if two bias potentials are applied at CG and PGS or PGD, as shown in Figure 10.3(d), then multi-V_T characteristic and dynamic control of the polarity is presented. In this case, the TIGFET transistor can be represented as two unipolar MOSFET devices in series.

10.3 Device-level considerations

In order to comply with BCB's methodology, the device-level performances of the considered technologies must be obtained using the 15-nm node technology assumed in [20–22]. In this section, the *I–V* curve, intrinsic capacitance, and intrinsic area of TIGFET's SiNW device, following the design rules set by BCB, are presented.

10.3.1 Electrical properties

First, the commercially available TCAD Sentaurus tool from Synopsys Inc. was used to determine the device-level electrical parameters [28]. These include extracting the OFF-current (I_{OFF}), ON-current (I_{ON}), and nominal voltage (V_{DD}) of the TIGFET device. The TIGFET structure, as used in this chapter, is shown in Figure 10.4 following the device parameters compliant to ITRS's 15 nm technology node [3]. The channel length is 15 nm, the separation between contacted gates is 15 nm, and the dielectric layer is HfO_2 with a thickness of 6.92 nm and an *equivalent oxide thickness* (EOT) of 1.08 nm. The channel is an 8-nm diameter GAA silicon nanowire with Si–NiSi Schottky contacts at source and drain. The materials used in this device were selected to ensure full compatibility with standard CMOS processes (e.g., CMOS HP).

The TCAD Sentaurus model, as used in [17,28], solved Poisson's equations and continuity equations for both electrons and holes. The models were fitted using experimental results obtained in [17]. The extracted model used Fermi statistics, the

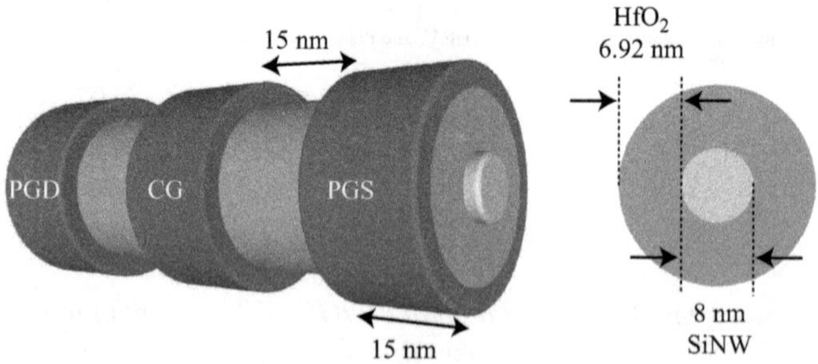

Figure 10.4 Experimental drawing of the silicon-based nanowire TIGFET

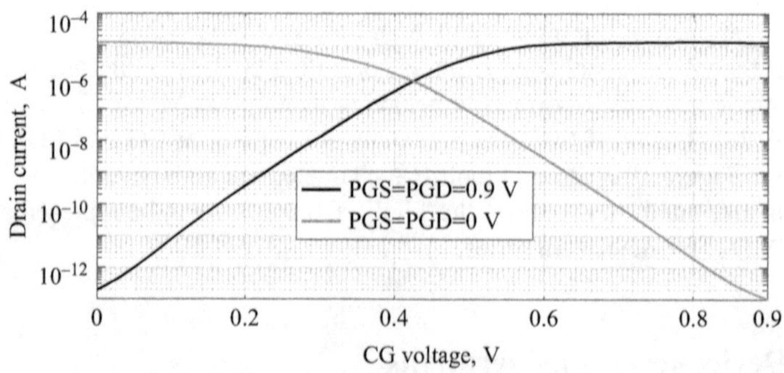

Figure 10.5 Drain current (I_D) versus CG voltage (V_{CG}) for a single TIG SiNWFET at $V_{DD} = 0.9$ V. The black and gray curves indicate the nMOS and pMOS realization, respectively [28].

Philips unified mobility model, the Slotboom bandgap narrowing model, the band-to-band tunneling model for electrons in the source electrode and holes in the drain electrode, and lastly the barrier lowering model for Schottky contacts.

The *I–V* curve presented in Figure 10.5 shows the results obtained from the TIGFET SiNW device under consideration. A supply voltage of 0.9 V is selected to ensure the device permits a large enough barrier on the OFF-state mode and, at the same time, lowers the effective Schottky barrier height enough at the source/drain terminals in the ON-state mode. This results in the largest possible I_{ON}/I_{OFF} metric. Note that lower bandgap channel materials can be used to reduce the voltage while maintaining similar current densities. In this model, the Schottky source/drain contacts are designed with mid-gap workfunction to achieve symmetric n- and p-type ON-currents. This is experimentally achieved using Si–NiSi contacts in [17]. The TCAD simulations extracted an ON-current of 12.6 μA, OFF-current of 0.135 pA, and V_{DD} of 0.9 V. Following the methodology from BCB 3.0 [22], the OFF-state current is the summation of drain current and leakage current coming from the MOS gate. Since the TIGFET device contains three-gated MOS terminals, the leakage current originating from the gates is multiplied by a factor of 3. This results in an OFF-current density of $I_{OFF} = 3.8$ nA/μm with an ON-current density of $I_{ON} = 501$ μA/μm [28].

10.3.2 Capacitance consideration

Following the BCB methodology, the total capacitance of the device must be determined. As a reminder, the ideal single gate capacitance per unit area (F/m^2) can be calculated as

$$C_g = \frac{\varepsilon_r \varepsilon_0}{EOT} L_g, \tag{10.2}$$

where L_g is the length of the gate and EOT is the electrical EOT that includes both dielectric and quantum capacitance—set to 1.08 nm in BCB. ε_r is the dielectric constant of the oxide and ε_0 is the permittivity of free space. The BCB methodology then assumes CMOS HP's parasitic capacitance is approximately 1.5× larger than the gate capacitance—estimated by ITRS for the 15-nm technology node [3]. It means that the considered CMOS HP's total capacitance, C_{tot}, is equal to $2.5 \times C_g$. The alternative technologies presented in the BCB methodology have a specific total capacitance associated with their intrinsic devices. To generalize these capacitances, the adjustment factor, M_{adj}, is introduced.

$$C_{tot} = \frac{\varepsilon_r \varepsilon_0}{EOT} L_g (M_{adj} + M_{par}), \tag{10.3}$$

where M_{par} is the parasitic capacitance of an ideal single gate (set to 1.5) and M_{adj} is normalized to 1 for CMOS HP. The adjustment factor cannot be set to a single value for TIGFET technology. It turns out that TIGFET devices can operate with multiple parasitic capacitances. To understand how this is possible, consider the case when PGS = PGD. This configuration should have a lower total capacitance than in the case of PGS $\neq$ PGD since the capacitance between the PGS and PGD electrodes would not be included. Therefore, the amount of inputs switching will lead to different parasitic capacitances—producing different adjustment factors (M_{adj}) for the BCB methodology.

To extract the multiple capacitance contributions of a TIGFET device, the authors in [28] used COMSOL Multiphysics to perform 3D Poisson electrostatic simulations. The geometric model follows the same structure as shown in Figure 10.4. In order to establish a baseline comparison between TIGFET and CMOS HP, the extracted capacitances coming from the TIGFET simulations were compared to a FinFET (MOSFET) structure simulated with the same parameters from the BCB methodology. Thus, the capacitance values between all TIGFET electrodes were calculated and multiplied by the corresponding baseline factor.

The circuit-level opportunities, as described in Section 10.4, can be expressed using four modes of operation as shown in Figure 10.6. These device configurations depend on the number of fixed bias potentials that lead to distinct switching capacitances. Figure 10.6 shows the total capacitances extracted with their corresponding adjustment factors.

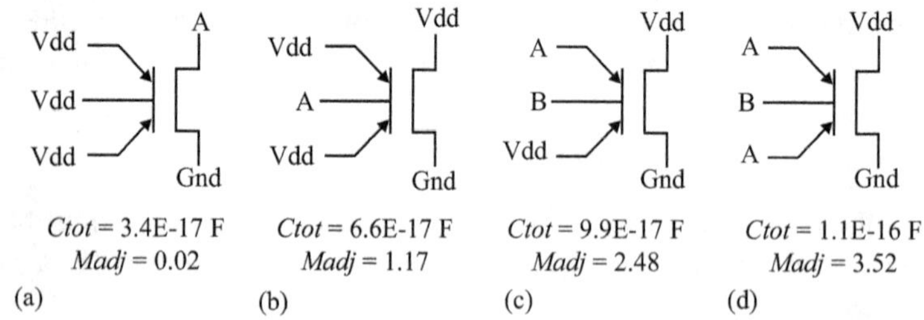

Figure 10.6 Total capacitance (C_{tot}) and adjustment factor (M_{adj}) for the multiple modes of operation. A and B correspond to different bias inputs. The transistors in (a), (b), and (c) are in n-type configurations. The constant bias gates can be set to G_{ND} for the p-type configurations.

10.3.3 Layout considerations

The BCB methodology follows the scalable design rules described in [21]. Taking these rules into consideration, the layout of the intrinsic TIGFET transistor can be constructed as shown in Figure 10.7(a). The feature size designated to be a DRAM's half pitch, F, of 15 nm was chosen by the BCB methodology. Using a maximum mask misalignment, $\lambda = (1/2)$F, the *Pitch* (*P*), and *Width* (*W*) of the TIGFET transistor were designed to be 8 F and 4 F, respectively [28]. The PGS, CG, and PGD gate lengths are set to 2 F (15 nm) and the separation between polysilicon lines are set to be 2 F. The transistor's width and contact size are set to the same values as the CMOS HP transistor described in [21].

The layout geometry implemented in this chapter uses the *sea-of-tile* (SoT) design [37] as implemented in [28]. This geometry allows for layout "tiles" to be placed side-by-side and surrounded by a grid of power rails as shown in Figure 10.7(b). The tiles are composed of two pairs of TIGFET transistors with common source and drain gates. This geometry provides less problematic routing congestions [38], while still allowing a large range of circuits to be mapped [39]. While TIGFET technology requires two additional contacted gates per device, the SoT design topology mitigates routing bottlenecks and achieves similar routing efficiency as CMOS technology [38]. The uniformity of the SoT architecture is expected to provide a homogeneous front-end production which can help achieve a high yield at advanced technology nodes. Now, since TIGFET's doping-free process allows for improved robustness [19], full integration of TIGFET devices into leading chip manufacturers is feasible as their yield and reliability are expected to be comparable to CMOS technology.

10.4 Circuit-level opportunities

The benefits of TIGFET technology in circuit architecture have been largely investigated in literature [40–45]. The capability to design basic logic gates is analogous to CMOS technology where the pull-up (PU) network represents the complement of

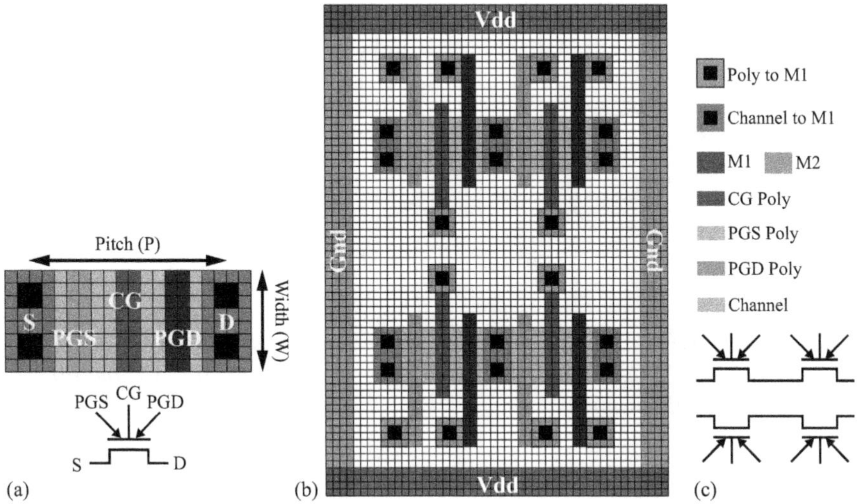

Figure 10.7 Layout view of the (a) intrinsic TIGFET device and (b) uncommitted 4-transistor TIGFET tile with their corresponding circuit representations. (c) Key with designated contact material naming.

the pull-down (PD) network. This is illustrated in Figure 10.8, where the MOSFET transistors are replaced with their corresponding TIGFET device representations as previously discussed in Figure 10.3. First, the realization of unipolar n- and p-type TIGFET devices is possible when two of its three MOS gates are tied to a constant bias. As an illustration, an inverter with an n-type device in the PD network and a p-type device in the PU network can be achieved in LVT mode (when both of the polarity gates are tied together) or in HVT mode (when the PGD and CG gates are tied together) as shown in Figure 10.8(a) and (b), respectively. Similar to CMOS FinFET design, several TIGFET transistors can be connected in parallel to increase the current drive without affecting the logic output. This has been accomplished with multiple stacked nanowires to lessen the extra layout space needed to route transistors in parallel [17]. Next, the ability to emulate two series n- or p-type MOSFETs with a single TIGFET transistor is possible by using two potential biases within a single TIGFET transistor. This is illustrated in the PD network of a two-input NAND gate with TIGFET technology as shown in Figure 10.8(c) and (d) for the LVT and HVT configurations, respectively. The reduction of transistors does not affect the drive strength of the device since both the n- and p-type devices are sized equally.

The TIGFET device-level representation to realize two n- or p-type MOSFETs in series can be generalized and used to create novel circuit architectures. As an example, by placing these TIGFET transistors in both the PU and PD networks, the two-level 4-to-1 static multiplexer [45] can be mapped as shown in Figure 10.9.

Finally, the TIGFET technology's richer set of operations, coming from two potential bias applied at the CG and at the polarity gates (PGS and PGD), allow

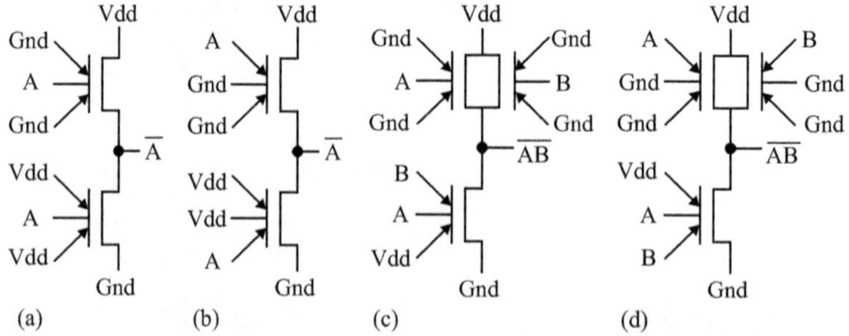

Figure 10.8 TIGFET circuit representations of an inverter with (a) LVT and (b) HVT configurations and a two-input NAND with (c) LVT and (d) HVT configurations

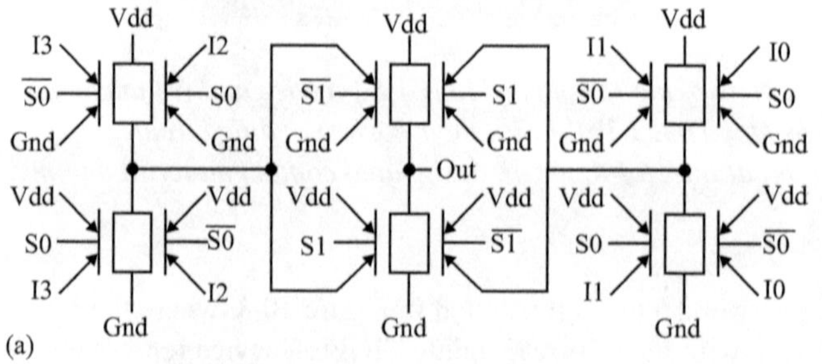

S1	S0	Out
0	0	I0
0	1	I1
1	0	I2
1	1	I3

(b)

Figure 10.9 (a) TIGFET circuit representations of a two-level 4-to-1 static multiplexer with (b) with corresponding X and Y values to map the functions

for very efficient exclusive OR (XOR) and the majority (MAJ) gates circuits. The schematics of a two-input XOR [40], a three-input XOR [43], and a three-input MAJ [43] are shown in Figure 10.10(a), (b), and (c), respectively. These circuit realizations are compacted within four transistors and some inverters—resulting in lower transistor count as compared with CMOS.

In addition to combinational logic, sequential logic circuits are often used in digital signal processing systems such as in the case of building arithmetic logic units (ALUs). These architectures commonly use flip-flops as memory elements. By employing the full control of a TIGFET transistor, i.e., applying a bias per MOS gate, a *true single-phase clock* (TSPC) flip-flop with TIGFET architecture can be constructed with eight transistors and two inverters as shown in Figure 10.11.

While not explicitly exploited in the rest of this chapter, TIGFETs also enable an interesting power gating scheme using *differential cascade voltage switch logic* (DCVSL) cells [44], as shown in Figure 10.12. The DCVSL is a logic style that implements a logic function f and its complement f' simultaneously. The removal of

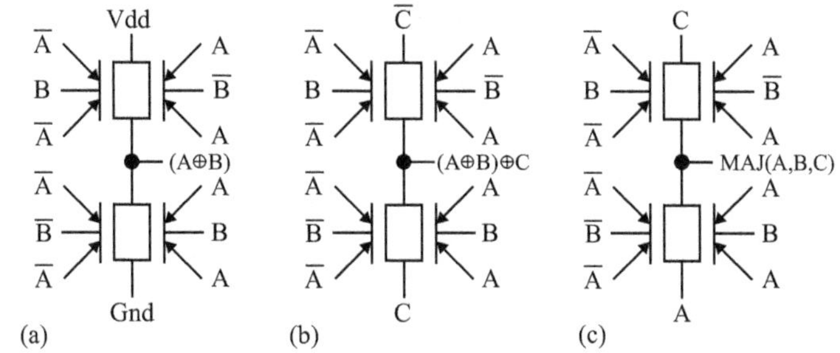

Figure 10.10 TIGFET circuit representations of (a) a two-input XOR gate, (b) a three-input XOR gate, and (c) a three-input MAJ gate

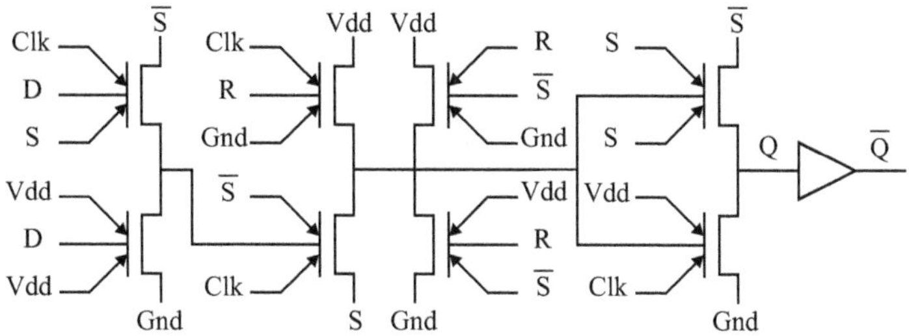

Figure 10.11 TIGFET circuit representations of a TSPC flip-flop

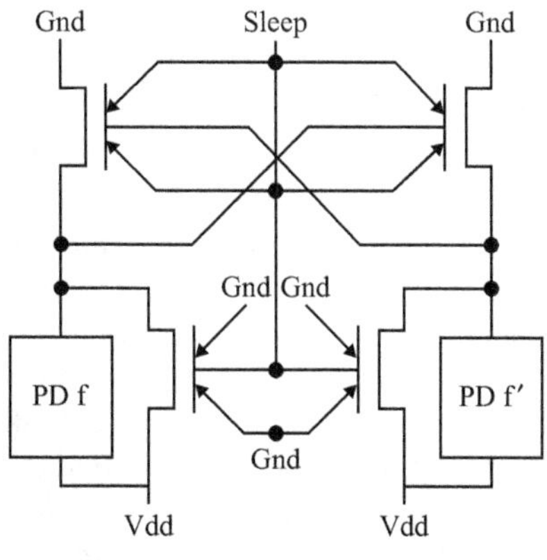

Figure 10.12 TIGFET circuit representations of a power-gated DCVSL [44]

the series device avoids major performance degradation compared to CMOS. The PU network contains p-type devices with the polarity modulated by the *Sleep* signal.

An efficient implementation of a ripple-carry adder can be derived from the presented primitives. The BCB methodology uses a 32-bit ripple-carry adder as a meaningful desire to demonstrate the performance of combinational logic circuits.

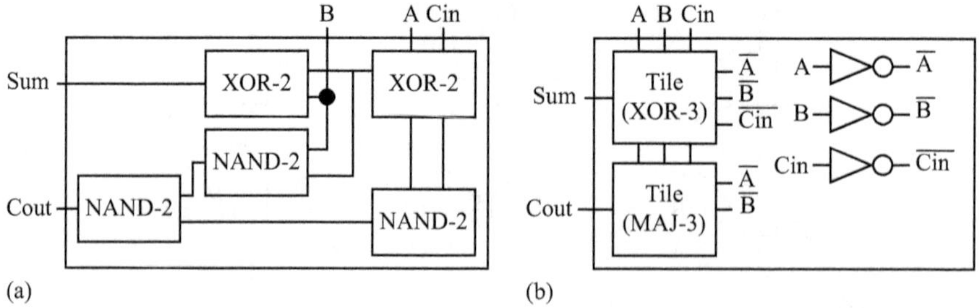

Figure 10.13 1-Bit full-adder (a) as used in the BCB methodology and (b) designed for TIGFET architecture

Since this circuit is composed of multiple 1-bit full adders, it is important to understand the circuit architecture in its most basic form. The 1-bit adder implemented in the BCB methodology is shown in Figure 10.13(a). It is composed of two two-input XOR gates and three two-input NAND gates. TIGFET technology does have the capability of representing this adder with a 1-to-1 replacement of transistors since two-input NANDs and two-input XORs can be easily implemented with TIGFET technology. However, the increased functionality of TIGFET transistors allows for very efficient three-input XORs and three-input MAJs to be used. The 1-bit adder, as shown in Figure 10.13(b), is designed with one three-input XOR gate to calculate the *Sum* signal $((A \oplus B) \oplus Cin)$, one three-input MAJ gate to calculate the *Cout* signal $(MAJ(A, B, Cin))$, and three inverters to invert the input signals.

By cascading the 1-bit adder in Figure 10.13(b), one can obtain a 32-bit adder. Nevertheless, the three-input XOR and the three-input MAJ gates are in parallel and thus output their signals simultaneously. Since the *Cin* signal is inverted at every stage, the critical signal contains 32 inverter components. In order to reduce the logic level of this circuit, both the regular and the complimentary carry signals may be computed in parallel. This operation is expensive in CMOS but is cheap with the TIGFET technology since the implementation of MAJ gates requires only a couple of additional transistors per stage. The adder stage, as shown in Figure 10.14(a), uses a three-input MAJ gate, with inputs $\overline{A}$, $\overline{B}$, and $\overline{Cin}$, to produce *Cout* and results in the simultaneous computation of the regular and inverted carry signals at each stage and removes the inverter contribution per stage. The stage diagram for the 32-bit adder, while using this circuit implementation, is shown in Figure 10.14(b). This circuit architecture is used for the 32-bit adder in the BCB methodology for TIGFET technology.

In a similar way, complex sequential circuits can be built from the logic functions presented in this work. The best sequential gate metric used in the BCB methodology is the 32-bit ALU. This circuit is able to realize the addition, subtraction, NAND, and NOR of two 32-bit numbers, as shown in Figure 10.15(a). The operation function is controlled by the four signals: *Ctrl*0, *Ctrl*1, *Clk*0, and *Clk*1 (offset by a half clock-cycle). Starting from the left of Figure 10.15(a), the two 32-bit numbers

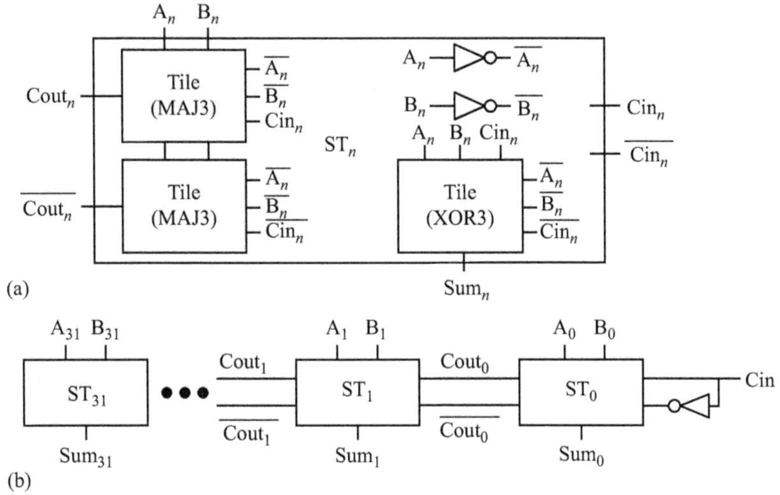

Figure 10.14 (a) Single stage, ST_n, where n denotes the stage number. (b) Stage diagram for the 32-bit ripple-carry adder

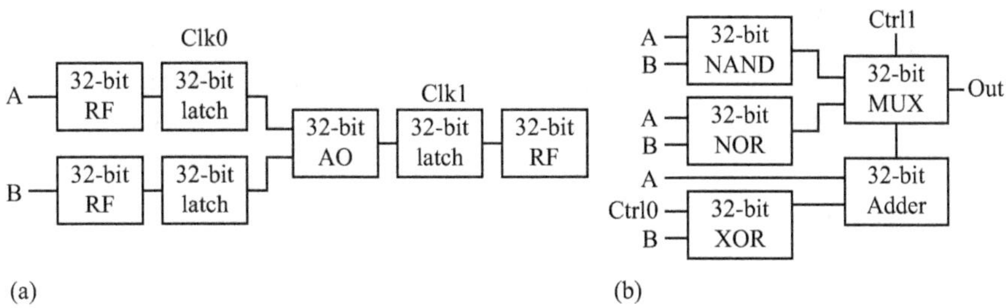

Figure 10.15 Scheme of (a) the ALU with clock signals Clk0 and Clk1 and of (b) the AO with control signals Ctrl0 and Ctrl1

(A and B) are stored in 32-bit register files (using 6-transistor SRAM cells). The 32-bit latch (using the flip-flops) transmit the output of the RF block into the *arithmetic operator* (AO) at the rising edge of the *Clk*0. The AO scheme is shown in Figure 10.15(b). At the rising edge of *Clk*1, the output of the AO is transferred and stored into the 32-bit registers. On the falling edge of *Clk*0, the inputs to AO are isolated from the RFs. Lastly, on the falling edge of *Clk*1, the inputs to the RFs are isolated from the outputs of AO. The control signal *Ctrl*0 enables the option between add or subtract and the control signal *Ctrl*1 selects the arithmetic operation to be performed. This ALU architecture is common for all alternative technologies in the BCB methodology .

10.5 Performance evaluation

The BCB methodology was developed as a guideline to evaluate the performances of various beyond-CMOS devices. In particular, the best circuit metrics performed in the BCB was the 32-bit adder and 32-bit ALU. As said earlier, these circuits are derived from circuit primitives, and thus, the area, delay, energy, and standby power for the inverter, two-input NAND, two-input XOR, three-input XOR, three-input MAJ gate, 4-to-1 multiplexer, TSPC flip-flop, and SRAM cell must be determined.

10.5.1 Area estimation

Let us first examine the area components of the circuits benchmarked in BCB 3.0 [22]. As discussed in Section 10.3.3, we consider a pragmatic layout assumption called SoTs that uses regular "tiles" containing four transistors (two pairs). This design efficiently realizes a variety of unate and binary logic functions. The area of a tile (A_T), expressed in F^2, is calculated to be

$$A_T = (2P + 7) \cdot (2W + 25), \tag{10.4}$$

where the *pitch* (P) of the intrinsic transistor is set to 8 F and the *width* (W) of the intrinsic transistor is set to 4 F. The first component of (10.4) is the width of the tile, while the second component is the length of the tile. The area of the 4-transistor tile (A_T) can be used to determine the area of the other circuits. Starting with the inverter, the area ($A_{inv\mathrm{FO}}$) is equal to half a tile times the *fan-out* (FO) number. This is the case since increasing the FO is equivalent to adding more transistors in parallel to increase the drive strength of the gate. The area of the two-input NAND (A_{nand2}) is simply the area of a tile.

$$A_{inv\mathrm{FO}} = 0.5A_T \cdot FO, \quad A_{nand2} = A_T. \tag{10.5}$$

Next, we derive the areas of the two-input XOR (A_{xor2}), three-input XOR (A_{xor3}), and three-input MAJ (A_{maj3}). As discussed in Section 9.4, these three logic gates require one tile (four TIGFET transistors) plus some inverters. The XOR gates require one inverter per input, while the three-input MAJ gates require only two inverters.

$$A_{xor2} = A_T + 2A_{inv1}, \quad A_{xor3} = A_T + 3A_{inv1}, \quad A_{maj3} = A_T + 2A_{inv1}. \tag{10.6}$$

The area of a two-level 4-to-1 static multiplexer (A_{mux}) is equivalent to three tiles to create the body logic plus two sets of inverters to invert the control signals. The area of the TSPC flip-flop (A_{ff}) is equivalent to two tiles plus two inverters to invert the output signal and the set signal. The 6-T SRAM cell has an area of 1.5 tiles to implement the six transistors.

$$A_{mux} = 4A_T M_g, \quad A_{ff} = 3A_T M_g, \quad A_{sram} = 1.5A_T, \tag{10.7}$$

where M_g is the gate overhead factor used in the BCB methodology to model the space required to route interconnects. This factor is set to 1.5 for CMOS and 1.3 for TIGFET technology. The overhead factor is lower for TIGFET technology since regular layouts require more space, and we do not want to increase the interconnect area needed to route the same number of signals. The XOR and MAJ gates do not require the overhead space to route interconnects since two adjacent tiles do not need extra space. The SRAM cell in BCB uses the optimized layout design and therefore also does not need overhead space.

Based on the above results, the area of the 1-bit full-adder (A_{1b}) can be calculated by adding the area of one three-input XOR, one three-input MAJ gate, and three inverters. This results in the area of two tiles plus three inverters. In order to take the area to route interconnects into consideration, the M_g gate overhead factor was added. The area of the 1-bit full adder is then shown to be

$$A_{1b} = (2A_T + 3A_{inv1})M_g = 3.5A_TM_g. \tag{10.8}$$

The most important metric for combinational circuits, as considered in this chapter, is the 32-bit adder. The unique TIGFET-based design used for this circuit is shown in Figure 10.14. Each stage in this architecture requires one three-input XOR, two three-input MAJ gates, and two inverters to generate the outputs *Sum*, *Cout*, and $\overline{Cout}$. An M_g factor is added to allocate space to route interconnects between the stages. Note that the design also includes an inverter for every five stages in order to restore the signal passing through the devices.

$$A_{32b} = 32(3A_T + 2A_{inv1})M_gM_g + 7A_{inv1}M_g = 131.5A_TM_g. \tag{10.9}$$

The sequential gate metric used in the BCB methodology is a 32-bit ALU. This circuit is capable of computing the addition, subtraction, NAND, and NOR of two 32-bit numbers. The area of the AO, A_{ao}, is first calculated. This circuit is composed of one 32-bit NAND, one 32-bit NOR (equivalent to a NAND in BCB), one 32-bit XOR, one 32-bit MUX, and one 32-bit adder as shown in Figure 10.15(b). The scheme for the 32-bit ALU consists of a 32-bit AO, three 32-bit flip-flops, and three 32-bit SRAMs. Similarly, an extra M_g factor is added to route interconnects between the circuits.

$$A_{ao} = A_{32b} + 32(A_{xor2} + 2A_{nan2} + A_{mux})M_g, \tag{10.10}$$

$$A_{alu} = (A_{ao} + 32(3A_{ff} + 3A_{sram})M_g)M_g. \tag{10.11}$$

10.5.1.1 Area summary

The results obtained from the area equations for the TIGFET and CMOS technologies are shown in Table 10.1. The TIGFET-based results are dependent on the area of a tile which, in turn, depends on the technology node and on the overhead factor. The logic gates that have a 1-to-1 replacement of TIGFET transistors, such as the inverter, two-input NAND, and SRAM, have larger total areas. This was predicted since the area of a single TIGFET device is 1.67× larger than a MOSFET device. However, the unique circuit opportunities brought on by TIGFET's functionality extension allowed some circuits to be highly compacted. This includes the two-input XOR, the two-level 4-to-1 MUX, the 1-bit full adder, and the 32-bit adder. Note that the results of the three-input XOR and the three-input MAJ gates are not shown. The BCB methodology did not benchmark these logic gates. The 32-bit ALU only saw a 6% reduction in layout area since it contained multiple inefficient NAND gates.

Table 10.1 Total area summary expressed in F^2

Logic function	CMOS HP and LV	TIGFET	Comparison
FO-1 Inverter	160	380	2.37×
Two-input NAND	288	759	2.64×
Two-input XOR	1,728	1,518	0.88×
4-to-1 MUX	4,368	3,962	0.91×
Flip-flop	2,448	2,972	1.21×
SRAM	320	759	2.37×
1-bit Adder	7,200	3,467	0.48×
32-bit Adder	345,600	168,920	0.49×
32-bit ALU	1,596,672	1,563,000	0.98×

10.5.2 Delay estimation

We now examine the switching delay of the circuits implemented in the BCB methodology. The switching delay is made up of (1) the intrinsic component and (2) the interconnect component. First, the RC delay of the intrinsic device (t_{int}) corresponds to the charging of the device's parasitic capacitance up to a time constant. This term uses the primitive lumped circuit equation to calculate the delay of a transistor. It is determined to be the total capacitance (C_{tot}) of the device times the supply voltage (V_{DD}) divided by the ON-current (I_{ON}).

$$t_{int,M_{adj}} = \frac{C_{tot}V_{DD}}{I_{\text{ON}}}. \tag{10.12}$$

However, the intrinsic delay for TIGFET technology is dependent on the capacitance adjustment factor (M_{adj}) since the total capacitance of the device (C_{tot}) is dependent on the mode of operation as discussed in Section 10.3.2. V_{DD} and I_{ON} are not dependent on M_{adj}.

The delay estimation of an interconnect (t_{ic}) can be similarly calculated by considering the charge of the interconnect capacitance (C_{ic}) that is equal to the capacitance of a wire per length of interconnect. This particular length is equal to the typical length of an interconnect and is assumed to be five times the pitch in BCB [22]. Note that the t_{ic} term is not dependent on the mode of operation. Since the capacitance of an interconnect is solely dependent on the length of the interconnect, a delay factor of 0.69 is added. This factor comes from the lumped circuit approximation, as done in [46], where the ON-resistance ($R_{\text{ON}} = V_{DD}/I_{\text{ON}}$) and interconnect capacitance term ($0.69R_{\text{ON}}C_{ic}$) sufficiently approximates the total interconnect delay for interconnect lengths below 100 times the gate pitch. This approximation is well suited for the BCB methodology.

$$t_{ic} = 0.69\frac{C_{ic}V_{DD}}{I_{\text{ON}}}. \tag{10.13}$$

Since the two-input NAND and the *fan-out-of*-1 inverter share the exact same structure for TIGFETs and MOSFETs, the delay expression will be the same. The only difference here is that the intrinsic delay of a TIGFET is dependent on the mode of operation. The use of subscripts (a)–(d) will identify the mode of operation as illustrated in Figure 10.6. The inverter has one input to CG, while PGS and PGD are fixed and thus mode (b) is used. The NAND gate, on the other hand, has an input to CG and PGS while PGD is fixed in the PD network and one input to CG while PGS and PGD are fixed in the PU network. The worst case delay is chosen here and thus the intrinsic delay uses mode (c).

$$t_{inv1} = 2M_{tinv}t_{int,b} + t_{ic}L_{inv}, \quad t_{nand2} = 2M_{tnand}t_{int,c} + t_{ic}L_{nand}, \tag{10.14}$$

where M_{tinv} and M_{tnand} are the intrinsic adjustment factors as modeled in [22]. The length factors L_{inv} and L_{nand} take into consideration the average length of the interconnects. They can be calculated using

$$L_{dev} = \max\left(1, \frac{\sqrt{A_{dev}}}{l_{ic}}\right), \tag{10.15}$$

where the subscript *dev* is the logic circuit under consideration. This equation states that if the dimensions of the logic gate are less than five times the pitch (l_{ic}), then assume the interconnect length is set to l_{ic}. If the dimensions are larger than l_{ic}, then the interconnect length increases and the capacitance grows to account for larger interconnect contributions.

For the following logic functions, it becomes convenient to derive the delay of a tile (t_T) that is dependent on the mode of operation (M_{adj}):

$$t_{T,M_{adj}} = 2M_{tT}t_{int,M_{adj}} + t_{ic}L_T, \tag{10.16}$$

where the adjustment parameter, M_{tT}, is considered to be equal to the adjusted parameters for a two-input NAND gate. This is chosen due to the similarities in the number of transistors used and in the number of inputs and outputs per gate, as done in the BCB methodology [22].

The delay consideration for the two-input XOR (t_{xor2}), three-input XOR (t_{xor3}), and three-input MAJ (t_{maj3}) gates can now be determined. All three circuits have the same critical path that lead to the same switching delay. This is possible since all three gates contain one tile with mode of operation (d) and the critical path goes through a single inverter.

$$t_{xor2} = t_{T,d} + t_{inv1}, \quad t_{xor3} = t_{T,d} + t_{inv1}, \quad t_{maj3} = t_{T,d} + t_{inv1}. \tag{10.17}$$

The delay of the 4-to-1 multiplexer (t_{mux}), the TSPC flip-flop (t_{ff}), and the 6-T SRAM cell (t_{sram}) can be similarly calculated. The delay of the 4-to-1 multiplexer (t_{mux}) can be approximated to be the delay of a tile with mode (c) plus two inverters. The TSPC flip-flop (t_{ff}) on the other hand contains one delay tile component and three delay inverters. The 6-T SRAM cell (t_{sram}) has two delay inverters as similarly stated in the BCB methodology.

$$t_{mux} = t_{T,c} + 2t_{inv1}, \quad t_{ff} = t_{T,c} + 3t_{inv1} + t_{ic}L_{ff},$$
$$t_{sram} = 2t_{inv1} + t_{ic}L_{sram}. \tag{10.18}$$

Finally, the delay of the 1-bit full adder and of the 32-bit adder are derived by examining the 1-bit adder's schematic in Figure 10.13(b). One can identify that the delay term of this circuit is quite simple since *Sum* and *Cout* are evaluated simultaneously and all three inverters are evaluated simultaneously. It means that the critical path of the 1-bit adder is composed of one tile with mode of operation (d), one inverter, and an interconnect component. The 32-bit adder is based on the schematic shown in Figure 10.14. The delay of the 32-bit adder can be broken down into two sections. First, the critical path goes through the initial stage. This delay component is equal to that of the delay of a 1-bit adder. It is important to note that the gates of all transistors have been charged at this point. So, when the critical path goes through the second stage, all the gates are pre-charged, and the tiles now act as the mode of operation (a). This mode of operation means that only the parasitic capacitances are being currently charged. For this reason, the delay of the tile is now dependent on mode (a) plus the delay component of the interconnects. Six delay components of the (restoration) inverters are added since they are connected series.

$$\begin{aligned} t_{1bit} &= t_{T,d} + t_{inv1} + t_{ic}L_{1bit}, \\ t_{32bit} &= t_{1bit} + 31(M_{tT}t_{int,a} + t_{ic}L_{1bit}) + 6t_{inv1}. \end{aligned} \tag{10.19}$$

10.5.3 Energy estimation

We now move to deriving the switching energy of the considered circuits. Similarly to the switching delay, the switching energy is made up of (1) the intrinsic component and (2) the interconnect component. The energy of the intrinsic device (E_{int}) is calculated first. This term is not dependent on the ON-current but is instead dependent on the supply voltage (V_{DD}) and on the total capacitance (C_{tot}) of the device. The mode of operation affects the device's capacitance contribution, and thus, the intrinsic energy is also dependent on the adjustment factor (M_{adj}).

$$E_{int,M_{adj}} = C_{tot}V_{DD}^2. \tag{10.20}$$

The switching energy of the interconnect (E_{ic}) is dependent on the capacitance of the interconnect (C_{ic}) and on the supply voltage (V_{DD}). The interconnect capacitance, C_{ic}, is equal to the capacitance of a wire per length of interconnect and is multiplied by the number of output signals. Since the energy consumed by the interconnect is not dependent on the gate's logic behavior, an energy associated factor of 1/2 is added to the switching energy—coming from the 50% probability of switching input signals.

$$E_{ic} = 0.5C_{ic}V_{DD}^2. \tag{10.21}$$

The switching energy of the two-input NAND and the *fan-out-of*-1 inverter with TIGFET transistors have similar expressions as compared with CMOS. The difference comes from the mode of operation that affects the total capacitance of a TIGFET device. We also reuse the mode of operation notation introduced in Section 10.3.2.

$$E_{inv1} = M_{EinvE}E_{int,b} + E_{ic}L_{inv}, \quad E_{nand2} = 2M_{Enand}E_{int,c} + E_{ic}L_{nand}, \tag{10.22}$$

where M_{Einv} and M_{Enand} are the intrinsic adjustment factors as modeled in [22]. Note that the same explanation is used for the length factor (L_{dev}), as written in (10.15), where *dev* is the logic circuit under consideration.

For convenience, in deriving the subsequent equations, we express the switching energy of a tile (E_T) as

$$E_{T,M_{adj}} = M_{ET}E_{int,M_{adj}} + E_{ic}L_T, \tag{10.23}$$

where the adjustment parameter M_{ET} is considered to be equal to the adjusted parameter for a two-input NAND gate. This is again chosen due to the similarities between CMOS-based two-input NANDs and the number of inputs that can be mapped onto a tile. The interconnect energy component ($E_{ic}L_T$) is multiplied by the number of output signals per gate as to approximate the amount of additional interconnect space needed. The interconnect components of the flip-flop and adders are therefore altered since they have higher number of output signals.

The formula for the switching energy of a tile can then be used to extract the expressions of various logic gates. For instance, the switching energy for the two-input XOR (E_{xor2}) and the three-input MAJ (E_{maj3}) gates contain a tile component, with adjustment parameter (*d*), and two inverter components. This is the case since both of these gates contain the same amount of transistors with the same mode of operation and the same amount of inverters. The switching energy of the three-input XOR (E_{xor3}), on the other hand, takes an extra inverter into consideration.

$$\begin{aligned} E_{xor2} &= E_{T,d} + 2E_{inv1}, \quad E_{xor3} = E_{T,d} + 3E_{inv1}, \\ E_{maj3} &= E_{T,d} + 2E_{inv1}. \end{aligned} \tag{10.24}$$

The switching energy of the 4-to-1 multiplexer (E_{mux}), the TSPC flip-flop (E_{ff}), and the 6-T SRAM cell (E_{sram}) are calculated as follows. The energy of the 4-to-1 multiplexer (E_{mux}) can be approximated by the energy of two tiles with mode (c) plus three inverters. The energy of the TSPC flip-flop (E_{ff}) is approximated to be one tile component plus four inverter components. Note that the interconnect component is multiplied by a factor of 2 in order to take the two output signals into consideration. Lastly, the 6-T SRAM cell's switching energy (E_{sram}) has three inverter components:

$$E_{mux} = 2E_{T,c} + 3E_{inv1}, \quad E_{ff} = E_{T,c} + 4E_{inv1} + 2E_{ic}L_{ff}, \tag{10.25}$$

$$E_{sram} = 3E_{inv1} + E_{ic}L_{sram}. \tag{10.26}$$

From there, we can derive the switching energy of the 1-bit full adder and the 32-bit adder. The 1-bit adder, as shown in Figure 10.13(b), contains energy components of one three-input XOR, one three-input MAJ, and three inverters. However, as included in the methodology of BCB, the probability of switching (activity factors) must be taken into account. The activity factor of the three-input XOR (α_{xor3}) is equal to $3/16$ and of the three-input MAJ (α_{maj3}) is calculated to be $1/4$. The probability of switching for the inverter is $1/2$. Furthermore, the switching energy of the 32-bit adder, as shown in Figure 10.14, is similarly derived by taking into consideration all of the logic gates in the circuit. A single stage contains a 1-bit full adder (with a three-input MAJ gate replacing one inverter) and an interconnect component. Since the circuit contains 32 stages, this term is multiplied by 32. The energy component of the initial inverter (to invert *Cin*) and of the six restoring inverters are added.

$$E_{1bit} = \alpha_{xor3}E_{T,d} + \alpha_{maj3}E_{T,d} + 3E_{inv1}/2 + 2E_{ic}L_{1bit}, \quad (10.27)$$

$$E_{32bit} = 32(E_{1bit} - E_{inv1}/2 + \alpha_{maj3}E_{T,d} + E_{ic}L_{1bit}) + 7E_{inv}/2. \quad (10.28)$$

10.5.4 Standby power estimation

The last benchmark variable of interest in the BCB methodology is the standby power. In logic circuits, the standby power is dissipated due to OFF-state leakage current between the power supplies. The BCB methodology assumes that the OFF-state standby power contains two components: (1) leakage current flowing from source to drain, S_{sd}, and (2) leakage current flowing from the gate dielectric, S_g.

$$S_{sd} = (J_{OFF}w_x)V_{DD}, \quad S_g = (J_gL_{ch}w_x)V_{DD}, \quad (10.29)$$

where J_{OFF} is the OFF-state leakage current per gate length, w_x is the channel width, L_{ch} is the channel length, and J_g is the gate leakage per unit area. Both of the standby power components come from the standard equation of power (current × voltage). The voltage is simply the supply voltage, V_{DD}, while the current is either the drain to source current ($J_{OFF}w_x$) to calculate S_{sd} or the gate leakage current ($J_gL_{ch}w_x$) to calculate S_g. The standby power of the intrinsic device can be written as the sum of S_{sd} and $3S_g$. The factor of 3, for TIGFET devices, comes from the three MOS gates on the semiconducting channel. The standby power of the interconnects, on the other hand, is set to zero since the BCB methodology assumes zero leakage current flowing through the interconnects.

$$S_{int} = S_{sd} + 3S_g, \quad S_{ic} = 0. \quad (10.30)$$

Following the methodology of BCB, the standby power of the inverter and two-input NAND are calculated as follows:

$$S_{inv1} = S_{int}, \quad S_{nand2} = M_{Snand2}S_{int}, \quad (10.31)$$

where the M_{Snand2} term is the adjustment factor for a two-input NAND gate as modeled in [22]. The intrinsic factor for the inverter is set to 1. Similarly to the delay and energy components, the standby power of a tile can be calculated to simplify the subsequent equations.

$$S_T = M_{ST}S_{int}. \quad (10.32)$$

Note that, as compared to the other metrics, the power is not dependent on the capacitance of the tile, and thus, the mode of operation does not affect the standby power. The M_{ST} term is equal to the adjustment parameter for the two-input NAND gate. This was similarly chosen since the two-input NAND contains the same number of transistors and the same number of inputs/output signals for the subsequent logic functions.

The formula for the standby power of a tile, in a similar manner, allows for a straightforward way to find the standby power of the other logic functions. The standby power metrics for the two-input XOR (S_{xor2}) and the three-input MAJ (S_{maj3})

gates are equal to the standby power of a tile plus the standby power of two inverters. The standby power for the three-input XOR (S_{xor3}) contains one tile component and three inverter components (by taking the extra inverter into consideration).

$$S_{xor2} = S_T + 2S_{inv1}, \quad S_{xor3} = S_T + 3S_{inv1}, \quad S_{maj3} = S_T + 2S_{inv1}. \tag{10.33}$$

The standby power of the TSPC flip-flop is similarly calculated by summing up all of the components in the circuit (equal to the standby power of a tile plus the standby power of four inverters). Since all transistors consume leakage power in the OFF-state, the standby power of the 4-to-1 static multiplexer contains two tile components and four inverter components. The only unique derivation here is the standby power for the SRAM cell which contains two inverter components. The BCB methodology assumes that the access transistors do not contribute to the standby power and the TIGFET-based SRAM cell uses a 1-to-1 ratio of TIGFET replacement.

$$S_{ff} = S_T + 4S_{inv1}, \quad S_{mux} = 2S_T + 4S_{inv1}, \quad S_{sram} = 2S_{inv1}. \tag{10.34}$$

Lastly, the standby power of the 1-bit adder and the 32-bit adder can easily be determined by taking into consideration all of the gates within the circuits. The 1-bit adder contains two tiles and three inverters so the standby power sums these five components. On the other hand, the 32-bit adder contains 32 1-bit adder standby power components. To take into consideration that the *Cout* is calculated using a three-input MAJ gate, the standby power of 32 tiles are added and the standby power of 32 inverters are subtracted. The standby power of seven inverters are also added (six restoring inverters plus the initial inverter to invert *Cin*).

$$S_{1bit} = 2S_T + 3S_{inv1}, \quad S_{32bit} = 32(S_{1bit} + S_T - S_{inv}) + 7S_{inv}. \tag{10.35}$$

10.6 Comparison of technologies

The plots obtained from the BCB methodology give a broad overview of how the multiple alternative technologies compare with each other and with CMOS HP/CMOS LV. In this section, the metrics and comparisons will be shown at the (1) device-level and (2) circuit-level.

10.6.1 Device-level performance

Let us first describe the metrics for the intrinsic devices in the BCB methodology. The different electronic-based technologies have four major differences: ON-current, OFF-current, voltage, and capacitance. With these four terms, the effective charge of the device ($Q = E_{int}/V_{DD}$), the effective resistance of the device ($R = V_{DD}/I_{ON}$), and the effective capacitance ($C = Q/V_{DD} = C_{tot}$) can be calculated. These metrics are summarized in Figure 10.16 with their corresponding optimum corners.

The ON/OFF current versus voltage plot in Figure 10.16(a) shows us that TIGFET devices have large ON/OFF current ratios, but they require the highest supply voltage. In particular, TIGFETs have 7.76× larger ON/OFF current and 1.23× larger voltage than CMOS HP. The supply voltage can be improved by utilizing low bandgap

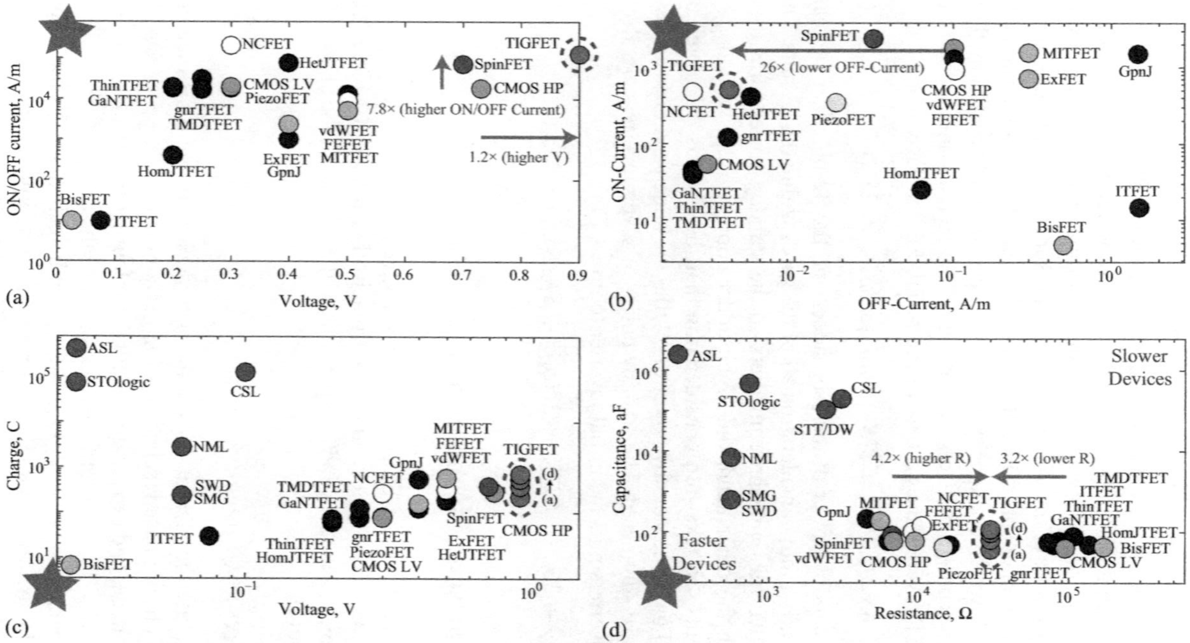

Figure 10.16 Intrinsic device characteristics. (a) ON/OFF current and voltage. (b) ON-current and OFF-current. (c) Charge and voltage. (d) Capacitance and resistance. The stars show the optimum corners.

materials that require less voltage to achieve similar current densities [47]. In the same manner, while TIGFET devices do not have the largest ON-current, as shown in Figure 10.16(b), it does have one of the smallest OFF-currents—as low as 26× lower as compared with CMOS HP. TIGFET's low leakage floor is possible by preventing carrier injection through the thick Schottky barriers at the source and drain electrodes. The charge versus voltage plot in Figure 10.16(c) shows the amount of charge accumulated in the capacitor per supply voltage, while the capacitance versus resistance plot in Figure 10.16(d) displays the speed of the intrinsic device. Faster devices are shown in the bottom left corner of Figure 10.16(d), while slower devices lie in the top right corner. The TIGFET devices in these two last plots show the different modes of operation and clearly present the increasing amount of capacitance and charge for higher number of varying potentials per gate. The spin-based devices require the largest charge and have increased capacitance. Electronic devices have very small amount of charges and thus have the lowest capacitances. At the device-level, TIGFETs have increased supply voltage when using a standard silicon channel and use relatively large capacitance values, as compared with the other electronic devices. While TIGFETs do perform under favorable high ON/OFF current ratios, the device-level metrics, in general, do not show promising results. However, in the following subsection, the circuit benchmarks will show advantageous results when using TIGFET technologies. To explain these results, the origins of benefit will be analyzed in Section 10.6.3.

10.6.2 Circuit-level performance

Let us first review the metrics that have become available by applying the switching delay (t), switching energy (E), and standby power (S) metrics. First, the standby energy ($E_{standby}$) is the energy consumed in the OFF-state and is calculated to be the standby power times switching delay ($E_{standby} = S \times t$). Next, the active power (P_{active}) is the power of the active circuit, expressed as the switching energy divided by the switching delay ($P_{active} = E/t$). This metric shows the rate at which energy is transferred and stored into the electric fields of the intrinsic devices as the output switches. If the circuit requires a long period of time to drain the same amount of energy, then the active power will decrease. Similarly, if the switching energy is lower for a given amount of switching time, then there is less active power. Finally, the power density ($P_{density}$) is the summation of active power and standby power divided by total area ($P_{density} = P_{active}/A + S/A$). This is a good representation of the total power that is consumed in the ON- and OFF-states. The authors of BCB 2.0 [21] set an upper cap on the power density at P_{cap} = 10 W/cm^2 since there is a practical limit of power that can be removed by a heat sink in chips. When the circuits run at a larger power density, the frequency must be decreased—leading to reduced throughput. The term throughput (T) refers to the rate of production per unit area, calculated to by the inverse of area times delay ($T = 1/A \times t$). The technologies that hit the power density cap have a throughput reduced by a factor of $P_{cap}/P_{density}$.

The benchmarking results for the 32-bit adder are shown in Figure 10.17. The first plot shows the switching energy (E) versus switching delay (t) metrics for the

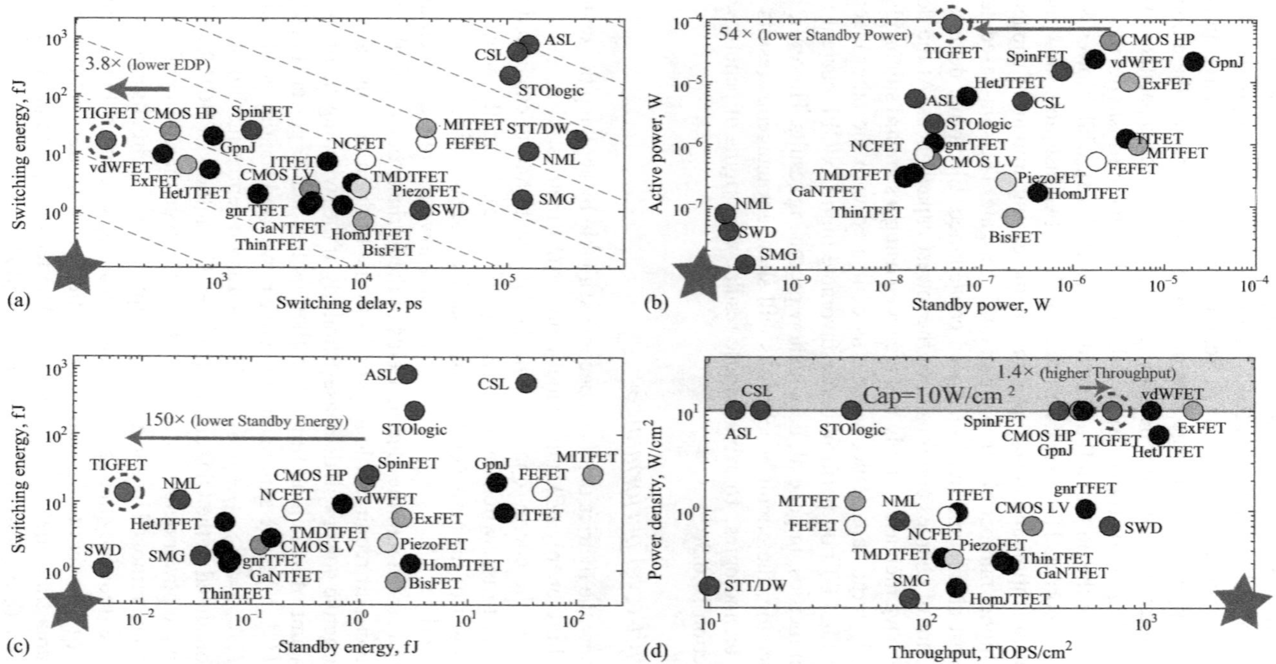

Figure 10.17 *Benchmark for the 32-bit adder. (a) Switching energy and delay. (b) Power density and throughput. (c) Switching energy and standby energy. (d) Active power and standby power. The stars show the optimum corners.*

various devices in BCB. The dashed lines show the constant EDP trends and helps to identify technologies with better performances. The preferable devices (low energy and low delay) are in the bottom-left "optimum" corner of Figure 10.17(a). The TIGFET technology has the best EDP metric as compared with all devices for the 32-bit adder. In particular, TIGFETs have 3.83× lower EDP as compared with CMOS HP and 4.18× lower EDP as compared with CMOS LV. One would not expect this from TIGFETs intrinsic performances (Figure 10.16). However, TIGFET technology is the only alternative device that can effectively use the three-input XOR and three-input MAJ gate as discussed in Section 10.4. Next, the active power (P_{active}) and standby power (S) are shown in Figure 10.17(b). The optimum corner is lower left since this corresponds with lower power consumption. The TIGFET technology has 53.8× lower standby power than CMOS HP and 2.42× higher standby power than CMOS LV. On the other hand, TIGFETs have the highest active power consumption (1.5× larger than CMOS HP). This comes from operating the transistors at full speed with similar energy consumption metrics. The switching energy (E) versus standby energy ($E_{standby}$) plot for the 32-bit adder is shown in Figure 10.17(c). The preferable devices, with lower energy and smaller delay, would lie in the bottom-left corner since this would signify lower energy consumed in the OFF-state and ON-state. The TIGFET technology sees a 147× lower standby energy as compared with CMOS HP and a 15.4× lower standby energy as compared with CMOS LV. The low standby power comes from TIGFETs low standby power at high switching speeds. Lastly, the power density ($P_{density}$) and throughput (T) are shown in Figure 10.17(d). TIGFET technology has the highest throughput of all the Beyond-CMOS devices—coming from the small area of the 32-bit adder and reduced switching delay. However, the cap on the power density reduces the possible throughput of TIGFETs, placing it behind the vdWFET, HetJTFET, and ExFET technologies. The resulting metric shows that TIGFET has 1.39× higher throughput than CMOS HP and 2.31× higher throughput than CMOS LV.

The same set of plots are shown in Figure 10.18 for the 32-bit ALU. The switching delay versus switching energy plot in Figure 10.18(a) shows that TIGFETs now have 1.54× higher EDP than CMOS HP and 1.75× higher EDP than CMOS LV. The power comparison plot (P_{active} versus S) in Figure 10.18(b) shows that TIGFETs have 36.6× lower standby power and 1.5× larger active power as compared with CMOS HP. The energy comparison plot (E versus $E_{standby}$) in Figure 10.18(c) shows that TIGFETs have 55.6× lower standby energy as compared with CMOS HP and 5.88× lower standby energy as compared with CMOS LV. In the final plot, as shown in Figure 10.18(d), shows the power density ($P_{density}$) versus throughput (T). The power density of TIGFETs and CMOS HP are similarly capped at 10 W/cm^2, while the throughput of TIGFETs reduce to 1.43% larger than CMOS HP. Note that all of the metrics here have seen a reduction in performance. This is explained since some of the circuit elements in the ALU do not take full advantage of the TIGFET's improved functionalities.

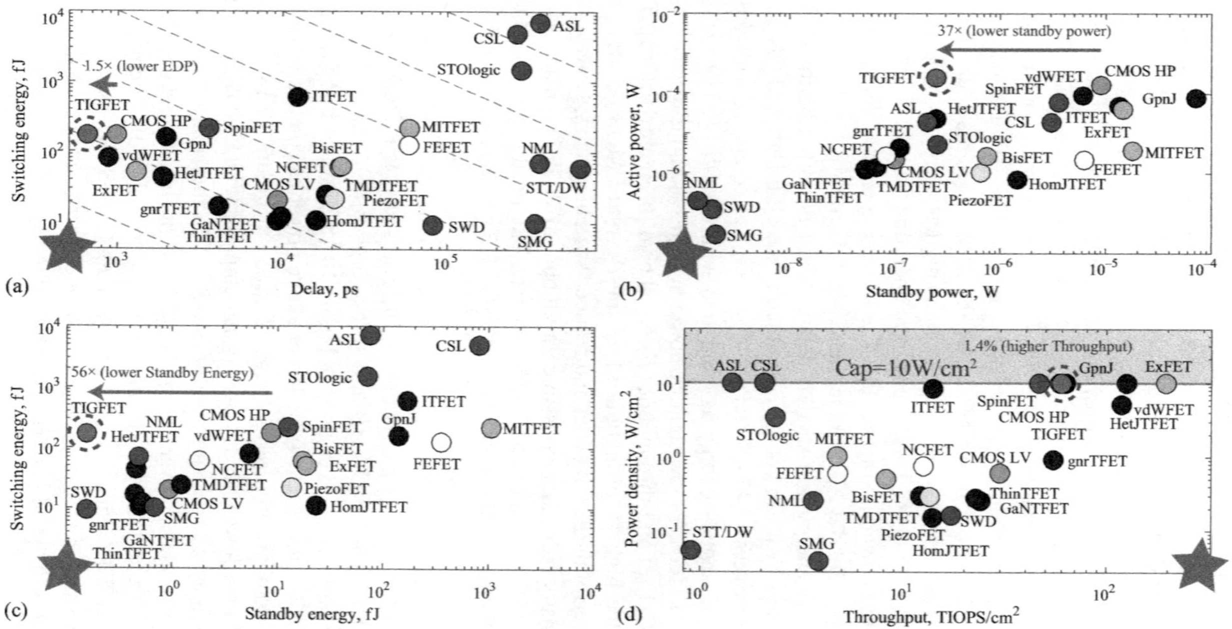

Figure 10.18 *Benchmark for the 32-bit ALU. (a) Switching energy and delay. (b) Power density and throughput. (c) Switching energy and standby energy. (d) Active power and standby power. The stars show the optimum corners.*

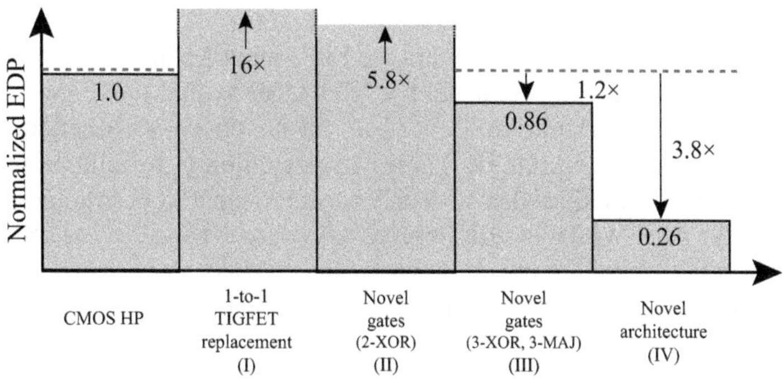

Figure 10.19 EDP gain origin for the 32-bit adder with the various TIGFET architectures (normalized to CMOS HP)

10.6.3 Origin of EDP results

To understand how the TIGFET architecture ends up having such advantageous circuit-based benchmarking results from poor device-level metrics, we identify the different contributions to the EDP gain (normalized to CMOS HP) of the 32-bit adder and illustrate how the different adder architectures improve the circuit performance as shown in Figure 10.19. Each segment has a corresponding EDP term with unique 32-bit adder architectures.

- *1-to-1 TIGFET replacement*
  First, the simplest adder architecture replaces a MOSFET with the considered SiNW TIGFET transistor and observes a large 15.8× EDP penalty. The EDP increase is primarily due to the TIGFET's two-input NAND (1) 2.64× larger device area, 1.59× larger parasitic capacitances—due to additional contacts—and (2) 3.6× smaller ON-current for a 23.3% increase in supply voltage—due to the Schottky barrier at source and drain terminals. While these parameters may impede further device development, it is important to consider the circuit-level innovations that this technology offers.

- *Novel gates (two-input XOR)*
  Next, the two-input XOR gates with TIGFET architecture, built using four TIGFET transistors and two inverters as shown in Figure 10.10(a), are used to replace the NAND-based two-input XOR gate in the 1-bit adder—observing a 5.84× EDP penalty for the 32-bit adder. While this circuit design allows for an improvement of 2.7× in EDP (compared to the 1-to-1 replacement), its switching delay and switching energy are still larger than CMOS HP due to the two-input XOR gate 3.5× larger parasitic capacitance, 4.4× larger V_{DD}/I_{ON}, and two inverters. Note that, at this stage, while the two-input XOR's are no longer NAND dominated, the full adders still contain the inefficient TIGFET-based two-input NAND gates.

- *Novel gates (three-input XOR and three-input MAJ)*
  Next, the three-input XOR gate and the three-input Majority (MAJ) gate with TIGFET architecture may remodel the 1-bit adder as shown in Figure 10.13(b). This circuit is used to build the 32-bit adder and thus observes a benefit of 1.17× in EDP, as compared to CMOS HP. These two gates allow us to build the previously NAND-driven 1-bit full adders with only one three-input XOR gate and one three-input MAJ gate. While we still observe 4.4× larger V_{DD}/I_{ON}, the reduction of contacts (1.89× lower number of gates) and a 63% decrease in overall average capacitance leads to a reduction in parasitic capacitance which leads to an overall EDP benefit.

- *Novel architecture*
  We have so far introduced XOR/MAJ-dominated adders into the methodology. However, there is one final 32-bit adder architecture that TIGFET technology can be used for. If the inverter to output $\overline{Cout}$ is replaced with a three-input majority gate, then this allows for the parallel computation of *Cout*, *Sum*, and $\overline{Cout}$. While this increases the energy consumption by 1.3×, the delay is reduced by 4.2×. The reduction in delay is explained by (1) the removal of most inverters from the critical path—since $\overline{A}$ and $\overline{B}$ can be executed simultaneously—and (2) the critical path from the tiles have reduced intrinsic capacitance—since all gates will be precharged.

10.7 Conclusion

While TIGFETs do not have advantageous *I–V* characteristics compared to CMOS HP (as shown by the 1-to-1 TIGFET replacement in Figure 10.19), the multiple gates of TIGFET transistors allow for unique circuit opportunities not possible with other technologies, and these circuit designs give the advantageous delay, energy, area, power, and throughput benchmarks. The results listed in this chapter show how the added functionality of FED technology may be a powerful tool to improve circuit performances. The presented SiNW TIGFET device illustrates the promises of the growing family of FEDs, where the technology's capabilities allow designers to create superior circuits and systems.

Acknowledgment

This material is based upon work supported by the National Science Foundation under Grant No. 1644592 and under CAREER Award No. 1751064.

References

[1] Moore GE. Cramming More Components Onto Integrated Circuits. IEEE Solid-State Circuits Society Newsletter. 2006 Sep;11(3):33–35.

[2] Collaert N, Alian A, Arimura H, *et al.* Beyond-Si materials and devices for more Moore and more than Moore applications. In: 2016 International Conference on IC Design and Technology (ICICDT); 2016 Jun. p. 1–5.

[3] International Technology Roadmap for Semiconductors [Online]; 2015. Available from: http://www.itrs2.net/.

[4] Zhirnov VV, Cavin RK, Hutchby JA, *et al.* Limits to Binary Logic Switch Scaling – A Gedanken Model. Proceedings of the IEEE. 2003 Nov;91(11): 1934–1939.

[5] Theis TN, Wong HSP. The End of Moore's Law: A New Beginning for Information Technology. Computing in Science Engineering. 2017 Mar;19(2):41–50.

[6] Seabaugh AC, Zhang Q. Low-Voltage Tunnel Transistors for Beyond CMOS Logic. Proceedings of the IEEE. 2010 Dec;98(12):2095–2110.

[7] Wolf SA, Awschalom DD, Buhrman RA, *et al.* Spintronics: A Spin-Based Electronics Vision for the Future. Science. 2001 Nov;294(5546):1488–1495.

[8] Banadaki YM, Srivastava A. Scaling Effects on Static Metrics and Switching Attributes of Graphene Nanoribbon FET for Emerging Technology. IEEE Transactions on Emerging Topics in Computing. 2015 Dec;3(4):458–469.

[9] Liu L, Lu Y, Guo J. On Monolayer MoS_2 Field-Effect Transistors at the Scaling Limit. IEEE Transactions on Electron Devices. 2013 Dec;60(12):4133–4139.

[10] Currivan JA, Jang Y, Mascaro MD, *et al.* Low Energy Magnetic Domain Wall Logic in Short, Narrow, Ferromagnetic Wires. IEEE Magnetics Letters. 2012 Apr;3:3000104–3000104.

[11] Sugahara S, Tanaka M. A Spin Metal–Oxide–Semiconductor Field-Effect Transistor Using Half-Metallic-Ferromagnet Contacts for the Source and Drain. Applied Physics Letters. 2004 Mar;84(13):2307–2309.

[12] Salahuddin S, Datta S. Use of Negative Capacitance to Provide Voltage Amplification for Low Power Nanoscale Devices. Nano Letters. 2008 Dec;8(2):405–410. PMID: 18052402.

[13] Jo J, Choi WY, Park JD, *et al.* Negative Capacitance in Organic/Ferroelectric Capacitor to Implement Steep Switching MOS Devices. Nano Letters. 2015 Jun;15(7):4553–4556.

[14] Khan AI, Chatterjee K, Duarte JP, *et al.* Negative Capacitance in Short-Channel FinFETs Externally Connected to an Epitaxial Ferroelectric Capacitor. IEEE Electron Device Letters. 2016 Jan;37(1):111–114.

[15] Seo J, Lee J, Shin M. Analysis of Drain-Induced Barrier Rising in Short-Channel Negative-Capacitance FETs and Its Applications. IEEE Transactions on Electron Devices. 2017 Apr;64(4):1793–1798.

[16] Zhang J, Marchi MD, Sacchetto D, *et al.* Polarity-Controllable Silicon Nanowire Transistors with Dual Threshold Voltages. IEEE Transactions on Electron Devices. 2014 Nov;61(11):3654–3660.

[17] Marchi MD, Sacchetto D, Frache S, *et al.* Polarity control in double-gate, gate-all-around vertically stacked silicon nanowire FETs. In: 2012 International Electron Devices Meeting; 2012 Dec. p. 8.4.1–8.4.4.

[18] Zhang J, Marchi MD, Gaillardon PE, *et al.* A Schottky-barrier silicon FinFET with 6.0 mV/dec subthreshold slope over 5 decades of current. In: 2014 IEEE International Electron Devices Meeting; 2014 Dec. p. 13.4.1–13.4.4.

[19] Gaillardon PE, Ghasemzadeh H, Micheli GD. Vertically-stacked silicon nanowire transistors with controllable polarity: a robustness study. In: 2013 14th Latin American Test Workshop – LATW; 2013 Apr. p. 1–6.
[20] Bernstein K, Cavin RK, Porod W, *et al.* Device and Architecture Outlook for Beyond CMOS Switches. Proceedings of the IEEE. 2010 Dec;98(12):2169–2184.
[21] Nikonov DE, Young IA. Overview of Beyond-CMOS Devices and a Uniform Methodology for Their Benchmarking. Proceedings of the IEEE. 2013 Dec;101(12):2498–2533.
[22] Nikonov DE, Young IA. Benchmarking of Beyond-CMOS Exploratory Devices for Logic Integrated Circuits. IEEE Journal on Exploratory Solid-State Computational Devices and Circuits. 2015 Dec;1:3–11.
[23] Pan C, Naeemi A. Beyond-CMOS non-Boolean logic benchmarking: Insights and Future Directions. In: Design, Automation Test in Europe Conference Exhibition (DATE), 2017; 2017 Mar. p. 133–138.
[24] Newns D, Elmegreen B, Liu XH, *et al.* A Low-Voltage High-Speed Electronic Switch Based on Piezoelectric Transduction. Journal of Applied Physics. 2012 Apr;111(8):084509.
[25] Khitun A, Wang KL. Nano Scale Computational Architectures with Spin Wave Bus. Superlattices and Microstructures. 2005 Sep;38(3):184–200.
[26] Nikonov D, Bourianoff G, Ghani T. Proposal of a Spin Torque Majority Gate Logic. IEEE Electron Device Letters. 2011 Jul;32(8):1128–1130.
[27] BCB Methodology with TIGFET Architecture; 2018. Available from: https://sites.google.com/site/pegaillardon/downloads.
[28] Romero-Gonzalez J, Gaillardon PE. BCB Evaluation of High-Performance and Low-Leakage Three-Independent-Gate Field Effect Transistors. IEEE Journal on Exploratory Solid-State Computational Devices and Circuits. 2018 Apr;4:1–9.
[29] Heinzig A, Slesazeck S, Kreupl F, *et al.* Reconfigurable Silicon Nanowire Transistors. Nano Letters. 2012 Nov;12(1):119–124.
[30] Nakaharai S, Iijima T, Ogawa S, *et al.* Electrostatically Reversible Polarity of Dual-Gated Graphene Transistors. IEEE Transactions on Nanotechnology. 2014 Nov;13(6):1039–1043.
[31] Resta GV, Sutar S, Balaji Y, *et al.* Polarity Control in WSe2 Double-Gate Transistors. Scientific Reports. 2016 Jul;6:29448. Article.
[32] Resta GV, Agarwal T, Lin D, *et al.* Scaling Trends and Performance Evaluation of 2-Dimensional Polarity-Controllable FETs. Scientific Reports. 2017 Mar;7:45556.
[33] Lin YM, Appenzeller J, Knoch J, *et al.* High-Performance Carbon Nanotube Field-Effect Transistor with Tunable Polarities. IEEE Transactions on Nanotechnology. 2005 Sep;4(5):481–489.
[34] Lundstrom MS, Antoniadis DA. Compact Models and the Physics of Nanoscale FETs. IEEE Transactions on Electron Devices. 2014 Feb;61(2): 225–233.
[35] Ionescu AM, Riel H. Tunnel Field-Effect Transistors as Energy-Efficient Electronic Switches. Nature. 2011 Nov;479(7373):329–337.

[36] Salvatore GA, Bouvet D, Ionescu AM. Demonstration of subthreshold swing smaller than 60 mV/decade in Fe-FET with P(VDF-TrFE)/SiO2 gate stack. In: 2008 IEEE International Electron Devices Meeting; 2008 Dec. p. 1–4.
[37] Bobba S, Gaillardon PE, Zhang J, *et al.* Process/design co-optimization of regular logic tiles for double-gate silicon nanowire transistors. In: 2012 IEEE/ACM International Symposium on Nanoscale Architectures (NANOARCH); 2012 Jul. p. 55–60.
[38] Zografos O, Gaillardon PE, Micheli GD. Novel grid-based power routing scheme for regular controllable-polarity FET arrangements. In: 2014 IEEE International Symposium on Circuits and Systems (ISCAS); 2014 Jun. p. 1416–1419.
[39] Gasnier C, Gaillardon PE, Micheli GD. SATSoT: a methodology to map controllable-polarity devices on a regular fabric using SAT. In: 2013 IEEE/ACM International Symposium on Nanoscale Architectures (NANOARCH); 2013 Jul. p. 46–51.
[40] Jamaa MHB, Mohanram K, Micheli GD. Novel library of logic gates with ambipolar CNTFETs: opportunities for multi-level logic synthesis. In: 2009 Design, Automation Test in Europe Conference Exhibition; 2009 Apr. p. 622–627.
[41] Raitza M, Kumar A, Vlp M, *et al.* Exploiting transistor-level reconfiguration to optimize combinational circuits. In: Design, Automation Test in Europe Conference Exhibition (DATE), 2017; 2017 Mar. p. 338–343.
[42] Zhang J, Tang X, Gaillardon PE, *et al.* Configurable Circuits Featuring Dual-Threshold-Voltage Design With Three-Independent-Gate Silicon Nanowire FETs. IEEE Transactions on Circuits and Systems I: Regular Papers. 2014 Oct;61(10):2851–2861.
[43] Gaillardon PE, Amaru L, Zhang J, *et al.* Advanced system on a chip design based on controllable-polarity FETs. In: 2014 Design, Automation Test in Europe Conference Exhibition (DATE); 2014 Mar. p. 1–6.
[44] Amar L, Gaillardon PE, Zhang J, *et al.* Power-Gated Differential Logic Style Based on Double-Gate Controllable-Polarity Transistors. IEEE Transactions on Circuits and Systems II: Express Briefs. 2013 Oct;60(10):672–676.
[45] Giacomin E, Gonzalez JR, Gaillardon PE. Low-power multiplexer designs using three-independent-gate field effect transistors. In: 2017 IEEE/ACM International Symposium on Nanoscale Architectures (NANOARCH); 2017 Jul. p. 33–38.
[46] Rakheja S, Naeemi A. Interconnects for Novel State Variables: Performance Modeling and Device and Circuit Implications. IEEE Transactions on Electron Devices. 2010 Oct;57(10):2711–2718.
[47] Trommer J, Heinzig A, Mhle U, *et al.* Enabling Energy Efficiency and Polarity Control in Germanium Nanowire Transistors by Individually Gated Nanojunctions. ACS Nano. 2017 Jan;11(2):1704–1711.

Chapter 11
Exploratory logic synthesis for multiple independent gate FETs

Luca Amarù[1], Pierre-Emmanuel Gaillardon[2], Subhasish Mitra[3], and Giovanni De Micheli[4]

11.1 Introduction

The use of *multiple independent gate field effect transistors* (MIGFETs) is a promising scaling path for digital electronics [1]. Originally introduced to achieve a better electrostatic control over a FET channel [2,3], MIGFETs have recently demonstrated the ability to enclose complex switching functions into a single device [4–8]. From a design perspective, enhancing the functionality of elementary components opens up new efficient logic implementations. For example, MIGFETs in [4] realize fast datapath circuits [9] as they switch based on the exclusive NOR (XNOR) operation between gate signals. Analogously, MIGFETs in [6] enable compact control logic circuits as they switch based on the AND/OR between gate signals. Other MIGFETs with expressive switching functions, e.g., gamble [5] and threshold [8] functions, advantageously fit other classes of circuits.

While a MIGFET functionality increases with the number of gates, also its physical implementation cost grows. For example, a three-independent gate FET ideally implements more complex switching functions than a two-independent gate FET but it requires the physical realization of an extra gate. Only MIGFETs enabling more system-level benefits than overhead are interesting to design next-generation integrated circuits. In such a scenario, the natural question that arises is *how many gates do we need?* In this chapter, we address this question from a logic synthesis standpoint. Our aim is to determine whether or not an increasing number of gates leads to more compact design implementations. We propose a logic synthesis methodology that exploits at a fine grain a switching function for a target MIGFET, potentially being any Boolean function. By using device and interconnect models, we estimate the characteristics of the synthesized circuits. In this study, we consider five promising classes

[1]Design Group, Synopsys Inc., USA
[2]Department of Electrical and Computer Engineering, LNIS Laboratory, University of Utah, USA
[3]Robust Systems Group, Stanford, USA
[4]Integrated Systems Laboratory, École Polytechnique Fédérale de Lausanne (EPFL), Switzerland

of MIGFET devices and the corresponding representative functions, with number of inputs ranging from 1 to 7. Physical device data are extrapolated from a 22-nm technology node [1]. Experimental results over Microelectronics Center of North Carolina (MCNC) benchmarks show nontrivial area/delay/energy minima, located between 1 and 4 gates, depending on a switching function class and MIGFET technology. We also conduct custom design experiments for two selected MIGFET devices, targeting cryptographic and arithmetic designs. These experiments confirm a tight interdependency between optimal number of physical gates and functionality of the circuit under synthesis. Such results can help technologists guiding their research efforts.

Our MIGFET synthesis tool, available online at [10], shows the flexibility to read any switching function and/or technology data. In this way, other researchers can evaluate the promises of their emerging devices.

The remainder of this chapter is organized as follows. Section 11.2 provides a background on contemporary MIGFET devices. Section 11.3 presents our MIGFET synthesis methodology. Section 11.4 describes the experimental setup and shows the synthesis results. Section 11.5 discusses our custom design experiments for two selected MIGFET devices. Section 11.6 concludes the chapter.

11.2 Survey on emerging MIGFETs

This section surveys emerging MIGFETs, with their associated switching functions, demonstrated up-to-date. We will use this review to determine some promising classes of switching functions realizable by prospective multi-gate devices.

A MIGFET is a switching device controlled by more than one independent physical gate. MIGFETs can be realized in different technologies, geometries and materials. The implementation choice of a MIGFET determines its physical and logic features. Each physical gate in a MIGFET can either (i) enhance the conduction properties or (ii) increase the device intelligence by enriching the switching function. We focus in this chapter on the latter case. We briefly report hereafter on five notable examples of such MIGFETs. Their sketch structures and switching functions are depicted by Figure 11.1. Note that many other MIGFETs exist but they are not reported here for the sake of brevity.

11.2.1 Double-gate silicon nanowire field effect transistor

Double-gate silicon nanowire FETs (DG-SiNWFET) [4] are emerging devices whose polarity can be configured online via the second gate, usually called the polarity gate. Figure 11.1 shows the conceptual structure of DG-SiNWFET fabricated in [4]. Owing to the online polarity configuration, the on/off state of such transistors is biconditional (XNOR) on both gates signals (Figure 11.1).

11.2.2 Independent-gate-FinFET-LV$_{th}$

An *independent-gate FinFET* (IG-FinFET) is a *Fin*-shaped transistor where the gate electrodes are isolated by a masked etch, allowing for separate biasing [2].

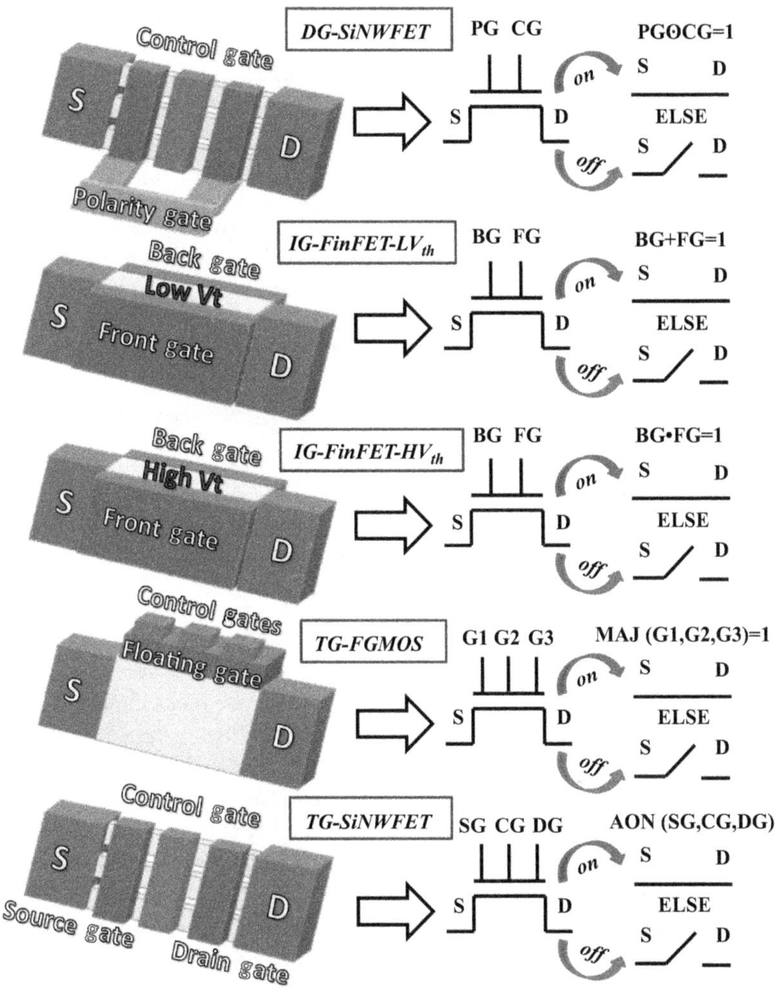

Figure 11.1 Structure and functionality of five notable MIGFETs, based on FinFET, SiNW and floating-gate technologies

Figure 11.1 sketches an IG-FinFET. When the IG-FinFET is a low-threshold (LV_{th}) device, the activation of just one of the two gates is sufficient to enable the channel formation [6]. Thus, the switching function of such transistor is a disjunction (OR) of the gates signals.

11.2.3 Independent-gate-FinFET-HV_{th}

Analogously to IG-FinFET-LV_{th}, high-threshold (HV_{th}) IG-FinFETs have an enhanced functionality. In this case, both gates must be activated to enable the channel formation due to the higher device threshold [6]. Here, the switching function becomes a conjunction (AND) of the gates signals.

Note that a similar IG-FinFET mechanism can be also exploited in UTBB FDSOI technology [11].

11.2.4 Triple-gate-floating-gate metal-oxide-semiconductor (MOS)

Floating-gate MOS (FGMOS) are transistors having multiple input gates that interact with an extra floating-gate capacitance [8]. Figure 11.1 depicts a *triple-gate* (TG) FGMOS. In such a transistor, the on/off state is controlled by a weighted sum (threshold function) of all input gates signals. In the particular TG-FGMOS of Figure 11.1, the switching function is a 3-input majority (MAJ).

11.2.5 Triple-gate-silicon nanowire field effect transistor

Recently introduced in [5], TG-SiNWFETs are an extension of DG-SiNWFET from [4]. TG-SiNWFET enables individual control of the gated regions (Figure 11.1) enriching the switching function. The *on* state of such a TG-SiNWFET is a gamble function (*All-Or-Nothing*—AON) of all the three gate signals.

11.2.6 Logic abstraction and discussion

From the aforementioned MIGFETs, we observe five classes of switching functions, namely, AND, OR, XOR, MAJ and AON. In this work, we focus mainly on the logic functionality of prospective multi-gate devices without a strong link to actual physical devices. Indeed, the final implementation technology for emerging MIGFETs is likely to evolve in time. Here, we want to estimate the optimal number of physical gates exploiting a class of switching functions, without highly precise physical information but still under conservative assumptions. The optima gate points happen where the enhanced functionality advantage exceeds (at its most) the interconnection and realization overheads deriving from the extra physical gates. In this context, we use logic synthesis to anticipately help technologists guiding their research efforts.

11.3 Logic synthesis for MIGFETs

In this section, we propose a synthesis framework enabling a fair comparative evaluation within a class of MIGFET switching functions. First, we give a brief overview on logic synthesis with useful notations and concepts. Second, we describe our circuit design considerations. Finally, we present a logic synthesis methodology capable to harness the expressive power of enhanced functionality switches, such as MIGFETs.

11.3.1 Brief overview on logic synthesis

Logic synthesis is the process by which virtually all digital integrated circuits are designed [12]. In its most general formulation, logic synthesis aims at transforming a general Boolean function description into its minimal circuit implementation.

Such a process consists of two phases: logic optimization and technology mapping. Logic optimization seeks for a concise Boolean representation on a given data structure. Technology mapping minimizes its physical implementation cost. Among the different techniques, *binary decision diagrams* (BDDs) are a canonical data structure [13] efficiently supporting both optimization and mapping techniques. On the optimization side, BDDs enable efficient logic circuit decomposition [14]. On the mapping side, BDDs simplify core operations such as cell-matching [15]. Among the several strengths of BDDs, it is worth noticing the efficient support of generalized cofactoring [16]. Such technique extends Shannon's circuit expansion $f(x,y,\dots,z) = x\cdot f(1,y,\dots,z) + x'\cdot f(0,y,\dots,z)$ over a set of orthonormal basis functions ϕ_i with $i = 1,2,\dots,k$ and $f = \sum_{i=1}^{k} \phi_i \cdot f_{\phi_i}$ [12]. In generalized cofactoring, the choice of the basis ϕ_i determines the efficacy of the expansion.

To assess the potential advantage of enhanced functionality devices in complex circuits, we make use of the design assumptions presented hereafter.

11.3.2 Circuit design considerations

Nowadays, complementary static is a popular style to design integrated circuits. In our study, we are not restricting to the subset of MIGFETs satisfying complementary static style requirements, e.g., presence of both carrier types, *self-dual* switching function,[5] etc. To cover all MIGFET devices, we decided to handle only the *pull-down network* (PDN) of logic cells in a complex circuit. Thus, we assume dynamic or pseudo-logic styles, where a pull-up device provides a conditional path between V_{dd} and the output. Note that different logic styles are expected to shift absolute circuit metrics but not to significantly drift relative minima points of our interest. Figure 11.2 depicts such a logic cell made of MIGFETs. Pull-up devices have the same cost of an elementary MIGFET. Each MIGFET has a switching function f_M of n variables, with one variable per each independent gate. We assume that all the MIGFETs in a logic circuit have the same number of physical gates, to enforce layout regularity at advanced technology nodes.

The compactness of a logic cell, in terms of device number and stack, depends on the expressive power of a MIGFET switching function f_M. We present hereupon a synthesis methodology to fully exploit a MIGFET logic expressiveness.

11.3.3 Synthesis methodology

Given an initial circuit description and a target k-gates MIGFET, with its characteristic function, our aim is to produce a netlist of logic cells utilizing as few devices as possible. Note that we do not target optimal results for a single MIGFET switching function but a fair comparative framework. Informally, we achieve this goal by two steps: (i) circuit optimization into a look up table (LUT) network and (ii) mapping of each LUT node into a compact logic cell. The optimization in step (i) is accomplished by state-of-art LUT-synthesis techniques. For the mapping in step (ii), we propose a

[5]A function $f(x,y,\dots,z)$ is *self-dual* if $f'(x,y,\dots,z) = f(x',y',\dots,z')$. For example, the MAJ is *self-dual* while the AND is *non-self-dual*.

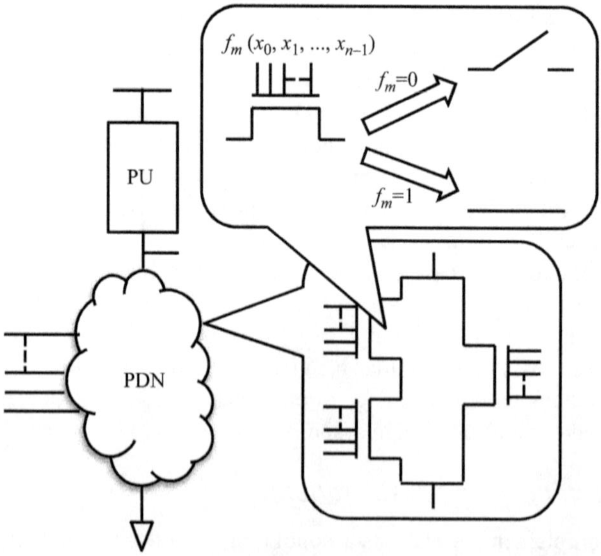

Figure 11.2 Logic cell showing hierarchical inclusion of MIGFETs

match/decompose strategy based on canonical *decision diagrams* (DDs). In this context, DDs serve as data structure for efficient logic representation and manipulation. For the sake of clarity, we introduce such mapping strategy by means of an example. Let us assume that we want to implement the function $f = abc' + bcd' + acd'$ in a logic cell. Let us also assume that the available MIGFETs have switching function $f_M = ab + ac + bc$ (MAJ function). Since we handle only the PDN of a logic cell, we want to make a connection from output to V_{ss} when $f = 0$, thus according to f'. Apparently, just one MIGFET is not enough to realize the PDN for this example. Intuitively, this is because f' depends on four variables, while f_M depends only on three. To deal with this occurrence, we use generalized and Shannon's expansions to decompose a logic function into simpler components. Between those two, the expansion reducing most height and size of a DD is chosen. The generalized expansion is specifically computed with respect to f_M, here $MAJ(a, b, c)$. Figure 11.3 depicts the transistor-level realization of these logic expansions into a generic logic cell. A circuit stratagem is used to obtain the negative basis by inverting the positive basis. In this way, we can generate f'_M without any assumptions on f_M *self-duality*. Back to our example, the Shannon's and generalized cofactors are $\{f'_a = b'd + cd + b'c',$ $f'_{a'} = b' + c' + d\}$ and $\{f'_{f_M} = c \cdot d, f'_{f'_M} = 1\}$, respectively. The details on their efficient computation with canonical DDs are omitted for the sake of brevity. In this case, the generalized cofactors w.r.t. f_M enable a larger simplification than Shannon's cofactors, in terms of logic representation size. Consequently, they are selected for the logic cell implementation, as depicted by Figure 11.4. There, the device on the right identifies the positive basis (f_M) while the device on the left identifies the negative basis (f'_M). The expansion procedure continues recursively with the obtained cofactors,

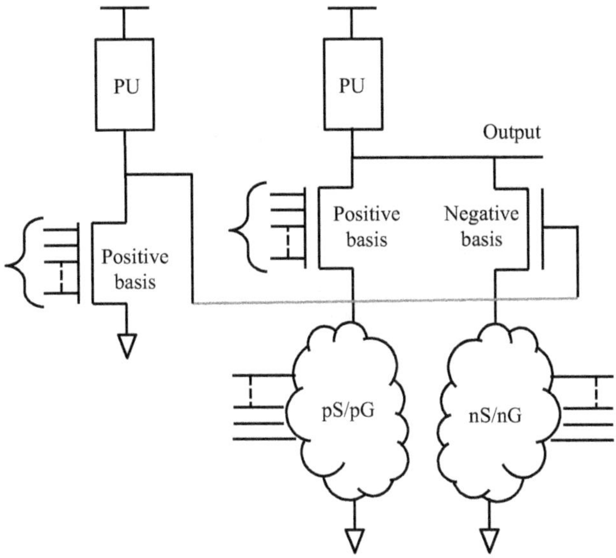

Figure 11.3 Logic expansions mapped into a logic cell. The negative basis is realized by inverting the positive basis

until simple matches are found. In this case, $f'_{f'_M} = 1$ just requires a direct connection to ground while $f'_{f_M} = c \cdot d$ is already included in $f_M = ab + ac + bc$ (mapped into a single device). Up to this point, the mapping is complete and valid. However, with a final redundancy check, it can be noted that the blue FET (positive basis) is removable as not essential to implement the desired function.

The LUT optimization and cell mapping steps form our proposed MIGFET synthesis methodology, sketched by Algorithm 1. The algorithmic procedure flows as follows. First, the initial circuit is optimized and mapped onto k-LUTs (Algorithm 1 $-\alpha$). The choice of the parameter k allows us to size the LUT nodes to match a MIGFET functionality. For this reason, k is usually set to the number of inputs of the f_M, or larger values. The switching function f_M (plus some of its NPN permutations) is represented and stored within a DD. Note that any canonical DD extension can be used here, as long as the representation uniqueness is preserved, together with its efficient manipulation properties.

After the LUT optimization step, the logic cell mapping begins (Algorithm 1 $-\beta, \gamma$). Each LUT node is considered in a *for loop* and mapped individually onto a logic cell. The complemented function[6] of each node is also represented with a DD, sharing the same data structure used for f_M. In this way, any logic match is identifiable in software by a simple pointer comparison. If the DD for the current logic node is contained in the DD for f_M, then the PDN can be implemented by a

[6]We operate on negated logic functions to directly handle the inverted PDN implementation polarity.

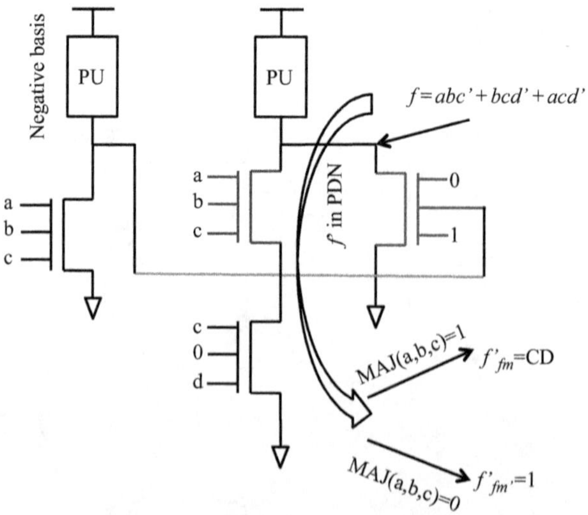

Figure 11.4 Logic cell mapping example. MIGFET switching function $f_M = ab + ac + bc$, target function $f = abc' + bcd' + acd'$

single MIGFET, with inputs assignment corresponding to DD variables (Algorithm 1 $-\beta$). Otherwise, logic decomposition is needed (Algorithm 1 $-\gamma$). Generalized (w.r.t. f_M) and Shannon's expansions are used for this purpose. The expansion reducing most a DD complexity cost metric is chosen. The circuit expansion continues iteratively, adding new cofactors to a queue, if not already included in the DD for f_M (else reduces to a simple match). Finally, inverters are mapped, if any, possible sharings between the PDN branches are identified and enforced and a redundancy removal routine eliminate superfluous FETs.

The ability of this synthesis flow to harness a MIGFET switching function is demonstrated in the next section.

11.4 Experimental results

In this section, we present first our experimental methodology and technology models. Then, we give details on the synthesis tool and comment on the results obtained.

11.4.1 Methodology

In our experiments, we test five promising classes of MIGFET devices and associated switching functions, with number of physical gates ranging from 1 to 7. The synthesis methodology and estimation models are embedded into a tool, which synthesizes combinational circuits onto such MIGFETs.

Algorithm 1: MIGFET logic synthesis

INPUT: Logic circuit C, MIGFET switching function f_M
OUTPUT: Netlist of pseudo-logic gates made of MIGFETs

```
k = |f_M|;                                          ┐
net ← k-LUT mapping (C);                            │
DD_fM ← canonical DD for f_M plus                   │ α
partial NPN configurations;                         ┘
for each LUT node i in net do
   create a new logic cell - add the pull-up device;
   DD_i ← canonical DD for node(i)';
   if DD_i ∈ DD_fM then                             ┐
      add a MIGFET to the PDN;                      │ β
      map inverters if any;                         ┘
   else                                             ┐
      f-to-map ← DD_i;                              │
      while f-to-map ≠ ∅ do                         │
         j=last function in the f-to-map queue;     │
         {pS, nS}= Shannon's cofactors of j;        │
         {pG, nG}= generalized cofactors of j w.r.t. f_M;
         if cost(pG)+cost(nG)<cost(pS)+cost(nS) then │
            map generalized expansion onto logic cell;
            add {pG, nG} ∉ DD_fM to f-to-map;       │
         else                                       │ γ
            map Shannon's expansion onto logic cell;│
            add {pS, nS} ∉ DD_fM to f-to-map;       │
         end if                                     │
      end while                                     │
      map inverters if any;                         │
      identify sharing - remove redundancy;         │
   end if                                           │
end for                                             ┘
```

11.4.1.1 Benchmarks MIGFETs

We consider the five classes of switching functions, associated with MIGFET devices, presented in Section 11.2, being XNOR, OR, AND, MAJ and AON. Different versions of such devices are studied, with variable number of physical gates. All of them have a single-gate version, which is a traditional MOSFET. The environment technology node is 22 nm [1], which defines basic device and interconnect properties. Note that MIGFET devices can be fabricated in different technologies (tech. nodes, materials, etc.), but in this study, we virtually assume a common technology for all devices. This assumption does not affect relative comparisons within the same class of devices. Instead, an interclass superiority assessment is not fair, but anyway out of the scope of this work. Table 11.1 shows a simplified set of technology parameters used for

Table 11.1 Technology parameters

Parameter	Value	Unit
Switch delay	1	ps
Device area	3,000	nm^2
Device capacitance	15	aF
Interconnection delay	9	fs/nm
Interconnection capacitance	0.2	aF/nm
Voltage	1	V

a single gate device. Increasing number of gates are considered, up to seven. As of now, only MIGFETs with up to 3/4 gates have been experimentally fabricated in such classes. With just two gates, many studies report very limited (or almost null) area/delay overhead, thanks to smart device geometries [17]. Also with three and four gates, compact device implementations exist, but the overhead in this case is not negligible. Averaging these considerations, we forecast a general area/delay increase by about 15% when adding one gate to a MIGFET. This law is consciously optimistic, especially when extended to 4–7 gates MIGFETs yet not developed. However, this is in line with our study: we want to predict an upper bound on the benefit deriving by future MIGFETs, in their best physical realization scenario. Nevertheless, the impact of extra gates on the interconnection complexity is taken into account, as detailed in the next subsection.

11.4.1.2 Estimation models

To estimate area, delay, energy and power metrics in a synthesized circuit, we combine device and interconnect models. In particular, the interconnects are important in presence of multiple gates. Indeed, a MIGFET with many physical gates requires more interconnects than a standard FET to wire a logic circuit. We use Rent's rule [18] to estimate the local/global number of terminals as function of the gates/cells number. We then use Donath's rule [19] to estimate average wirelenghts. Using these rules in conjunction, it is possible to assess what are the extra delay and capacitance deriving from the interconnects. Together with device technology parameters, the overall circuit metrics can now be computed. The area is just the sum of all devices physical occupation on the circuit. The delay is the sum, over the logic cells on the critical path, of (i) maximum stack of switching devices times their switching delay plus (ii) estimated wirelength times its unit delay. The energy *consumption* is computed as $C_{sw-tot}V_{dd}^2$, where C_{sw-tot} is the switching capacitance, consisting of both devices and interconnects contribution. Finally, the power *consumption* is the energy *consumption* times the maximum switching frequency achievable by a circuit. Buffering and sizing impacts are omitted in this work but currently under study for integration in future tool releases.

11.4.1.3 Synthesis tool

We implemented in C language the logic cell synthesis method in Algorithm 1 supplied with the aforementioned estimation models. The associated tool is available online [10]. We use the *Biconditional BDD (BBDD)* [20] logic manipulation package available online at [21] to accomplish canonical DD tasks. The inputs to our MIGFET synthesis tool are (i) a logic circuit in Berkeley logic interchange format (BLIF) format already optimized/decomposed in *k*-LUT, (ii) a switching function in BLIF format and (iii) a technology file with device/interconnect information. The outputs are (i) the overall area/delay/energy/power estimated metrics and (ii) a netlist of logic cells with detailed synthesis information. In our experiments, we considered the largest MCNC and additional arithmetic benchmarks pre-optimized with ABC academic tool [22], using the synthesis command *if -K 7*. Such LUT-mapping command is kept untouched for all gates numbers, as it experimentally demonstrated good results. Still, a custom mapping is possible for each gate number.

Note that many other experimental scenarios can be tested, as the MIGFET synthesis tool [10] is designed to be flexible.

11.4.2 Results

Table 11.2 shows average synthesis results, normalized with respect to the single gate FET case. Delay, area and energy values are reported, with their corresponding minimum highlighted in bold. Power values are omitted for the sake of brevity. Anyhow, they can be obtained using energy and delay values. For AND and XOR MIGFETs, the best delay happens with two gates. For AON MIGFETs, the best delay is at three gates, while for MAJ MIGFETs is at four gates. The best area is obtained with two gates considering AND and MAJ MIGFETs, three gates for AON MIGFETs and one gate for the others. The energy for all MIGFETs is at its best for single gate. This is because, for multiple gates, the energy is increased by interconnects deriving from extra gates.

Figure 11.5 shows an average *area* × *delay* figure of merit for the MIGFETs on test. We see optimal gates numbers being one (OR, XOR types), two (AND type), three (AON type) and four (MAJ type) for different device classes. Also, the increase on the average energy is reported in the bottom-right plot.

11.4.3 Discussion

The synthesis experiments showed that designing ICs with expressive but complex devices is not always beneficial. While the logic functionality increases when considering many gate FETs, the capabilities of contemporary synthesis techniques, or the natural features of a circuit, limit the corresponding savings. In this unfortunate case, the overhead deriving from interconnects and device physical implementation offsets the enhanced functionality advantage. We highlighted optimal points for some MIGFET devices where the increased functionality benefit is maximized. After those points, the design complexity grows, frequently exceeding also a traditional single gate FET implementation cost.

Table 11.2 MIGFET synthesis average results normalized values for delay, area and energy

Number of gates	AND			AON			XOR			OR			MAJ		
	D	A	E	D	A	E	D	A	E	D	A	E	D	A	E
1	1.00	1.00	**1.00**	1.00	1.00	**1.00**	1.00	**1.00**	**1.00**	**1.00**	**1.00**	**1.00**	1.00	1.00	**1.00**
2	**0.96**	**0.98**	1.03	0.97	1.11	1.13	**0.97**	1.11	1.13	1.02	1.08	1.10	0.96	**0.98**	1.03
3	0.98	1.04	1.13	**0.84**	**1.00**	1.10	1.01	1.22	1.27	1.10	1.20	1.24	1.10	1.18	1.23
4	1.06	1.15	1.25	0.91	1.10	1.24	1.10	1.35	1.42	1.18	1.33	1.39	**0.94**	1.00	1.17
5	1.14	1.29	1.39	0.97	1.24	1.37	1.19	1.51	1.57	1.27	1.50	1.55	1.27	1.46	1.51
6	1.21	1.43	1.52	1.04	1.38	1.50	1.26	1.68	1.71	1.36	1.67	1.70	1.09	1.28	1.41
7	1.29	1.54	1.55	1.11	1.49	1.63	1.34	1.82	1.87	1.42	1.77	1.69	1.44	1.75	1.80

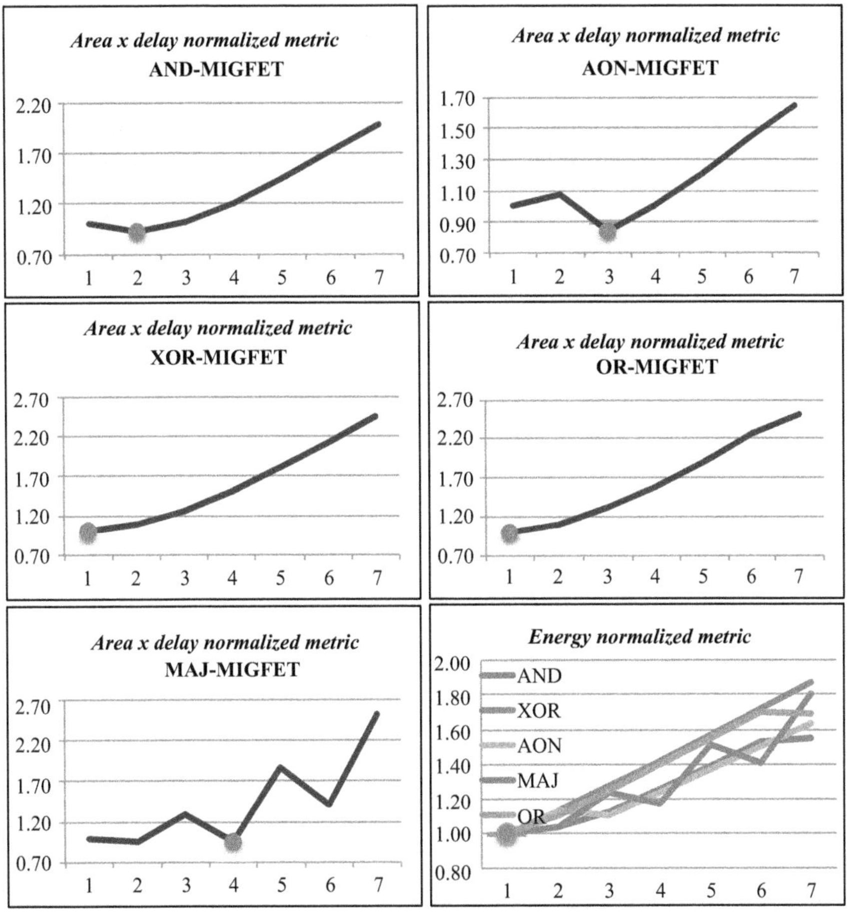

Figure 11.5 *Implementation quality metric versus number of gates. Average results over MCNC benchmarks. Different types of MIGFETs are individually presented*

11.5 Custom exploration of special MIGFET classes

Note that custom designs and/or *ad hoc* synthesis methodologies may offer a better exploitation for a specific technology, as compared to the previous results. For example, some advanced *BBDD* [20] techniques fit natively with XOR MIGFETs, enabling larger improvements for XOR-intensive circuits. Similarly, logic synthesis methodologies based on majority voting unlock the full potential of MAJ MIGFETs. In the following subsection, we study in more detail the opportunities of XOR and MAJ MIGFETs, with the help of custom logic tools.

11.5.1 Potential of XOR MIGFETs

When focusing on XOR-intensive functions, XOR MIGFETs show the potential for compact circuit implementations. One prime example here regards circuits for cryptography. They are dominated by biconditional logic, thus taking advantage of any native implementation of XOR operations at the device level. We consider as case study the *twine-sbox* block in the lightweight cipher recently introduced in [23]. To highlight the logic structure to be mapped directly onto XOR MIGFETs, we use a monolithic BBDD representation [20]. The resulting BBDD netlist of logic operations is depicted by the following listing:

```
module twine-sbox (
            x0, x1, x2, x3,
            t4, t5, t6, t7);
input x0, x1, x2, x3;
output t4, t5, t6, t7;
wire one, node14, node1, node8, node4, node15, node7, node13, node10,
node6, node5, node12, node11, node2, node9;
assign node1 = (x0 ^ x2) ? ~x2 : one;
assign node4 = (x0 ^ x2) ? ~one : x2;
assign node8 = (x3 ^ x0) ? ~node1 : node4;
assign node14 = (x1 ^ x3) ? node1 : node8;
assign node7 = (x3 ^ x0) ? x2 : one;
assign node15 = (x1 ^ x3) ? node4 : node7;
assign node10 = (x3 ^ x0) ? one : node4;
assign node5 = (x0 ^ x2) ? x2 : one;
assign node6 = (x3 ^ x0) ? node5 : x0;
assign node13 = (x1 ^ x3) ? ~node10 : node6;
assign node2 = (x0 ^ x2) ? one : x2;
assign node11 = (x3 ^ x0) ? ~node2 : one;
assign node9 = (x3 ^ x0) ? ~one : node1;
assign node12 = (x1 ^ x3) ? ~node11 : node9;
assign one = 1;
assign t5 = ~node14;
assign t6 = node15;
assign t4 = node13;
assign t7 = ~node12;
endmodule
```

In this example, the BBDD circuit consists of 14 nontrivial biconditional if–then–else operations. Each one can be mapped into a 2-input XOR MIGFET. The logic depth is three levels, which makes necessary to create two levels of XOR MIGFET gates, designed as detailed in Section 11.3.3. Using the previously introduced estimation models and technology parameters, the custom mapping into 2-input XOR MIGFET is 9.52%/4.32% smaller in area and 21.14%/8.56% faster than 2-input AND/XOR MIGFET realization (obtained through our automated flow). We attempted at grouping more XORs in the custom mapping, to evaluate wider XOR MIGFETs, but results—for this particular crypto circuit—indicated that 2-input XOR operations are the most flexible choice. This result is in line with the AND MIGFET trend depicted in Table 11.2. Nevertheless, specific circuits dominated by wide XOR operations may take more advantage in 3- and 4-input XOR MIGFETs.

11.5.2 Potential of MAJ MIGFETs

Inspired by computer theory results on majority logic [24], we evaluate the potential of MAJ MIGFETs to effectively implement special classes of Boolean functions, that are compactly represented in terms of majority operators. We know that majority networks can realize complex arithmetic functions [24], such as bit counters, adders, multipliers, etc., with a small number of majority gates and a constant number of logic levels. In particular, for the scope of this study, we focus on bit counters functions as they are primitive blocks for building various types of IPs in modern digital designs. We make use of the *Majority Arithmetic Compiler* (MAC) software tool [25] to synthesize netlist descriptions of bit counters, implemented by (wide) majority gates in a constant number (3) of levels. Such netlist is the starting point for mapping onto gates of MAJ MIGFETs. The following listing of majority logic operators represents a 7-input, parallel bit counter.

```
module bitcounter (xin, bitcount);
input [6:0] xin;
output [2:0] bitcount;
wire gelle1, ge3le3, ge5le5, ge7le7, ge2le3, ge6le7,
ge4le7, ge1, ge2, ge3, ge4, ge5, ge6, ge7, le1, le3, le5, le7;

majority #(13) inst0 ({xin[0], xin[1], xin[2], xin[3], xin[4], xin[5],
xin[6], 1'b1, 1'b1, 1'b1, 1'b1, 1'b1, 1'b1}, ge1);
majority #(11) inst1 ({xin[0], xin[1], xin[2], xin[3], xin[4], xin[5],
xin[6], 1'b1, 1'b1, 1'b1, 1'b1}, ge2);
majority #(9) inst2 ({xin[0], xin[1], xin[2], xin[3], xin[4], xin[5],
xin[6], 1'b1, 1'b1}, ge3);
majority #(7) inst3 ({xin[0], xin[1], xin[2], xin[3], xin[4], xin[5],
xin[6]}, ge4);
majority #(9) inst4 ({xin[0], xin[1], xin[2], xin[3], xin[4], xin[5],
xin[6], 1'b0, 1'b0}, ge5);
majority #(11) inst5 ({xin[0], xin[1], xin[2], xin[3], xin[4], xin[5],
xin[6], 1'b0, 1'b0, 1'b0, 1'b0}, ge6);
majority #(13) inst6 ({xin[0], xin[1], xin[2], xin[3], xin[4], xin[5],
xin[6], 1'b0, 1'b0, 1'b0, 1'b0, 1'b0, 1'b0}, ge7);
majority #(11) inst7 ({~xin[0], ~xin[1], ~xin[2], ~xin[3], ~xin[4],
~xin[5], ~xin[6], 1'b0, 1'b0, 1'b0, 1'b0}, le1);
majority #(7) inst8 ({~xin[0], ~xin[1], ~xin[2], ~xin[3], ~xin[4],
~xin[5], ~xin[6]}, le3);
majority #(11) inst9 ({~xin[0], ~xin[1], ~xin[2], ~xin[3], ~xin[4],
~xin[5], ~xin[6], 1'b1, 1'b1, 1'b1, 1'b1}, le5);
majority #(15) inst10 ({~xin[0], ~xin[1], ~xin[2], ~xin[3], ~xin[4],
~xin[5], ~xin[6], 1'b1, 1'b1, 1'b1, 1'b1, 1'b1, 1'b1, 1'b1, 1'b1}, le7);
majority #(3) inst11 ({ge1,le1,1'b0},gelle1);
majority #(3) inst12 ({ge3,le3,1'b0},ge3le3);
majority #(3) inst13 ({ge5,le5,1'b0},ge5le5);
majority #(3) inst14 ({ge7,le7,1'b0},ge7le7);
majority #(3) inst15 ({ge2,le3,1'b0},ge2le3);
majority #(3) inst16 ({ge6,le7,1'b0},ge6le7);
majority #(3) inst17 ({ge4,le7,1'b0},ge4le7);
majority #(7) inst18 ({gelle1,ge3le3,ge5le5,ge7le7,1'b1, 1'b1, 1'b1},
bitcount[0]);
majority #(3) inst19 (ge2le3,ge6le7,1'b1, bitcount[1]);
assign bitcount[2]=ge4le7;

endmodule
```

The switching function of each majority operator in this netlist can be implemented by a single MAJ MIGFET of the corresponding input size. Indeed, one can build a gate for each majority operator, using the style described in Section 11.3.3, with only one pull-down MAJ MIGFET with inputs properly connected. Inverted inputs need to be implemented separately.

In our evaluation, we assume that any MAJ MIGFET size is available, but decomposition of large majority operators into smaller devices is always possible. Using the previously introduced estimation models and technology parameters, the custom mapping into MAJ MIGFET is 15.46%/6.79% faster than 2-input AND/XOR MIGFET realization (obtained through our automated flow), at expenses of area overhead, being about 5.34%/7.01%. As bit counters have some XOR operations embedded in their functionality, XOR MIGFETs are more advantageous than AND MIGFETs, in this case.

Please note that, if larger MAJ MIGFETs are available with low switching delay, the advantages in speedup for this experiment would grow more marked. Indeed, we could implement larger bit counters: the MAC software can realize arbitrary bit counters in majority logic requiring wider fanin but still fitting in three logic levels [24,25]. This directly translates into faster bit counter implementations with MAJ MIGFETs.

11.5.3 Summary

Focusing on the classes of MIGFETs studied in this work through automated synthesis, we notice that majority (MAJ) and gamble (AON) functions lead to the largest relative benefits. This is thanks to their intrinsic expressiveness. Considering the custom exploration of XOR and MAJ MIGFET classes, we identified 2-input XOR MIGFET as most advantageous, targeting cryptographic circuits, and very wide MAJ MIGFET desirable for arithmetic-intensive circuits such as bit counters. Note that these results are based on a common virtual technology among MIGFETs, created for comparison and projection purposes. Once precise technology information becomes available for an emerging MIGFET device, all synthesis experiments need to be repeated to confirm the expected trend.

11.6 Conclusions

MIGFETs with enhanced logic functionality are promising *more than Moore* technologies. In such FETs, extra gates increase the device intelligence by enriching the switching function. However, each additional gate also introduces a side implementation cost. In this chapter, we addressed, from a logic synthesis standpoint, the naturally arising question: *How many gates do we need?* We proposed a logic synthesis methodology that exploits at a fine grain a MIGFET switching function. Using simplified design assumptions and device/interconnect models, we synthesized MCNC benchmarks on five promising MIGFET devices, with number of physical gates ranging from 1 to 7. Experimental results evidenced nontrivial area/delay/energy minima, located between 1 and 4 physical gates, depending on a MIGFET switching function and technology. We also conducted custom design experiments for two selected MIGFET devices, targeting cryptographic and arithmetic designs, which confirmed a tight interdependency between optimal number of physical gates and functionality of the circuit under synthesis. The proposed methodology can help technologists in guiding their research effort on the most advantageous devices.

References

[1] ITRS available online at *http://www.itrs.net.*

[2] J.-P. Colinge, Editor. *FinFETs and Other Multi-Gate Transistors*, Vol. 73. New York: Springer, 2008.

[3] S. Cristolovenau, *How many gates do we need in a transistor?*, International IEEE Semiconductor Conference, 2007. CAS 2007.

[4] M. De Marchi, M. Sacchetto, D. Frache, *et al.*, *Polarity control in double-gate, gate-all-around vertically stacked silicon nanowire FETs*, Electron Devices Meeting (IEDM), 2012 IEEE International, 8–4, 2012.

[5] J. Zhang, P.-E. Gaillardon, G. De Micheli, *Dual-threshold-voltage configurable circuits with three-independent-gate silicon nanowire FETs*, Proc. ISCAS, 2013.

[6] M. Rostami, K. Mohanram, *Novel dual-Vth independent-gate FinFET circuits*, Proc. ASPDAC, 2010.

[7] T. Oya, T. Asai, T. Fukui, Y. Amemiya, *A majority-logic device using an irreversible single-electron box*, IEEE Transactions on Nanotechnology, 2(1): 15–22, 2003.

[8] T. Shibata, T. Ohmi, *A functional MOS transistor featuring gate-level weighted sum and threshold operations*, IEEE Transactions on Electron Devices, 39(6): 1444–1455, 1992.

[9] L. Amaru, P.-E. Gaillardon, G. De Micheli, *Efficient arithmetic logic gates using double-gate silicon nanowire FETs*, Proc. NEWCAS, 2013.

[10] *MIGFET Synthesizer* available online at *https://lsi.epfl.ch/MIGFET.*

[11] E. Beigne, A. Valentian, B. Giraud, *et al.*, *Ultra-wide voltage range designs in fully-depleted silicon-on-insulator FETs*, Design, Automation & Test in Europe Conference & Exhibition (DATE), 613–618, 2013. IEEE.

[12] G. De Micheli, *Synthesis and Optimization of Digital Circuits*, McGraw-Hill, New York, 1994.

[13] R.E. Bryant, *Graph-based algorithms for Boolean function manipulation*, IEEE Transactions on Computers, C-35(8): 677–691, 1986.

[14] C. Yang, M. Ciesielski, *BDS: A BDD-based logic optimization system*, IEEE Transactions on Computer-Aided Design of Integrated Circuits and Systems, 21(7): 866–876, 2002.

[15] L. Benini, G. De Micheli, *A survey of Boolean matching techniques for library binding*, ACM Transactions on Design Automation of Electronic Systems, 2(3): 193–226, 1997.

[16] O. Coudert, J.C. Madre, *A unified framework for the formal verification of sequential circuits*, Proc. ICCAD, 1990
[17] O. Zografos, P.-E. Gaillardon, G. De Micheli, *Novel grid-based power routing scheme for contr.-pol. FET arrangements*, Proc. ISCAS, 2014.
[18] B. Landman, R. Russo, *On a pin versus block relationship for partitions of logic graphs*, IEEE Transactions on Computers, C-20: 1469–1479, 1971.
[19] W.E. Donath, *Placement and average interconnection lengths of computer logic*, IEEE Transactions on Circuits and Systems, 26(4): 272–277, 1979.
[20] L. Amaru, P.-E. Gaillardon, G. De Micheli, *An efficient manipulation package for biconditional binary decision diagrams*, Proc. 2014.
[21] *BBDD package available at: http://lsi.epfl.ch/BBDD.*
[22] ABC synthesis tool – available online.
[23] T. Suzaki, K. Minematsu, S. Morioka, E. Kobayashi, *Twine: A lightweight, versatile block cipher*, ECRYPT Workshop on Lightweight Cryptography, Vol. 2011, 2011.
[24] M. Krause, P. Pudlak, *On the computational power of depth-2 circuits with threshold and modulo gates*, Theoretical Computer Science, 174: 137–156, 1997.
[25] L. Amarú, P.-E. Gaillardon, G. De Micheli, *A sound and complete axiomatization of majority-n logic*, IEEE Transactions on Computers, 65(9): 2889–2895, 2016.

Chapter 12
Ultrafine grain FPGAs with polarity controllable transistors

Xifan Tang[1,2]*, Pierre-Emmanuel Gaillardon*[2]*, Ian O'Connor*[3]*, and Giovanni De Micheli*[1]

Abstract

In the quest to push further the Moore's scaling laws, intensive development efforts have been invested on seeking for alternatives to planar CMOS transistors at advanced technology nodes. In particular, recent years have witnessed the massive commercialization of the *fin-based field effect transistors* (FinFETs) at the 10, 14 and 22-nm technology nodes [1–3]. In addition to FinFET technology, several devices are currently investigated by following the trend toward 1D structures. Among them, carbon nanotubes FETs [4] and vertically stacked *silicon nanowires* FETs (SiNWFETs) [5] are promising extensions to current tri-gate FinFETs technology, which exploit the 1-D properties of their channels to exhibit superior performances. Moreover, these novel transistor technologies employ the gate-all-around (GAA) structure which can improve the electrostatic control of the channel, leading to a higher I_{ON}/I_{OFF} ratio and reduced leakage current [6].

More than the performance improvement, these novel transistor technologies present another possible direction in pushing the Moore's scaling laws: functionality-enhanced device. Especially at advanced technology nodes, transistors are strongly affected by Schottky contacts at the source and drain interfaces. As a result, transistors may operate with an ambipolar behavior, i.e., the device can exhibit n- and p-type characteristics simultaneously. Indeed, to achieve pure n- and p-type polarity, the ambipolar behavior of the devices is typically suppressed through additional process steps. However, new design methodologies [7–9] have shown attractive opportunities in controlling the ambipolar phenomenon through programmable polarity devices. By engineering the source and drain contacts and by constructing independent double-gate structures, the device polarity can be electrostatically programmed to be either n- or p-type. Such functionality-enhanced devices have been demonstrated using silicon

[1] School of Computer Science and Communication, École Polytechnique Fédérale de Lausanne (EPFL), Switzerland
[2] Department of Electrical and Computer Engineering, University of Utah, USA
[3] Department of Electronic, Electrical and Control Engineering, Ecole Centrale de Lyon, France

[10,11] and carbon electronics [12,13]. In this chapter, we focus on a *double-gate* SiNWFET (DG-SiNWFET), built using a top-down fabrication flow [14].

Combining both performance improvement and device-level reconfigurability, the functionality-enhanced devices has been exploited to build ultrafine grain reconfigurable logic blocks [7]. This technology shift allows a reconfigurable *logic cell* (LC) which is realized with only seven transistors to perform many different two-input Boolean operations [7]. The cell compactness opens an opportunity in rethinking the standard reconfigurable architectures. Take the example of the traditional *field-programmable gate arrays* (FPGAs) architectural; only less than 15% of the area is dedicated to the logic computation, while the other resources are used for the structure reconfigurability [15]. Such a large imbalance results in a large cost in terms of area, routing delay and power consumption.

In this chapter, we exploit the ultrafine grain LCs based on the DG-SiNWFET to improve the FPGA architecture:

1. We introduce a novel architectural organization where the standard *look-up tables* (LUTs) are replaced by ultrafine grain LCs. While it seems counterintuitive to further reduce the size of the logic blocks, therefore worsening the logic/routing imbalance, we will see that the ultrafine grain elements can be organized in compact matrix arrangements of LCs, called MClusters that become competitive as compared with the typical combinational elements of the FPGAs. To prevent the reconfigurable interconnection overload within the MClusters, we interconnect the cells through a layered, fixed and incomplete interconnection pattern, i.e., that an output on a given layer is permanently connected to a subset of inputs on the next layer.
2. To support the novel architecture, we propose a complete benchmarking tool flow suited to this organization. In particular, we develop a new tool, called MPack, aiming at mapping logic functions to MClusters and clustering MClusters into logic blocks of FPGA architectures. We integrate MPack into a state-of-the-art FPGA flow, enabling a completed evaluation on area, delay, and power consumption of the novel FPGA architectures.

Architecture-level results show that an FPGA structure using 2-by-2 MClusters in replacement of the traditional LUTs gives an average area and delay reduction of 43% and 23%, respectively, compared with a conventional CMOS FPGA at 22-nm technology node. This large performance gains are directly accounted from the efficient use of ultrafine grain LCs. In addition, the proposed approach opens a promising path to correct two intrinsic limitations of the FPGA circuits: (1) an 8% reduction in the logic/routing imbalance and (2) a better control in the distribution of the routing delay.

The rest of this chapter is organized as follows. Section 12.1 surveys the necessary background on FPGA architecture, functionality-enhanced devices, and ultrafine grain LC designs. Section 12.2 presents the novel architectural scheme based on ultrafine grain reconfigurable LC, by highlighting the concept and the organization at each architectural level. Section 12.3 introduces the specific benchmarking tool MPack, with an emphasis on the MCluster packing. Section 12.4 evaluates the performance of

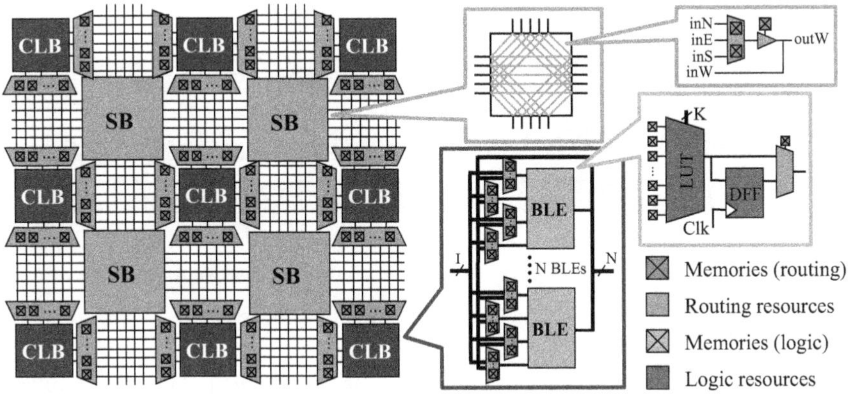

Figure 12.1 Baseline field-programmable gate arrays architecture [15]

the targeted architecture, in terms of granularity and compares them with its CMOS FPGA counterpart. Section 12.5 concludes this chapter.

12.1 Background

In this part, we first introduce the necessary background on FPGA architecture (see Section 12.1.1). Then, we provide a brief review on controllable-polarity devices (see Section 12.1.1) and associated ultrafine grain computation topology (see Section 12.1.3).

12.1.1 FPGA architecture

As shown in Figure 12.1, FPGAs are highly regular circuits typically consisting of an array of identical *configurable logic blocks* (CLBs) surrounded by reconfigurable routing resources [15]. Each CLB contains a set of N *basic logic elements* (BLEs) and rich local routing resources. A BLE is formed by a K-input LUT, a flip-flop and a two-input reconfigurable routing multiplexer to select either a combinational or a sequential output. The local routing resources, consisting of reconfigurable routing multiplexers, interconnect the I CLB inputs and the CLB outputs to the inputs of BLEs.

All design parameters N, K, and I can be set by the FPGA architect depending on the targeted system granularity. The reconfigurable routing resources outside CLBs are called global routing architecture, which is organized by large channels of width W, interleaved between the CLBs. The channels cross each other and the signals can be routed within the FPGA using *switch boxes* (SBs). An SB is a matrix at the intersection of channels, which is made of reconfigurable routing multiplexers.

As we see in Figure 12.1, routing resources appear in every hierarchy of FPGA architectures, being a dominating factor on the total area, delay and power

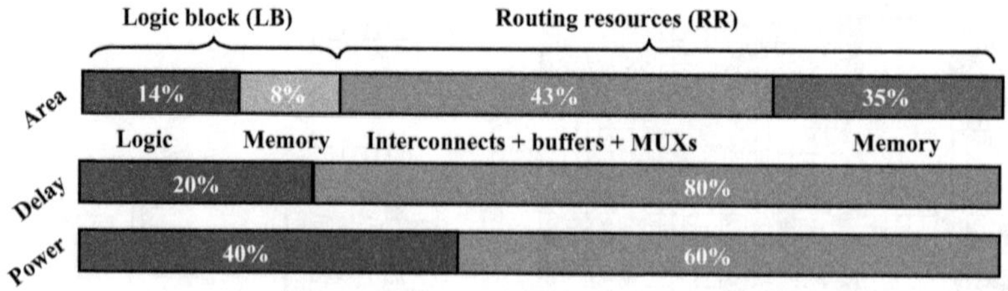

Figure 12.2 Field-programmable gate arrays area/delay/power repartition per block [16]

consumption. Figure 12.2 shows the area, delay and power breakdown of the various components of a baseline *static random access memory* (SRAM)-based FPGA. The heavy use of reconfigurable routing resources leads to the imbalance between logic and routing. Only 14% of the total area is used for actual computation elements, i.e., LUTs. In particular, roughly half of the area in both the logic blocks and the routing resources are consumed by configuration memories. In addition to occupying most of the die area, routing resources significantly contribute to FPGAs hurdles. Interconnect delays are reported to account for roughly 80% of the total path delay [16]. They also contribute to the high-power consumption of FPGAs, with more than 60% of the total dynamic power consumption.

In this chapter, we will develop novel logic block structures, by exploiting the reconfigurability at the device level, offered by novel device technologies, in the purpose of reducing the imbalance between logic and routing.

12.1.2 Transistors with controllable polarity

At advanced technology nodes (45 nm and below), the superposition of n-type and p-type carriers is observable even under normal bias conditions in several nanoscale FET devices. The phenomenon, called ambipolarity, occurs in many different materials, such as silicon [17], carbon nanotubes [18] and graphene [19]. The control of this ambipolarity enables device-level reconfigurability, allowing us to adjust the device polarity online. Transistors with a controllable polarity have been experimentally fabricated in several novel technologies, such as carbon nanotubes [12], graphene [13] and silicon nanowires (SiNWs) [10,11]. The operation of these FETs is enabled by the regulation of Schottky barriers at the source/drain (S/D) junctions thanks to an additional gate.

In this chapter, we consider here vertically stacked SiNWFETs [14], as illustrated in Figure 12.3. Featured by vertically stacked nanowires and *GAA* structures, the SiNWFETs can be regarded as a natural evolution of FinFET structures. As depicted in Figure 12.3(a), a SiNWFET has four electrodes: (1) the control gate; (2) the source node; (3) the drain node; and (4) the polarity gate (PG). Similar to conventional transistors, the control gate acts conventionally by turning *on* and *off* the device, while the source and drain nodes are connected to the channels. The key difference lies in the fourth electrode, the PG, acts on the side regions of the device, in proximity

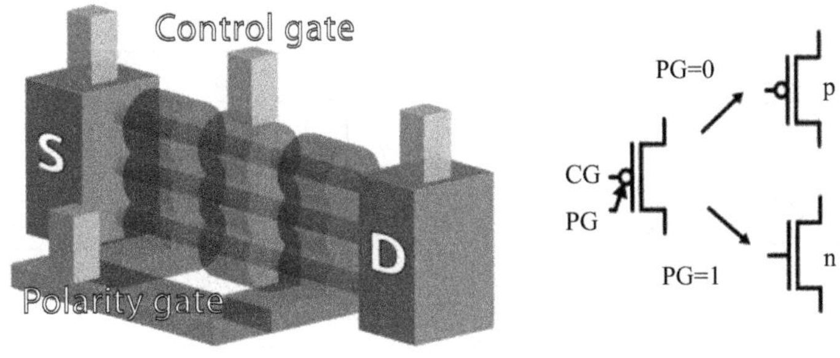

(a) (b)

Figure 12.3 (a) 3D sketch of the SiNWFETs featuring two independent gates and (b) its associated symbol [14]

of the S/D Schottky junctions, switching the device polarity dynamically between n- and p-type. As illustrated in Figure 12.3(b), when the PG is configured to logic 0, the SiNWFET behaves a p-type transistor, while its functionality can be switched to an n-type transistor by applying logic 1 to the PG. In addition, the control gate and the PG are fabricated with GAA structure, providing better electrostatic control over the channel and consequently superior scalability properties [14]. Compared to planar transistors, transistors with GAA structure can drive a much lower leakage current, below $4fA$ [14]. Note that the input and output voltage levels are compatible, logic gates can be cascaded as CMOS technology [14]. For a complete review on the design opportunities brought by these transistors, we refer the reader to [8].

12.1.3 Ultrafine grain reconfigurable logic gates

The in-field reconfigurability of ambipolar transistors has been exploited in [7] to build a compact reconfigurable cell. As shown in Figure 12.4(a), the reconfigurable cell can realize any of the eight Boolean logic functions Y of the two inputs A and B, which is detailed in Figure 12.4(b). The cell, consisting of only seven transistors, is implemented in the principle of the dynamic logics [20], arranged in two stages: logic function and follower/inverter. Signals PC_1, EV_1, PC_2, and EV_2 are the global precharge and evaluation signals of the two stages, respectively. The reconfiguration of the cell depends on the signals applied to the PGs, V_{BA}, V_{BB} and V_{BC}. Each of these signals configures the related transistors to either n- or p-type, thereby customizing the gate internal circuit. For a detailed description of the reconfigurable cells, we refer the reader to [7].

In this chapter, for practical reasons, we consider only binary configuration signals (either V_{DD} or GND), limiting the gate reconfigurability to eight functions. Previous work [7] considered ternary configuration signals, increasing the logic capability of the cell to 14 functions. We qualify the circuit as operating at ultrafine grain. The terminology of ultrafine grain is twofold. First, it covers the size of the cell, which is highly compact, as opposed to its CMOS equivalent. Second, it also

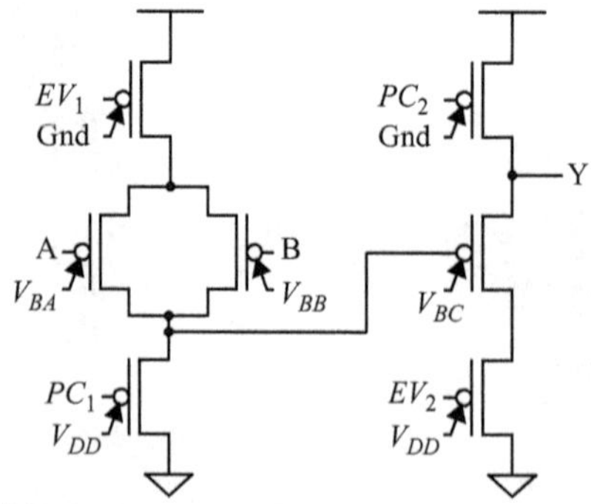

(a)

V_{BA}	V_{BB}	V_{BC}	Y
0	0	0	$\overline{A.B}$
0	0	1	A.B
0	1	0	$\overline{A}$+B
0	1	1	A.$\overline{B}$
1	0	0	A+$\overline{B}$
1	0	1	$\overline{A}$.B
1	1	0	A+B
1	1	1	$\overline{A+B}$

(b)

Figure 12.4 Ultrafine grain reconfigurable cell schematic [7] (a) and associated configurations (b)

describes the granularity of the covered logic functions, as compared with a traditional four-input LUT.

12.2 Leveraging the ultrafine granularity at the architecture level

In Section 12.1.3, we showed the circuit-level implementation of the new class of ambipolar devices. In this section, we will study, at the architectural level, the impact of the ultrafine gain cells onto the FPGA architecture.

In the vision of our FPGA architecture, the traditional LUTs are replaced by an ultrafine grain LC. In a conventional FPGA architecture, such a modification will require each cell to be directly connected to a full connectivity unit. As the typical cell size is comparable with a 1-b SB [7], this projection would cause a large overhead in terms of connections and also would worsen the already significant imbalance between routing and logic resources at the FPGA level. To correct the granularity issues, we propose to pack the cells into an intermediate structure, compatible in terms of size with the traditional levels of FPGAs, and called MClusters for Matrix Clusters.

In the rest of this section, we will detail the organization of MClusters from three aspects: multilayer structure (see Section 12.2.1), inter-MCluster interconnection (see Section 12.2.2) and integration to FPGA architectures (see Section 12.2.3).

12.2.1 Multilayer organization

To increase the logic coverage of the MClusters, we propose a 2-D matrix assembly of ultrafine grain logic. The LCs are arranged in a layered structure, with connections only existing between adjacent layers, as depicted in Figure 12.5. As such, long

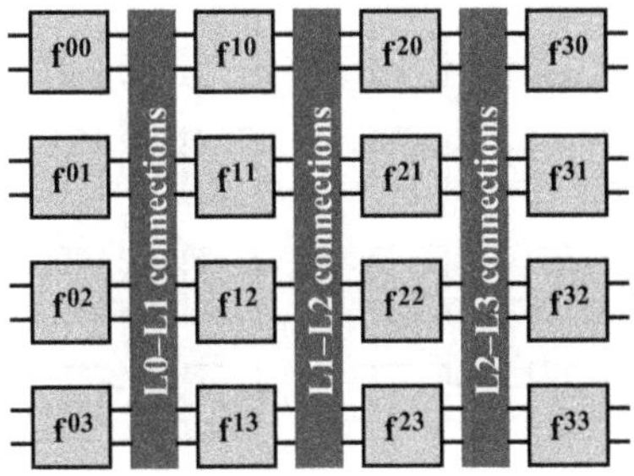

Figure 12.5 MClusters_4_4 arrangement for reconfigurable architectures

connections inside MCluster can be avoided and local connectivity can be maximized. Note that each layer is simply an array of LCs without pipeline or signal restoration.

In the rest of this chapter, MClusters will be defined by the d (depth) and w (width) parameters and named as *MClusters_d_w*. Note that, in Figure 12.5, each LC has two inputs, one is the output Y of the LC [Figure 12.4(a)], while the other is a duplication output Y. As such, we keep the same number of the input and output terminals per layers. The matrix patterns exhibit a fundamental structural regularity, being easily transposable to regular fabrics [8]. Note that the architectural concept is not limited to the presented architecture. Indeed, it is possible to consider diamond-shaped or triangular matrices and to have interconnects crossing different layers. However, to keep the focus of this chapter, only squared matrices, i.e., $d \times w$ LCs and $d = w$, and layer-to-layer connections are considered.

12.2.2 Intramatrix interconnecting

In relaxing the imbalance between logic and routing resources, the intramatrix interconnect will avoid introducing additional reconfigurable routing multiplexers. Therefore, any interconnecting topology giving a full interconnectivity is prohibited, due to the associated wiring complexity and intensive use of routing multiplexers. Similar to computer networks, our approach adapts incomplete interconnection sets to the matrix architecture. We employ *multistage interconnection networks* (MINs) to interconnect matrix layers. In computer engineering, MINs are used to interconnect layers of *SBs* to route information packets only. This concept has been applied in interconnect strategies for VLSI [21] and also FPGAs in order to reduce the wiring complexity between the logic blocks [22]. Note that the main difference lies in that *SBs* in the network context are replaced by LCs in our MCluster, thus introducing computing within the network itself. Furthermore, the use of very local MIN-style interconnect has a drastic impact on the size and the wire length. Among many MIN topologies and combinations [23,24], we report four typical permutations in Figure 12.6:

1. Banyan [Figure 12.6(a)],
2. Baseline [Figure 12.6(b)],

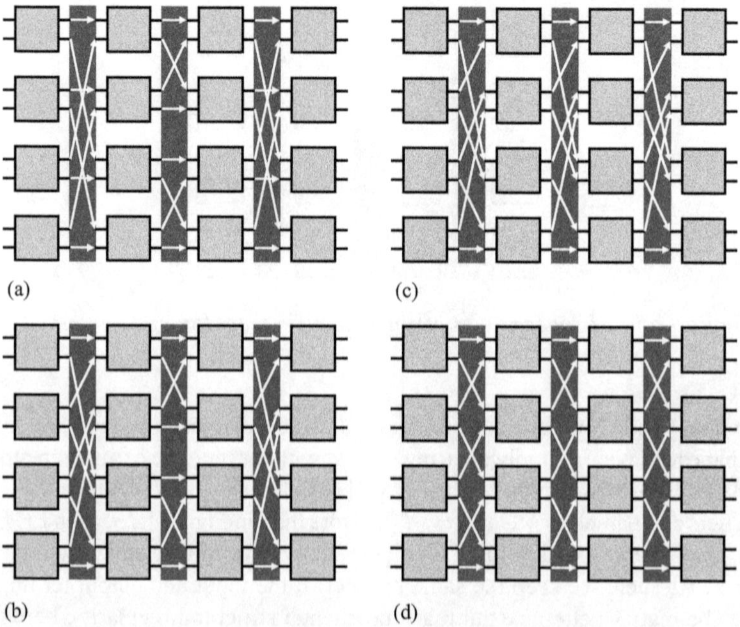

Figure 12.6 Matrix of 16 reconfigurable gates with fixed interconnect topology—(a) banyan, (b) baseline, (c) flip and (d) modified omega

3. Flip [Figure 12.6(c)] and
4. Modified omega [Figure 12.6(d)], where the modifications to standard omega maximize the shuffling between the layers.

Since the interconnect topology is fixed and static, the choice of topology is made by considering the application context. In the context of matrix-based logic, the performance of the different topologies has been investigated in [25]. It has been shown that a modified omega topology has the best performance in terms of mapping efficiency. Therefore, in the rest of this chapter, we consider only modified omega interconnect topology.

12.2.3 Integration into FPGA architecture

As MClusters contain rich reconfigurable LCs, they can potentially perform a large set of combinational logic functions, being a competitive candidates for substituting the original LUTs. Figure 12.7 shows the organization of the logic blocks at the CLB level. The CLB includes N MCluster-based BLEs, each of which consists of an *MClusters_d_w*, whose outputs can be individually latched on demand. Similar to conventional CLB organization, all the BLE outputs, i.e., $w \times N$ signals, are connected to the local routing resources. Each BLE input can be reached by every BLE outputs and every CLB inputs, thus achieving a full connectivity pattern.

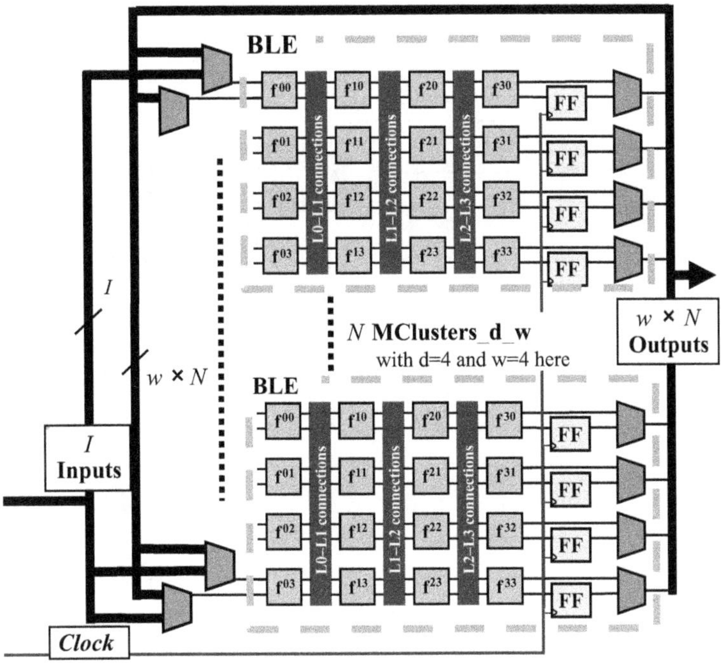

Figure 12.7 MCluster-based CLB proposal

At the top architectural level, we consider a standard FPGA organization shown in Figure 12.1, where the CLBs are interconnected by a global routing architecture.

12.3 MCluster CAD flow

To analyze the performance gain of the proposed FPGAs from an application perspective and to compare it to conventional counterparts, the benchmarking of standard circuits has to be carried out. The conventional FPGA computer-aided design (CAD) flow is developed for LUT-based FPGA architecture. To unlock the potential of MCluster, in this section, we develop novel CAD tool MPack supporting the MCluster-based FPGA architectures. And MPack is integrated into a complete benchmarking tool flow based on the well-known *verilog-to-routing* (VTR) flow [26].

In the rest of this section, we first give an overview of the developed benchmarking flow strategy in Section 12.3.1. Then, we introduce the principles of a novel CAD tool, called MPack, which is specifically designed to handle the MCluster packing step with fixed interconnect topologies (see Section 12.3.2). We detailed the mapping algorithms and clustering algorithms of MPack in Sections 12.3.3 and 12.3.4, respectively.

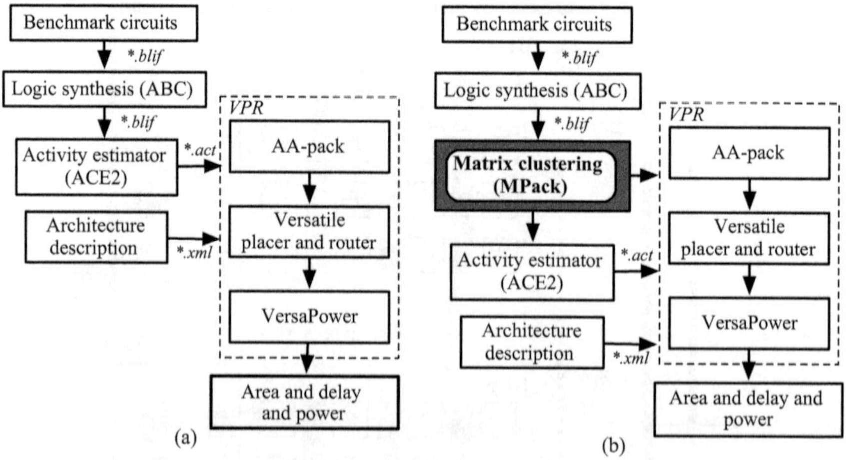

Figure 12.8 Comparison between (a) VTR flow for conventional FPGA architectures and (b) proposed MCluster-compatible benchmarking flow

12.3.1 General overview of the flow

For conventional FPGAs, architecture-level evaluations are conducted with a traditional FPGA tool flow, i.e., VTR flow [26], as illustrated in Figure 12.8(a). A set of logic circuits, called benchmarks, are first optimized at logic level using the ABC synthesis tool [27]. Subsequently, the logic packing of the synthesized circuit into CLBs is performed, using AA-PACK [28]. Finally, the placement and routing are carried out using *versatile place and route* (VPR) [15]. Power results are estimated by VersaPower [29] with the net activities computed by ACE2 [30].

However, the MCluster architecture introduces new logic circuit different from LUTs, which is not handled by traditional FPGA tools. In particular, the layered structure with various interconnect topology is not supported by traditional packing tools, such as AA-PACK [28]. Therefore, a specific tool, called MPack, is developed, and a novel design flow is proposed to leverage the potential of MPack. As shown in Figure 12.8(b), MPack maps the logic functions into LCs and MClustering, by considering the specific interconnect topologies as well as the associated buffering generation required by the layered structure. Details of the tool are further described in Sections 12.3.2–12.3.4.

12.3.2 MPack: the matrix packer

The matrix packer, MPack, is developed to pack netlists of LCs into MClusters, whose internal organization in shown in Figure 12.9. MPack consists of two principal algorithms: the mapper and the clusterer. The mapper maps simple logic functions on a unique MCluster. The clusterer splits the input logic network into subgraphs that can be subsequently mapped on individual MClusters thanks to the mapper. For a seamless

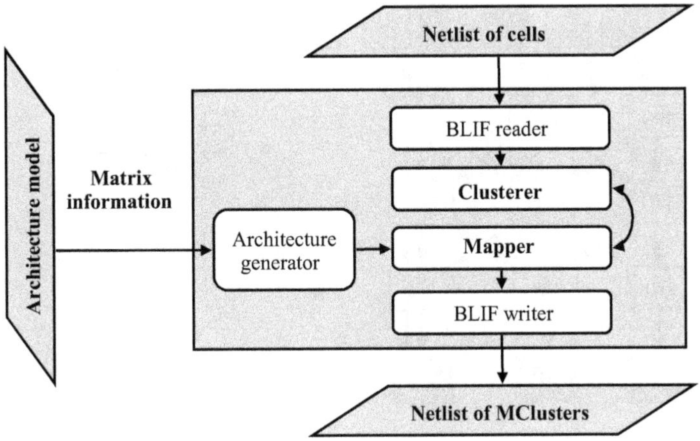

Figure 12.9 MPack model flow

integration with VTR flow, the input and output logic network are represented in the *Berkeley logic interchange format* (BLIF) file format.

MPack supports customized MCluster structures. A built-in Architecture Generator can automatically generate the MCluster template, with user-specified depth, width and interconnecting topology scheme. Figure 12.10(a) shows an example of a MCluster featured by $d = w = 3$ and modified omega interconnecting topology. In Figure 12.10(b), the nodes f^{ij} are uniquely addressed by the indices i and j, where i indicates the row while j denotes the column. MPack employs the adjacency matrix of the graph to represent the connectivity between nodes (LCs), where a 1 at the position (i,j) means that the node i is connected to the node j. The local adjacency matrices X_{nm} between the LC stages at depth n to m are defined as the individual cross-connectivity matrices. For instance, Figure 12.10(b) shows two examples: X_{01} and X_{12}, corresponding to the MCluster shown in Figure 12.10(a). In the example, the first row of the matrix X_{01} shows the connectivity of the cell f_{10}, while f_{10} is connected to f_{00} and f_{01}.

12.3.3 Matrix mapping algorithm

The mapping algorithm of MPack aims at fitting a netlist of LCs onto an MClusters architecture. The module is split into two parts: (1) an architecture optimization is performed to adapt the netlist to the physical architectural scheme (Section 12.3.3.1) and (2) the logic function netlist is mapped to the architecture netlist (Section 12.3.3.2).

12.3.3.1 Architectural optimization

The architectural optimization is a preprocessing stage, which aims at adapting the logic netlist to the physical architectural scheme. The layered structure forces the logic to be pipelined. The input netlist requires a processing where its connections is adapted to conform to the physical topology. For instance, a typical post-synthesis netlist

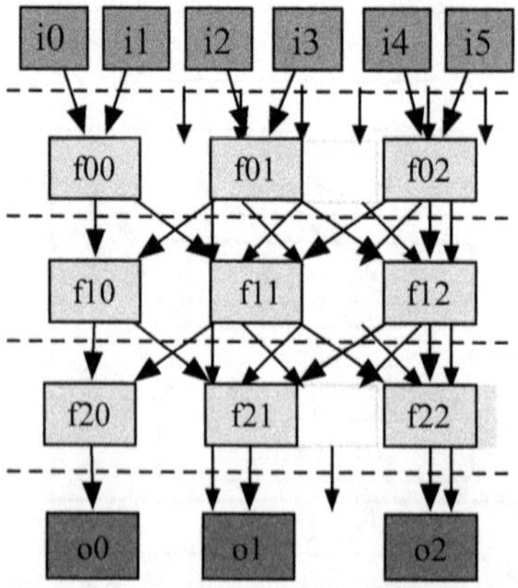

(a)

$$X_{01}=\begin{bmatrix}1&1&0\\1&0&1\\0&1&1\end{bmatrix}\Rightarrow\begin{bmatrix}f^{10}\\f^{11}\\f^{12}\end{bmatrix}=X_{01}\cdot\begin{bmatrix}f^{00}\\f^{01}\\f^{02}\end{bmatrix}$$

$$\Rightarrow\begin{matrix}f^{00}\rightarrow f^{10},\ f^{00}\rightarrow f^{11}\\f^{01}\rightarrow f^{10},\ f^{01}\rightarrow f^{12}\\f^{02}\rightarrow f^{11},\ f^{00}\rightarrow f^{12}\end{matrix}$$

$$X_{12}=\begin{bmatrix}1&1&0\\1&0&1\\0&1&1\end{bmatrix}\Rightarrow\begin{bmatrix}f^{20}\\f^{21}\\f^{22}\end{bmatrix}=X_{12}\cdot\begin{bmatrix}f^{10}\\f^{11}\\f^{12}\end{bmatrix}$$

$$\Rightarrow\begin{matrix}f^{10}\rightarrow f^{20},\ f^{10}\rightarrow f^{21}\\f^{11}\rightarrow f^{20},\ f^{11}\rightarrow f^{22}\\f^{12}\rightarrow f^{21},\ f^{10}\rightarrow f^{22}\end{matrix}$$

(b)

Figure 12.10 (a) MClusters_3_3 with a modified omega topology (logic cells are labeled by "f", virtual input nodes are labeled by "i", virtual output nodes are labeled by "o") and (b) associated cross-connectivity matrices

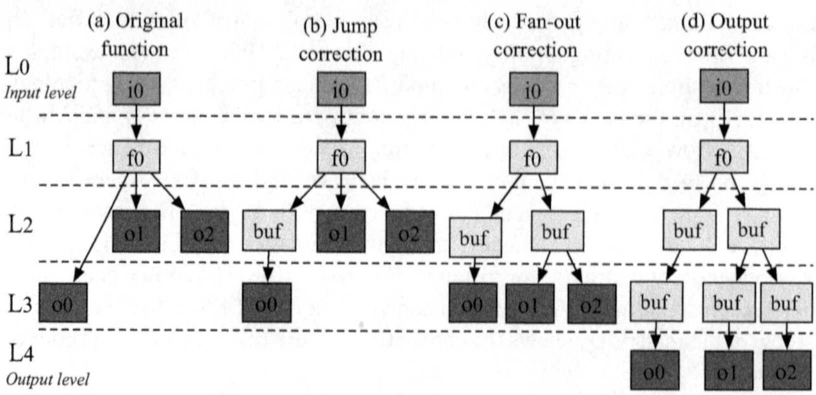

Figure 12.11 Architectural optimization of the function graph. (a) Original function before formatting. (b) Jump correction. (c) Fan-in–fan-out correction. (d) Output layer correction by buffer insertion

allows any logic gate output to be connected to several gate inputs, while MClusters have a fixed fan-in and fan-out. Therefore, to be compatible with MClusters, buffering nodes are added to the netlist, in order to limit the fan-out. Figure 12.11 provides an optimization example, where we see that the netlist is processed in three steps:

1. **Jump correction:** Since the layered topology, a connection prohibits jumps at least one logic layer, a connection between two layers must pass through an LC.

Consequently, instead of a jump, it is necessary to create a path by the addition of a buffer cell, in order to handle such a situation, as shown in Figure 12.11(a).

2. **Multiple output correction:** In the MClusters organization, the LCs are connected to a limited number of other cells due to a fan-in/fan-out limitation. Larger fan-in/fan-out requires duplicating the signals using buffers. Figure 12.11(a) shows an example case, where node f_0 drives three different nodes. Since each LC has only two outputs, a buffer is added to satisfy the constraint [Figure 12.11(b)].
3. **Output layer correction:** In this final optimization step, all the outputs of the logic functions are placed on the physical output layer. In Figure 12.11(b), we considered a logic function where the outputs are on layer $L3$. To meet the constraints of the MCluster architecture (here a MCluster with a depth of three and a width of three), buffers are added to place the outputs on the layer $L4$ [Figure 12.11(c)].

12.3.3.2 Mapping algorithm

After the architecture optimization steps, the logic function netlist is mapped to the MCluster structure, using the recursive algorithm reported in Algorithm 1. The mapping algorithm consists of two steps:

1. In the first step, the adapted netlist is analyzed in depth, where child branches are identified and recursively explored for each node in the MCluster structure. During this depth exploration, any edge orientation in the graph is not considered. As such, the connections between the nodes are identified and the mapping sequence can be initialized.
2. In the second step, logic nodes are assigned to ultrafine grain LCs, with respects to the physical interconnections between layers. The connectivity matrices of each layer are compared with the relevant interlayer architectural connectivity allowing (or not) the assignment of functions to cells.

Branching, i.e., the exploration of the immediately preceding alternative, is used when the arbitrary choice leads to a dead-end. The process is iterated until all functional nodes are assigned to physical cells.

For instance, let us consider a matrix with a depth of three and a width of three (*MClusters_3_3*) using a modified omega interconnect topology [Figure 12.10(a)]. Figure 12.12(a) shows a simple function to be mapped on a MCluster. With architecture optimization methodology explained in Section 12.3.3.1, the original function graph is modified to maximize the matching with the targeted MCluster. The architecture optimization engine will correct the position of the inputs/outputs and interlayer jumps, resulting in the graph as shown in Figure 12.12(b). The graph exploration is then triggered and the following sequence is obtained:

$$\begin{aligned} \text{pi0} \;\; &-\text{n1} - \text{pi1} - \text{buf} - \text{n4} - \text{n3} - \text{n2} - \text{pi2} - \text{pi3} \\ &-\text{buf} - \text{n5} - \text{po1} - \text{po0} \end{aligned} \tag{12.1}$$

where buf represents synchronization nodes. Subsequently, the graph assignment is performed.

Take the example shown in Figure 12.12(b), the first point n_1 is assigned to the ultrafine grain LC f^{00}. According to the path defined in the previous step, a buffer is

Algorithm 1: MPack mapping algorithm (pseudo-code)

```
Function Mapping_procedure(Current node):
    if (all children/parents of current node placed) then
        for (empty cell connected to children or parents) do
            Store the potential positions;
        end
    else
        Store all the positions on the current level
    end
    if (No positions have been found) then
        MAPPING FAILURE;
    end
    for (Each potential position) do
        Place the current node at the position;
        Mapping_procedure(Next node);
        if (Return is MAPPING FAILURE) then
            Unplace the current node from the position;
        end
        if (Return is MAPPING SUCCESS AND No more nodes) then
            MAPPING SUCCESS;
        end
    end
    if (No potential positions have been correct) then
        MAPPING FAILURE;
    end
end
```

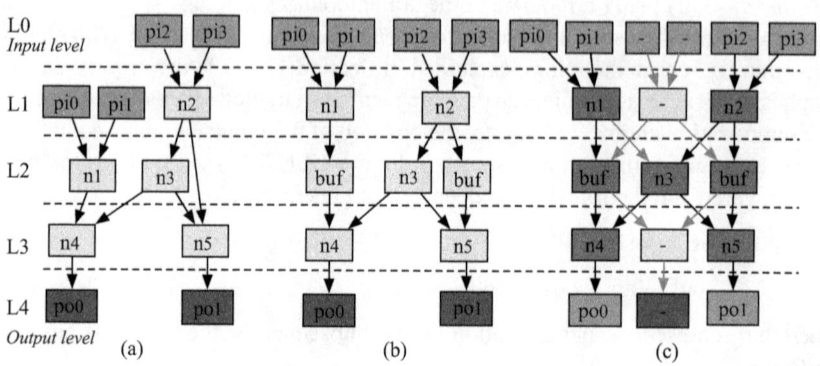

Figure 12.12 Mapping example. (a) Original function graph to map onto an MClusters_3_3 with a modified omega interconnect topology. (b) Function graph after correction step. (c) Mapped function

the next node to assign to a ultrafine grain LC in the matrix. Since f^{00} is physically connected to f^{10} and f^{12}, the LC with lower index (here f^{10}) is arbitrarily chosen for the buffer assignment, and the other possibility is memorized. In our example, the final programmed matrix is shown in Figure 12.12(c), where the nodes of the logic function graph and the nodes added for synchronization purposes are correctly placed on the cell matrix.

12.3.4 Clustering algorithm

The clustering algorithm is the hardcore of the cluster shown in Figure 12.9. It aims at splitting the initial logic network into subgraphs that will subsequently be mapped onto MClusters. Such functionality has been widely used in many CAD tools. The MPack tool borrows the clustering algorithms from VPACK [15]. Each MCluster is constructed sequentially, and the ultrafine grain LCs are greedily packed into MClusters. The pseudo-code for the algorithm is shown in Algorithm 2.

It begins by choosing a seed LC for the current MCluster. The clustering algorithm selects the unclustered LC with the most used inputs as the seed, which has been proved as an efficient way [15]. This approach prioritizes early placement of cells using the largest number of cluster inputs, which are a scarce resource. Next, the algorithm selects the LC with the highest attraction to the current MCluster and checks if it could legally be added to the current cluster. The mapping algorithm described in

Algorithm 2: Clustering algorithm (pseudo-code)

```
Remain_Cells: set of unclustered logic cells.
C: set of cells packed in the current MCluster.
MClusters: set of MClusters.
while (Remain_Cells != NULL) do
    /* More Cells to cluster                          */
    C = Seed(Remain_Cells);
    /* Most Used Inputs and legal Cell                */
    while (|C| < (MCluster_w.MCluster_d)) do
        /* Cluster is not full                        */
        Best_Cell = MaxAttractionLegalCell(C,Remain_Cells);
        if (Best_Cell == NULL) then
            BREAK;
        end
        Remain_Cells = Remain_Cells - Best_Cell;
        C = C ⋃ Best_Cell;
    end
    MClusters = MClusters ⋃ C;
end
```

Section 12.3.3.2 evaluates the legality of adding new LCs. If the cluster is mappable, then the LC is definitively added to the current cluster and the mapping result is recorded. Similar to [15], we define the attraction between an LC and the current MCluster (MC) as a function related to the number of the inputs and outputs they have in common:

$$Attraction(LC) = |Nets(LC) \cap Nets(MC)|. \tag{12.2}$$

This greedy procedure continues until either

1. the MCluster is full or
2. adding any additional cell would make the current cluster illegal.

If the current cluster is full, a new seed is selected and the packing starts for another MCluster.

12.4 Experimental results

In this section, we perform architecture evaluation for the novel FPGA architecture based on MClusters introduced in Section 12.2 with the novel benchmarking flow presented in Section 12.3. We first introduce our methodology of comparing the performance of FPGA architecture in Section 12.4.1. Then, we perform an architectural exploration to identify the best MClusters sizing in Section 12.4.2. With best MCluster sizing, we compare the novel FPGA architecture to its traditional CMOS-based counterpart in Section 12.4.3.

12.4.1 Methodology

In this chapter, our major objective is to study the impact on replacing standard CMOS LUTs by SiNWFET-based MClusters in FPGA architecture. We evaluate the performances of a set of logic circuits taken from the *Microelectronics Center of North Carolina* (MCNC) [31] and ISCAS'89 [32] mapped on simple homogeneous FPGA architectures. Our baseline FPGA architecture is a fully optimized homogeneous FPGA architecture, as shown in Figure 12.1. The CMOS reference architecture is composed of four-input non-fractionable LUTs with a CLB organization of ten BLEs and 22 external inputs ($K = 4, N = 10, I = 22$). The CLB architecture is optimal for homogeneous CMOS-based FPGAs [33]. We then proceed to a one-to-one replacement of the LUTs by MClusters according to Figure 12.7, i.e., keeping $N = 10$. The size of MCluster under evaluation is swept from *MClusters_1_1* to *MClusters_4_4*. To ensure a proper routability of the logic cluster, the number of BLEs inputs and outputs are determined by considering the size of the MCluster. The number of CLBs external inputs is set to

$$I = \frac{(N + 1) \times nb\_MCluster\_inputs}{2} \tag{12.3}$$

This relation typically ensures that 98% BLEs are utilized on average [33]. For example, an *MClusters_2_2*-based CLB has 22 external inputs ($I = 22$), while an

MClusters_4_4 has 44 external inputs ($I = 44$). For these two examples, the local routing circuits consist of a set of 42- and 84-b multiplexers, respectively. SRAM circuits are used to store the configuration information for both LUTs and MClusters.

For CMOS LUT-based FPGA architecture, we evaluate by using the benchmarking flow described shown in Figure 12.8(a). For SiNWFET-based MCluster FPGA architecture, we consider the benchmarking flow described shown in Figure 12.8(b). The physical parameters of the two FPGA architectures are extracted for a 22-nm technological node [1]. MClusters and LUTs area evaluation estimates the size of the blocks as a function of the elementary transistors area [1,14] and the transistor sizing. For instance, a 2-by-2 MCluster costs an area of 2.22 μm^2, while a four-input LUT costs 5.45 μm^2. To extract the basic average delay and power consumption of LUTs and MClusters, electrical analysis are conducted by using HSPICE simulations [34]. For CMOS circuits, the SPICE netlists are built with FinFET high performance at 22-nm model took from [35]. For SiNWFET-based circuit, the SPICE netlists are constructed with a simple table-based model for DGSiNWFETs took from [14]).

The architectural evaluation considers four metrics: (1) the area, (2) the critical path delay, (3) the dynamic power consumption and (4) the leakage power. These metrics are computed during the place and route iterations of the flow. The area corresponds to the sum of the logic area, i.e., the area of used CLBs, and the routing area, i.e., the area of the used routing resources. The critical path delay corresponds to the most constrained delay through the implemented structures. Finally, the power consumptions include both the contribution of logic blocks and the contribution of routing structures. All the metrics are normalized with respect to the most constrained CMOS design.

12.4.2 Impact of the granularity

We study the impact of MCluster sizes on the area, the delay, and the power consumptions averaged over the considered benchmark circuits. In this part, we sweep the dimensions of squared-shaped MClusters from *MClusters_1_1* to *MClusters_4_4*. Note that averaged over the different benchmarks and different cluster sizes, the utilization rate of the MClusters, i.e., the ratio between the employed LCs and the total number of LCs, stays at 90%, thereby indicating a good level of performances of MPack.

Figure 12.13(a) presents the area estimation of an FPGA based on MClusters. Small MCluster sizes, i.e., *MClusters_1_1* and *MClusters_2_2*, lead to best area results. Note that, the area increases drastically when a higher granularity (*MClusters_3_3* and *MClusters_4_4*) is applied. For large MClusters, the area overhead comes from the incomplete interconnectivity in their internal routing. In these cases, a large number of LCs are used only for buffering purposes, leading to a low mapping efficiency.

Figure 12.13(b) and (d) presents the critical path delay, the dynamic power consumption, and the leakage power consumption, respectively, under different MCluster sizes. Similar to the area results, the MClusters with small granularities, i.e., *MClusters_1_1* and *MClusters_2_2*, guarantee best delay and power consumption. Note that,

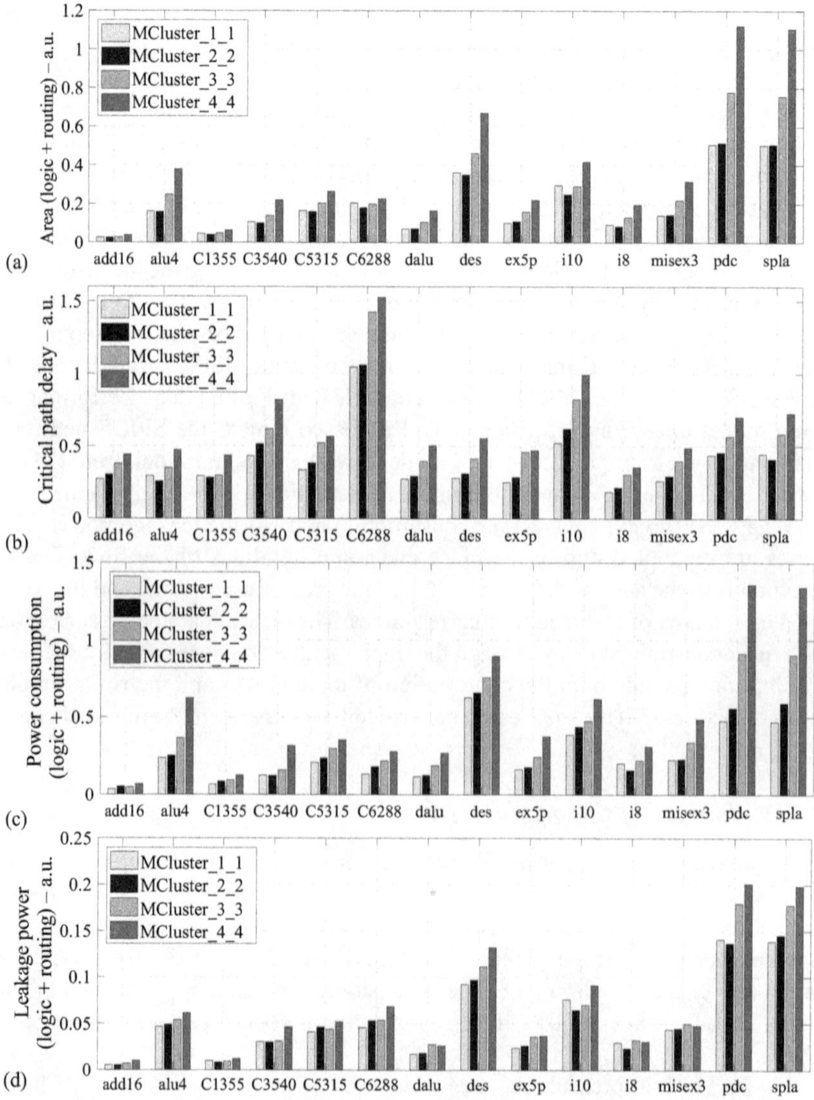

Figure 12.13 Impact of various granularities on MCluster-based FPGAs (a) global area, (b) critical path delay, (c) dynamic power consumption, (d) leakage power consumption. MClusters are sized as squares ranging from 1 to 4 reconfigurable cells

the delay and the power consumption are almost linear to the size of the MClusters. As large MClusters contain deeper logic pipeline, many LCs are used as buffers in order to fit the interlayer interconnecting topology. Therefore, the mapping inefficiency directly reduces the performances and increases the power consumption.

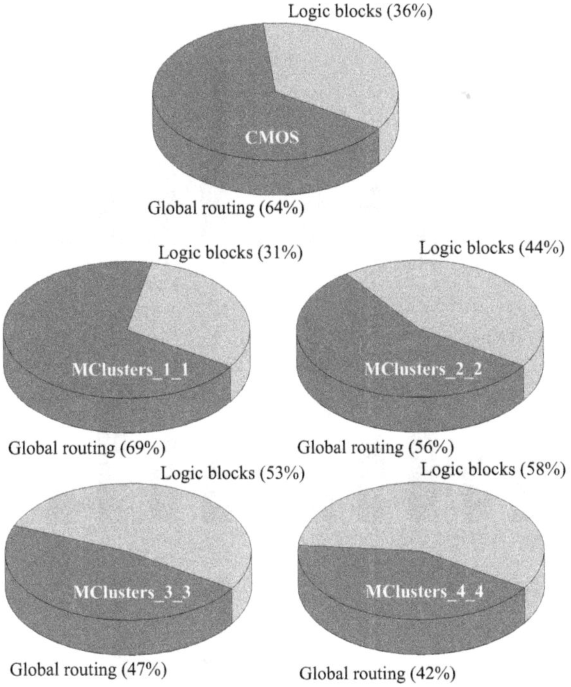

Figure 12.14 *Area breakdown of an interconnection intensive benchmark (i10) for 2-by-2 MCluster-based and CMOS-based FPGAs*

In terms of performance levels, *MClusters_1_1* overperforms slightly *MClusters_2_2* with 5%, 4%, 12%, and 1% gains in the area, the delay, the dynamic power and the leakage power, respectively. However, when determining the best MCluster size, we also consider more factors, i.e., the impact of the different MCluster granularities on the area breakdown between the logic blocks and global routing. Figure 12.14 shows the area breakdown for an interconnection intensive benchmark (i10). As discussed in Section 12.1.1, a fundamental drawback of the traditional FPGA architecture is its strong imbalance between the routing resources and the LCs. As shown in Figure 12.14, the use of *MClusters_1_1* worsens the natural imbalance. This violates our initial hypothesis as explained in Section 12.1, where the direct transposition of an FPGA architecture to ultrafine grain would help alleviate interconnection overhead. On the other side, large clusters can significant increase the logic share, but they will also strongly limit the performances of the architecture, as shown Figure 12.13.

Overall, *MClusters_2_2* provides a best trade-off between performance (Figure 12.13) and logic/routing balance (Figure 12.14). In the rest of this chapter, we consider *MClusters_2_2* as the ultrafine grain FPGA architecture when comparing to CMOS FPGAs.

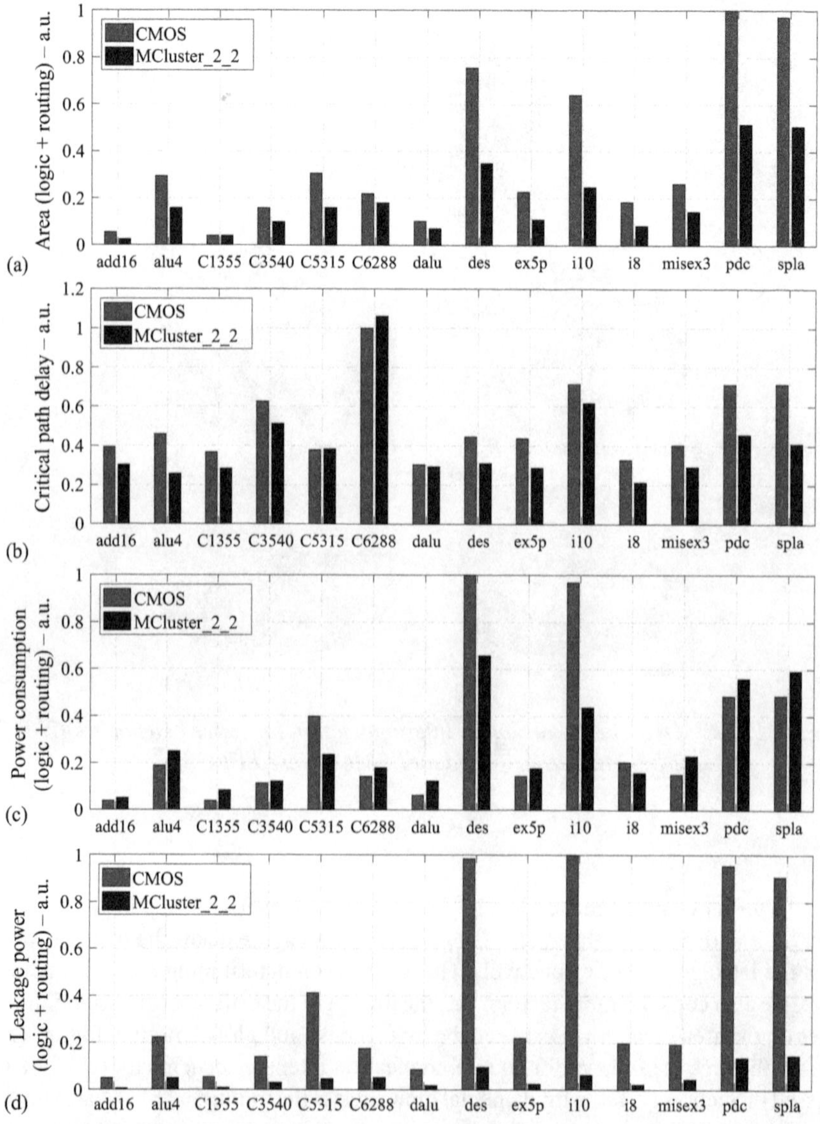

Figure 12.15 Performance comparisons between CMOS-based and MClusters_2_2-based FPGAs (a) global area, (b) critical path delay, (c) dynamic power consumption and (d) leakage power consumption

12.4.3 Performance comparison with CMOS

Figure 12.15 compares the area, delay and power consumption of *MClusters_2_2*-based FPGA and its CMOS counterpart.

In terms of area, Figure 12.15(a) shows that *MClusters_2_2*-based FPGA brings a reduction of up to 61%, with an average of 43%. The area benefits can be accounted: (1) to the performance of a logic-gate-based computation (as compared to the LUT approach) and (2) to the low area impact of ultrafine grain LCs, compared with the rather larger area required by a CMOS LUT. As mentioned in Section 12.4.1, a 2-by-2 MCluster is 2× smaller than a four-input LUT. Furthermore, in terms of implementing multi-output functions, *MClusters_2_2* is potentially more capable than its equivalent LUT. A four-input LUT can realize any function with four inputs and a single output, but two LUTs are required for a two-output function. While *MClusters_2_2* can only cover a subset of the single-output functions reachable by a LUT, they can produce four-input and two-output functions, beyond the capability of a LUT. Hence, MClusters can be regarded as outputting 2× more results for roughly 2× less area. Correlated to the efficiency of MPack, this demonstrates a clear advantage of our proposed FPGA architecture as compared with the CMOS approach.

Similar conclusion can be drawn for the delay [Figure 12.15(b)], the dynamic power consumption [Figure 12.15(c)] and the leakage power consumption [Figure 12.15(d)]. MCluster-based FPGAs can improve the critical path delay by up to 44% with an average value of 23%, with regard to the standard FPGA counterpart. The performance improvement comes from the higher performance of the MCluster structure as compared with LUTs. A dynamic power consumption reduction of up to 54% is observed, though, on average, we observe an increase of 19% in the dissipated power. Nevertheless, we emphasis that large dynamic power consumptions reductions are achieved on the larger benchmarks, indicating a promising trend toward the power reduction of larger circuits. Finally, leakage power can be reduced by up to 94% with an average reduction of 83%. This large reduction is due to the ultralow power capabilities of the controllable-polarity transistors, coming from the GAA structure, coupled to the resources reduction coming from the MCluster approach. Note that even if the granularity chosen in the previous section is not optimal for delay and power numbers, the improvements as compared with traditional FPGAs are significant and can be further pushed by choosing an appropriate granularity.

In addition to the performance improvement, we extend the study to the critical net routing delay. This metric strongly relates to the FPGA architecture efficiency as well. Figure 12.16 analyzes the distribution of the critical routing delay. In addition to improving the average routing delay by 43%, the use of MClusters allows us to reduce the standard deviation of the delay distribution by 45%. We can remark that while the CMOS distribution is quite large, the use of a *MClusters_2_2* implies a lowering of the extremes and tends to globally improve the performance of the mapped circuits. This behavior can be explained through the global impact of ultrafine granularity on the benchmarking tool flow. The ultrafine granularity induces a predominance of local inter-CLB interconnect instead of long wire connections, thus leading to the reduction of long critical paths. Such results are interesting from an architectural perspective as it tends to homogenize the results of mapped circuits and makes them less dependent to the route phase.

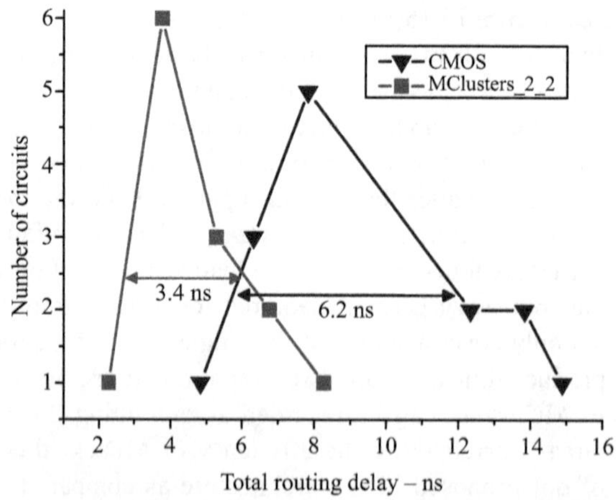

Figure 12.16 Critical net routing delay repartition for CMOS-based and 2-by-2 MCluster-based FPGAs

12.5 Conclusion

This chapter presents a new architectural scheme for FPGAs, which is based on the use of controllable-polarity transistors, i.e., SiNWFETs. Thanks to the device-level reconfigurability, ultrafine grain reconfigurable logic gates can be exploited at both circuit-level and architecture-level. We propose to replace LUTs in FPGAs by MClusters, where the ultrafine grain reconfigurable logic gates are organized in matrices and interconnected through a fixed interconnection pattern. To evaluate this approach objectively, we developed a novel tool MPack and its associated complete benchmarking flow, which is capable of packing logic circuits onto the considered architecture. We evaluated the potential of the MCluster-based architecture and compared it to a well-optimized homogeneous CMOS FPGA. We showed that 2-by-2 MClusters give an average improvement of 43% and 23% area and delay, respectively, with respect to CMOS LUT-based FPGAs at 22-nm technology node. Finally, we observed that the proposed approach opens a promising path to correct two intrinsic limitations of the FPGA circuits, with a reduction in the logic/routing imbalance, and a better control in the distribution of the routing delay.

References

[1] Auth C, Allen C, Blattner A, *et al.* A 22 nm High Performance and Low-Power CMOS Technology Featuring Fully-Depleted Tri-Gate Transistors, Self-Aligned Contacts and High Density MIM Capacitors. In: 2012 Symposium on VLSI Technology (VLSIT); 2012. p. 131–132.

[2] Intel's 10 nm Technology: Delivering the Highest Logic Transistor Density in the Industry Through the Use of Hyper Scaling. 2017. Available from:

https://newsroom.intel.com/newsroom/wp-content/uploads/sites/11/2017/09/10-nm-icf-fact-sheet.pdf.
[3] Natarajan S, Agostinelli M, Akbar S, *et al.* A 14 nm Logic Technology Featuring 2nd-Generation FinFET, Air-Gapped Interconnects, Self-Aligned Double Patterning and a 0.0588 μm^2 SRAM Cell Size. In: 2014 IEEE International Electron Devices Meeting; 2014. p. 3.7.1–3.7.3.
[4] Wong HSP, Mitra S, Akinwande D, *et al.* Carbon Nanotube Electronics – Materials, Devices, Circuits, Design, Modeling, and Performance Projection. In: 2011 International Electron Devices Meeting; 2011. p. 23.1.1–23.1.4.
[5] Suk SD, Yeo KH, Cho KH, *et al.* High-Performance Twin Silicon Nanowire MOSFET (TSNWFET) on Bulk Si Wafer. IEEE Transactions on Nanotechnology. 2008;7(2):181–184.
[6] Bangsaruntip S, Cohen GM, Majumdar A, *et al.* High Performance and Highly Uniform Gate-All-Around Silicon Nanowire MOSFETs with Wire Size Dependent Scaling. In: 2009 IEEE International Electron Devices Meeting (IEDM); 2009. p. 1–4.
[7] O'Connor I, Liu J, Gaffiot F, *et al.* CNTFET Modeling and Reconfigurable Logic-Circuit Design. IEEE Transactions on Circuits and Systems I: Regular Papers. 2007;54(11):2365–2379.
[8] Gaillardon PE, Amarù LG, Bobba S, *et al.* Nanowire Systems: Technology and Design. Philosophical Transactions of the Royal Society of London A: Mathematical, Physical and Engineering Sciences. 2014;372(2012). Available from: http://rsta.royalsocietypublishing.org/content/372/2012/20130102.
[9] Jamaa MHB, Gaillardon PE, Frégonèse S, *et al.* FPGA Design with Double-Gate Carbon Nanotube Transistors. ECS Transactions. 2011;34(1):1005–1010.
[10] Appenzeller J, Knoch J, Tutuc E, *et al.* Dual-Gate Silicon Nanowire Transistors with Nickel Silicide Contacts. In: 2006 International Electron Devices Meeting; 2006. p. 1–4.
[11] Heinzig A, Slesazeck S, Kreupl F, *et al.* Reconfigurable Silicon Nanowire Transistors. Nano Letters. 2011;12(1):119–124.
[12] Lin YM, Appenzeller J, Knoch J, Avouris P. High-Performance Carbon Nanotube Field-Effect Transistor with Tunable Polarities. IEEE Transactions on Nanotechnology. 2005;4(5):481–489.
[13] Harada N, Yagi K, Sato S, Yokoyama N. A Polarity-Controllable Graphene Inverter. Applied Physics Letters. 2010;96(1):012102.
[14] Marchi MD, Sacchetto D, Frache S, *et al.* Polarity Control in Double-Gate, Gate-All-Around Vertically Stacked Silicon Nanowire FETs. In: 2012 International Electron Devices Meeting; 2012. p. 8.4.1–8.4.4.
[15] Betz V, Rose J, Marquardt A, editors. Architecture and CAD for Deep-Submicron FPGAs. Norwell, MA, USA: Kluwer Academic Publishers; 1999.
[16] Lin M, Gamal AE, Lu YC, *et al.* Performance Benefits of Monolithically Stacked 3-D FPGA. IEEE Transactions on Computer-Aided Design of Integrated Circuits and Systems. 2007;26(2):216–229.

[17] Colli A, Pisana S, Fasoli A, *et al.* Electronic Transport in Ambipolar Silicon Nanowires. Physica Status Solidi (B). 2007;244(11):4161–4164. Available from: http://dx.doi.org/10.1002/pssb.200776154.

[18] Martel R, Derycke V, Lavoie C, *et al.* Ambipolar Electrical Transport in Semiconducting Single-Wall Carbon Nanotubes. Physical Review Letters. 2001;87(25):256805.

[19] Geim AK, Novoselov KS. The Rise of Graphene. Nature Materials. 2007;6(3):183–191.

[20] Rabaey, JM, Anantha PC, Nikolic B. Digital Integrated Circuits. Vol. 2. Englewood Cliffs, NJ: Prentice Hall, 2002.

[21] Lewis DL, Yalamanchili S, Lee HHS. High Performance Non-Blocking Switch Design in 3D Die-stacking Technology. In: VLSI, 2009. ISVLSI'09. IEEE Computer Society Annual Symposium on. IEEE; 2009. p. 25–30.

[22] Marrakchi Z, Mrabet H, Masson C, *et al.* Efficient Mesh of Tree Interconnect for FPGA Architecture. In: Field-Programmable Technology, 2007. ICFPT 2007. International Conference on. IEEE; 2007. p. 269–272.

[23] Wu CL, Feng TY. On a Class of Multistage Interconnection Networks. IEEE Transactions on Computers. 1980;100(8):694–702.

[24] Adams GBI, Agrawal DP, Siegel HJ. A Survey and Comparison of Fault-Tolerant Multistage Interconnection Networks. Computer. 1987;20(6): 14–27.

[25] Gaillardon PE, Clermidy F, O'Connor I, *et al.* Interconnection Scheme and Associated Mapping Method of Reconfigurable Cell Matrices based on Nanoscale Devices. In: Nanoscale Architectures, 2009. NANOARCH'09. IEEE/ACM International Symposium on. IEEE; 2009. p. 69–74.

[26] Rose J, Luu J, Yu CW, *et al.* The VTR Project: Architecture and CAD for FPGAs from Verilog to Routing. In: Proceedings of the ACM/SIGDA International Symposium on Field Programmable Gate Arrays. ACM; 2012. p. 77–86.

[27] Berkeley Logic Synthesis and Verification Group. ABC: A System for Sequential Synthesis and Verification; Release 20160717. Available from: https://people.eecs.berkeley.edu/\ ~alanmi/abc/.

[28] Luu J, Anderson JH, Rose JS. Architecture Description and Packing for Logic Blocks with Hierarchy, Modes and Complex Interconnect. In: Proceedings of the 19th ACM/SIGDA International Symposium on Field Programmable Gate Arrays. FPGA '11. New York, NY, USA: ACM; 2011. p. 227–236. Available from: http://doi.acm.org/10.1145/1950413.1950457.

[29] Goeders JB, Wilton SJE. VersaPower: Power Estimation for Diverse FPGA Architectures. In: 2012 International Conference on Field-Programmable Technology; 2012. p. 229–234.

[30] Lamoureux J, Wilton SJE. Activity Estimation for Field-Programmable Gate Arrays. In: 2006 International Conference on Field Programmable Logic and Applications; 2006. p. 1–8.

[31] Yang S. Logic Synthesis and Optimization Benchmarks User Guide: Version 3.0. Microelectronics Center of North Carolina (MCNC); 1991.

[32] Brglez F, Bryan D, Kozminski K. Combinational Profiles of Sequential Benchmark Circuits. In: IEEE International Symposium on Circuits and Systems (ISCAS). IEEE; 1989. p. 1929–1934.

[33] Ahmed E, The Effect of Logic Block Granularity on Deep-Submicron FPGA Performance and Density. MASc Thesis, University of Toronto. 2001;101.

[34] Synoposys Inc. HSPICE: The Gold Standard for Accurate Circuit Simulation; 2010. Available from: https://www.synopsys.com/content/dam/synopsys/verification/datasheets/hspice-ds.pdf.

[35] Sinha S, Yeric G, Chandra V, *et al.* Exploring Sub-20 nm FinFET Design with Predictive Technology Models. In: Proceedings of the 49th Annual Design Automation Conference. ACM; 2012. p. 283–288.

Chapter 13
Tunnel FET-based security primitive design

Yu Bi[1]*, Pierre-Emmanuel Gaillardon*[2]*, X. Sharon Hu*[3]*, Michael Niemier*[3]*, and Yier Jin*[4]

13.1 Introduction

The Internet of Things (IoT) is certain to impose new demands on modern cryptographic systems. As many IoT nodes and remote sensors are driven by batteries, power consumption is a critical design constraint. As a result, conventional encryption algorithms such as Advanced Encryption Standard (AES) [1] may not be suitable for these resource-constrained applications because of the high power and area associated with the hardware implementations. As such, the development of light-weight cryptographic algorithms has become a high priority, with various light-weight encryption algorithms being developed [2–5]. Meanwhile, the wide distribution of IoT devices gives attackers physical access to these devices, making side-channel attacks easier to apply. Therefore, countering side-channel attacks, such as differential power analysis, is an important design consideration even in these resource-constrained designs.

Ever since differential power analysis was first proposed by Kocher *et al.* [6], researchers have been working to develop solutions to counter DPA attacks. Countermeasures are generally classified into two categories: (i) hardware level solutions and (ii) algorithm level solutions. Kocher suggested that the cryptographic algorithm should be designed in a way that withstands a certain amount of information leakage [7]. For example, when using a hashing algorithm for generating new keys, the frequently changing keys will make it difficult for attackers to capture a sufficient amount of a power trace to mount a successful DPA attack. Another technique was presented by adding a transformed mask to S-box, where the masking methods can be applied to the nonlinear part of encryption algorithm [8]. After the substitution operation, the multiplicative mask is replaced with the original mask. Yang *et al.* proposed randomly varying voltage and frequency to prevent side-channel attacks at the gate level [9]. Therefore, the time and power consumption of the intermediate operations are more random, which minimizes the leakage of information through the side

[1]AMD, USA
[2]Department of Electrical and Computer Engineering, University of Utah, USA
[3]Department of Computer Science and Engineering, University of Notre Dame, USA
[4]Department of Electrical and Computer Engineering, University of Florida, USA

channels. A more practical circuit-level method on preventing DPA attack leveraged a sense amplifier-based logic style for cryptographic algorithm implementations [10]. The strength of this approach is the constant power consumption of differential logic which can counter power-based attacks as operation power is independent of processed data. However, traditional circuit-level protection schemes such as current mode logic (CML) trade power efficiency for security. When considering the IoT applications, embedded system designers are presented with a dilemma in which they can choose either high security or low-power consumption.

Orthogonal to current approaches of circuit-level optimization, in this chapter, we consider how emerging transistor technologies could help mitigate risks of side channel attacks while maintaining low-power consumption. Emerging devices have been proven to have unique applications in the hardware security domain [11–13]. In this work, we further extend research in this direction to use emerging devices to preserve low-power consumption but achieve the goal of DPA resilience. More specifically, we will demonstrate that by implementing CML with emerging tunnel transistors (TFETs) for lightweight encryption algorithms, one can significantly improve the circuit security at a fraction of the power when compared to CMOS equivalents.

13.2 Background

In this section, we briefly introduce frequently employed lightweight ciphers as well as CML. Both form the underlying basis of this work.

13.2.1 Light-weight cipher

Conventional encryption algorithms such as AES may not be suitable for applications where power and area are strictly limited. For example, devices used in components for the IoT, wireless sensor networks (WSN), etc. consist of various low-power and low-cost nodes to support communication over wireless channels. Furthermore, many IoT devices and wireless sensor nodes support military applications where security is a critical request to be addressed. Understanding how to secure communication channels at a lower cost is an important question for the security community. Thus, the development of light-weight cryptographic algorithms has become an important research area [2–5]. Following are the two important design principles that guide lightweight cipher design:

1. **Security:** In general, light-weight ciphers are resistant to typical attacks such as algebraic attacks and brute force attacks.
2. **Implementation efficiency:** Light-weight ciphers are small in terms of area and power consumption, compared to hardware required for standard cryptography algorithms.

One main difference distinguishing the light-weight from the conventional block cipher is that the block size of light-weight ciphers is usually less than 64 b—when compared to at least 128 b for conventional block ciphers. As examples:

- The DESL and DESXL ciphers represent the first attempt at light-weight encryption [3]. Since the algorithm is a derivative of DES, the key idea is to replace the DES round function by only one S-box and to remove the initial and final permutation.
- The PRESENT cipher is one of the most popular light-weight ciphers proposed in 2007 [2]. It is executed in 31 rounds and composed of a 64-b length block and keys of 80 or 128 b. A simple substitution–permutation network is adopted for the round function.
- Beaulieu *et al.* proposed a new cryptographic algorithm—the SIMON block cipher—where the block size covers a wide range from 32 to 128 b as well as key sizes ranging from 64 to 256 b [4]. Different key schedules can lead to different rounds of encryption varying from 32 to 72 rounds. The SIMON cipher uses the conventional Feistel network with a further reduction of hardware cost.
- Aiming to further lower the gate equivalent (GE, a measure of hardware complexity) of the block cipher, the KATAN and KTANTAN ciphers were proposed in 2009 [5]. The design of the KATAN cipher is a block cipher based on stream cipher design, which iterates 254 rounds to output the ciphertexts. The 80-b keys can be applied for three different variants, 32, 48 and 64 b.

13.2.2 Current mode logic

CML represents a differential digital logic family [10,14–17]. A CML gate includes a tail current source, a current steering core and a differential load. The working mechanism of a CML gate is to switch the constant current through the differential network of input transistors, utilizing the reduced voltage swing on the two load devices as the output.

Although CML is not widely used in mainstream circuit design, its unique features, namely, low latency and stable power consumption, can be leveraged for specific applications, such as DPA countermeasures—i.e., serving as a countermeasure against a DPA attack.

To some extent, employing CML primitives may be more efficient than other gate-level DPA countermeasures, such as gate masking [18] and dynamic management of voltage and frequency [9,15]. For example:

- Badel *et al.* formalized the generic standard cells for differential logic styles, including layout characterization and library generation methodologies [16].
- In order to further reduce the power of CML gates, Cevrero *et al.* leveraged the power gating technique for the differential logic design, where a standard cell was generated, and DPA analysis was launched on the power-gated CML AES design [17]. The implementation achieved both goals of reduced power consumption and DPA resilience.
- Macé *et al.* raised a potential connection between binary decision diagram (BDD) and the CML for cryptography [19]. It is promising since BDD is applied to optimize the Boolean representation and CML gates tend to be energy-hungry. The exploited isomorphism between CML and BDD can help design very efficient differential logic gates that optimize the performance in terms of power and area.

- Furthermore, the authors of [20] proposed a subthreshold CMOS CML design to reduce the power consumption associated with conventional MOS CML. Not surprisingly, a large PMOS load device and low drive currents ($I_{SD} = 1$ nA) lead to a design with a large area and low speed. These drawbacks limit the application of block ciphers, especially lightweight block ciphers, where both area and speed are two critical criteria.

TFET-based CML gates have also recently been introduced [13,21]. Initial CML gate designs based on the newly developed GaSb–InAs heterojunction TFET, which has improved on-state current with hetero-band alignment. Two logic gates, a buffer and a multiplexer, were studied and evaluated [21]. TFET-based CML design exhibited lower power consumption when compared to CMOS equivalents. However, (i) only two TFET CML gates were presented, (ii) the authors of [21] did not discuss how to leverage TFET CML gates for circuit-level designs and (iii) TFET CML was not applied at all in the hardware security domain. In order to fully evaluate TFET-based logic—not only from the perspective of traditional metrics such as delay and power but also with respect to new metrics such as security—in this chapter, we will construct a TFET CML gate library using a systematic approach and will demonstrate its applications in the hardware security area.

13.3 Tunnel FET

In this section, we briefly discuss the underlying technology (TFET devices) and modeling assumptions which are used to build the TFET CML gate library in this chapter.

13.3.1 Device description

Different types of tunneling FETs (TFETs) have been developed and fabricated [24,25]. Among them, III–V TFETs appear more promising due to their higher conduction current. More specifically, InAs homo-junction TFETs [26] and GaSb–InAs heterojunction TFETs [27] have been the subject of much study. Considering that the InAs homo-junction is more mature among these two devices, we will employ it as our TFET transistor model. FinFET 20 nm technology is also adopted for comparison. The physical structures (used in Synopsys TCAD simulation) of both the homo-junction TFET and FinFET are depicted in Figure 13.1 [22,23].

It is apparent that TFETs have asymmetrical doping where source and drain are p- and n-type doping, respectively. A gate voltage can induce band-to-band tunneling at the channel to control the tunneling current. In contrast, in a conventional CMOS transistor, current conduction occurs via electron carriers with enough energy to surmount the channel thermal barrier. The Fermi–Dirac distribution limits the subthreshold slope (SS) to 60 mV/decade. However, the high energy carriers in TFETs can be filtered by the gate-voltage-controlled tunnel such that a sub-60 mV/decade subthreshold swing is achievable at the room temperature [24]. With improved steep slope and high on-current at a low supply voltage, TFETs could enable supply voltage

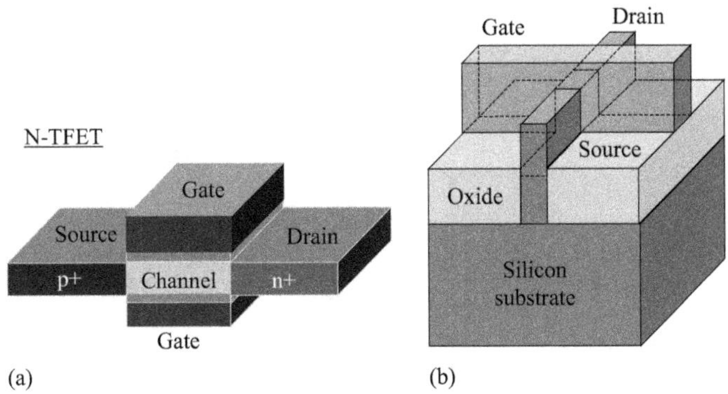

Figure 13.1 3-D physical structure of (a) a tunnel FET [22] vs. (b) a FinFET [23]

Table 13.1 InAs homo-junction TFET device parameters [26]

Gate length (L_G)	**20 nm**
Body thickness (T_ch)	5 nm
Dielectric thickness (HfO_2)	5 nm
Source doping (p+)	4×10^{19} cm^{-3}
Drain doping (n+)	6×10^{17} cm^{-3}
Si FinFET S/D doping	1×10^{20} cm^{-3}

scaling to further address challenges such as undesirable leakage currents, threshold voltage reduction, etc.

The device parameters assumed for the InAs homo-junction TFET (that we will employ in our circuit simulations) are listed in Table 13.1. A Si FinFET is also included as the baseline.

13.3.2 Device modeling

While a compact SPICE model has been recently developed for TFETs [28,29], in this work, we employ a look-up table based Verilog-A model derived from TCAD Sentaurus for our simulations as this model has been widely used and validated [21]. Figure 13.2(a) depicts the structure of the TFET Verilog-A model [30]. It is composed of three parts: gate-drain capacitance C_{GD}, gate-source capacitance C_{GS} and the transfer characteristics $I_{DS}(V_{GS}, V_{DS})$. The current models of different paths

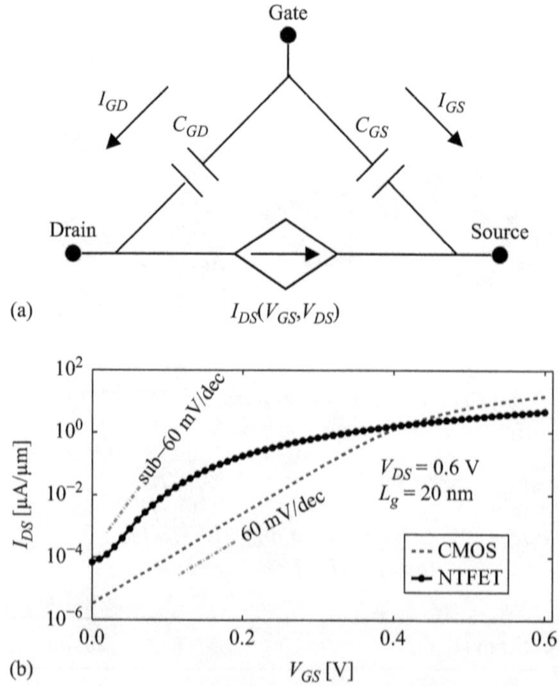

Figure 13.2 TFET device modeling: (a) TFET Verilog-A model (b) I_{DS} vs. V_{GS} [30]

are also listed in (13.1). The calculation of three current models refers to the look-up table that includes a range of fine-step voltage bias and capacitance.

$$\text{Look up table} = \begin{cases} I_{GD} = \dfrac{d}{dt}(C_{GD} \times V_{GD}) \\ I_{GS} = \dfrac{d}{dt}(C_{GS} \times V_{GS}) \\ I_{DS} \rightarrow (V_{GD}, V_{GS}) \end{cases} \tag{13.1}$$

By employing the TFET Verilog-A model, we evaluate the DC performance of an n-type TFET as shown in Figure 13.2(b), where the on-current I_{DS} varies with gate-source voltage V_{GS}. CMOS data are also included for comparison. Both CMOS and TFET devices assume 20 nm technology with V_{DS} = 0.6 V. A TFET's SS is improved when compared to CMOS. Notably, when the gate-source voltage is less than 0.4 V, the conducting current of TFETs outperforms the CMOS counterpart. (However, when $V_{GS} > 0.4$ V, the CMOS device exhibits a better on-current.) As a result, TFETs represent promising ultralow-power features that provide further V_{DD} scaling in integrated circuit designs.

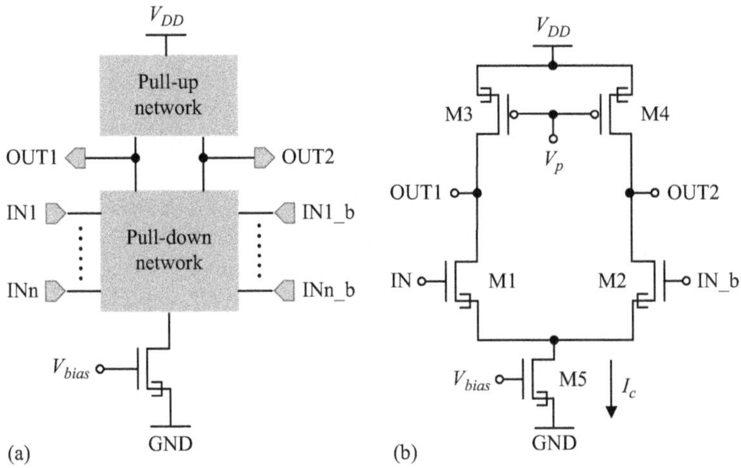

Figure 13.3 (a) The universal diagram of CML circuits and (b) schematic of the TFET-based CML inverter

13.4 Tunnel FET in hardware security

13.4.1 TFET-based current mode logic

One major difference between CML circuits and single-ended circuits is that the voltage swing of CML is smaller than that of static logic. Thus, differential logic styles were originally designed for high-speed communication. Due to invariant power consumption, researchers adopted this circuit-level method as a countermeasure against differential power analysis [15–17]. A "generic" TFET-based CML circuit is shown in Figure 13.3(a). The schematic is divided into two parts: a pull-up network and pull-down network.

For TFET CML, the pull-up network is constructed by either two resistors or two p-type TFETs (PTFETs). Since the consumption of power and area of the resistor are dramatically larger than a FET using modern technology, the FET-based pull-up network dominates. In CML, the pull-up network mainly works as the load device to manage the DC voltage drop on the output. By simply tuning the gate bias of a p-type FET, the on-resistance of PTFETs can be adjusted, thereby altering output voltage accordingly. At the bottom of Figure 13.3(a), one n-type FET (NTFET) is included to serve as a current source, which can determine the value of output voltage swing. On the other hand, the pull-down network that is composed of NTFETs mainly serves as the major functional unit in the CML circuit. The different logic functions can be achieved by distinct combinations of a group of NTFETs. Note that the inputs of the pull-down network are required to be differential pairs.

Figure 13.3(b) shows a schematic of a TFET-based current mode inverter/buffer. One pair of transistors is controlled by the differential inputs, IN and IN_b, respectively. The constant driving current is provided by the transistor M5, which is also tunable by the gate bias voltage V_{bias}. Together with M5, transistors M3 and M4 are employed to charge and discharge the output pair, OUT1 and OUT2. When IN is logic 1, M1 is turned on, and the constant current I_C flows through the left-handed path. Thus, OUT1 discharges to a certain value between VDD and GND, and OUT2 alternatively charges to quasi-VDD. Note that in the CML design, logic 0 is commonly defined as half VDD, and logic 1 is close to VDD. In this case, OUT1 voltage is less than logic 1, which is treated as logic 0. If OUT1 is extracted as the output pin and the inverted OUT2 is extracted as complementary output pin, the schematic achieves the inverter function. On the contrary, if OUT1 is treated as the complementary output pin and OUT2 is treated as the output pin, the circuit performs the buffer function.

13.4.2 TFET-based CML standard cells

The above analysis suggests that CML can perform various functions based on different configurations. In fact, three levels of CML implementations are introduced in [31]. By observing the stacked levels and different pairs, the delay of a gate with more than three-levels exceeds the delay of an equivalent three-level, static multiplexer. That is, the level of differential pairs is limited to three for the optimization in the CML implementation. Figure 13.4 depicts four two-input TFET-based CML functions with a two-level structure. Each of the gates has three differential pairs as inputs. A set of four functions (including AND, NAND, OR and NOR) can be derived from Figure 13.4(a) with different input/output configurations. The MUX, XOR/XNOR and D latch are also distinguished by wiring and the input/output selection shown in Figure 13.4(b)–(d), respectively.

As discussed in the previous section, we attempt to maintain the voltage swing of input and output between 0.15 and 0.3 V for TFET CML gates. The configuration of the supply voltage and voltage swing sets the baseline for the other parameters, such as transistor size and biasing voltages. Here, we configure the TFET width to be the same size as the technology length to minimize the area. The 20-nm technology nodes are used for our evaluations. Consequently, it is important to tune V_{bias} and V_p to achieve the necessary voltage swing for the entire standard logic cells. After voltage sweeping, the basic CML logic gates function best when $V_{bias} = 0.18$ V and $V_p = 0.14$ V.

Standard cells are characterized and simulated under the same biasing condition. Table 13.2 shows the area, delay and power for the standard cells of TFET-based CML. Only ten cells are described, but more CML logic functions can be derived from the standard cells proposed in Table 13.2. For instance, if we define OUT1 as the output pin, then a CML-based inversion function is possible per Figure 13.3(a). However, if we choose OUT2 as the output pin, the CML schematic works as a buffer. Moreover, a standard cell library usually accounts for the different driving strengths of each individual function. In CML gates, a simple solution is to increase the constant current by the tail biasing transistor [16].

The area of CML and static TFET gates is also provided in Table 13.2. With the exception of a CML buffer and a four-input AND gate, all other CML standard cells consume less area compared to static counterparts. This feature may also be a major

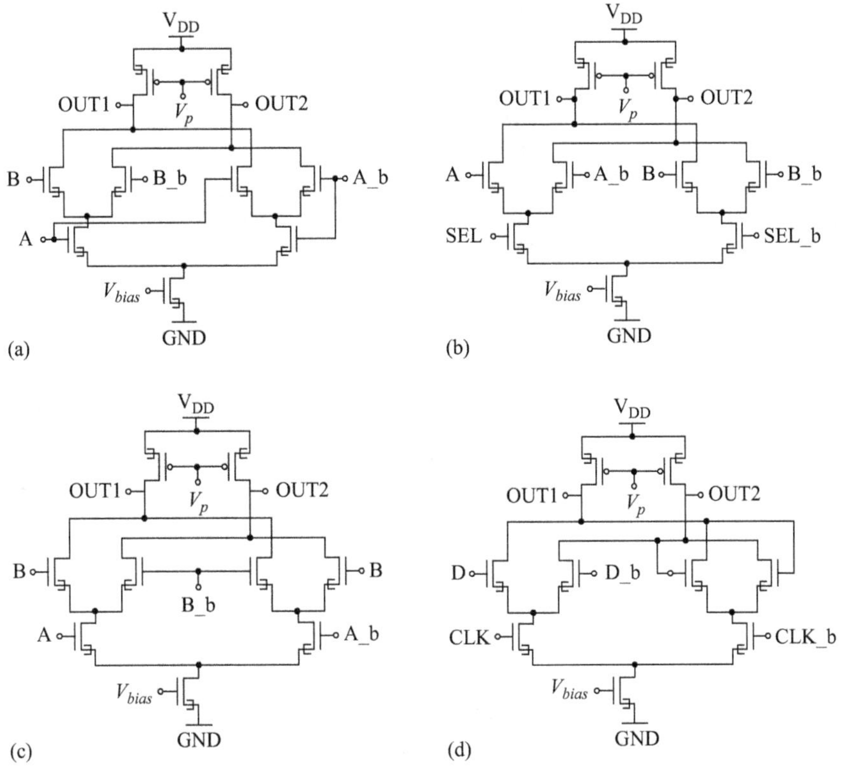

Figure 13.4 The universal schematics structure of four different CML circuits: (a) AND, (b) multiplexer (MUX), (c) exclusive-OR (XOR) and (d) D latch

advantage for cryptographic systems, especially light-weight ciphers such as KATAN, where majority of the hardware is composed of D flip flops and multiplexers.

13.4.3 CML implementation on KATAN

Due to large area and high power consumption, using CML to implement cryptographic hardware is not common—especially in lightweight cryptographic systems. To protect cryptographic circuits against DPA attacks, researchers often employ other techniques [32,33]. These solutions incur significant computation cost where the cryptography already involves massive computation and consumes relatively large power and area. As such, lower power, TFET-based CML could be especially valuable when considering devices for the IoT, WSN nodes, etc. Lacking an effective defense mechanism, hardware in these spaces can be substantially more vulnerable/susceptible to hardware attacks such as DPA.

To address these challenges, in the following sections, we consider the impact of TFET-based CML on a 32-b KATAN cipher. More specifically, (i) the KATAN cipher is a hardware-oriented block cipher with a low GE—even among other

Table 13.2 Area, delay and power of the TFET-based CML standard cells

Cells	Transistor counts	Area [μm^2]	Rising delay [ps]	Falling delay [ps]	Avg delay [ps]	Power [nW]	PDP [nW×ps]	CML area/ static area
Buffer	5	0.0022	90	124	107	30.588	3,272.916	1.833
OR2	9	0.0036	99	124	111.5	24.032	2,679.568	1
AND2	9	0.0036	75	165	120	22.97	2,756.52	0.818
AND4	27	0.011	476	644	560	70.828	39,663.68	1.8
MUX2	9	0.0036	71	115	93	24.183	2,249.019	0.5
XOR2	9	0.0039	99	105	102	25.848	2,636.496	0.817
D-Latch	9	0.0037	102	168	135	23.122	3,121.47	0.341
DFF	18	0.0074	100	200	150	45.500	6,825	0.341
1-b FA	45	0.0186	416	591	503.5	233.928	1.178×10^6	0.847
4-b FA	180	0.744	654	591	622.5	939.150	5.846×10^6	0.847

lightweight ciphers, (ii) applications that employ lightweight ciphers are typically power constrained—and thus could benefit from TFET technology—and (iii) the limit for the application of CML on conventional block ciphers is the large power overhead, but power consumption in a lightweight cipher is typically much less. In subsequent sections, we will briefly discuss the working mechanism of the KATAN cipher. Implementations of the 32-b KATAN cipher are provided in different circuit-level structures, where a table is presented to compare the TFET-based implementation with the CMOS implementation. We will then present the correlation power analysis on KATAN32 with experimental results through design simulations.

We now discuss how different transistor technologies could impact the power/performance of KATAN32 by using the Synopsys Design Compiler using 20 nm InAs homo-junction TFET [34] and the predictive technology model 20 nm FinFET technology [35]. In order to minimize the area consumption of KATAN32, the driving-strength-one library is employed for the synthesis. The synthesized transistor-level netlist is further converted into both the single-ended and differential modes. Synopsys FineSim is adopted for the gate-level simulation with less simulation time compared to the HSPICE simulator. The operating frequency of KATAN32 is set to 100 MHz to ensure its functional correctness.

Area and power data for four different implementations are summarized in Table 13.3. More specifically, we consider TFET and CMOS static implementations as well as CMOS CML with a 0.6 V supply, as well as TFET CML with a 0.3 V supply. A two-input NAND gate is assumed when comparing equivalent gate numbers. It is worth noting that the number of the synthesized static GEs is more than what is reported in [5], mainly because we simplify our library for both TFET and CMOS by using our own driving-strength-one and two-input standard cells. Complex logic gates, such as D flip flops and multiplexers, are not fully optimized and consume a relatively larger number of gates. (Future work will be performed to further optimize all TFET CML based logic gates.)

Table 13.3 Power consumption comparison among different implementations on KATAN32

	Voltage supply [V]	**Gate equivalent [#]**	**Area [μm²]**	**Average current [μA]**	**Power [μW]**	**Area change [%]**	**Power change [%]**
CMOS Static	0.6	1,013	3.534	16.09	9.96	–	–
CMOS CML	0.6	393	1.415	283.65	170.19	−59.96%	+1,608.73
TFET Static	0.6	1,013	3.536	3.14	1.89	+0.057%	−81.02
TFET CML	0.3	393	1.441	32.53	9.76	−59.22%	−2.01

13.4.4 Correlation power analysis on KATAN32

By observing the KATAN algorithm, it is apparent that the two nonlinear functions f_a and f_b are able to connect the plaintext/ciphertext with partial keys (or more precisely, subkeys). We can then select the 2 b each round generated by the nonlinear functions as our intermediate values or points of attack. Besides those two arithmetic functions, the majority of KATAN32 hardware is made up of D flip flops such that the overall power consumption mainly depends on the operation of shifting registers. As a result, it is important to come up with an attack mechanism that maximizes the power profile of two nonlinear operations.

In single-ended logic gates, power consumption only occurs during state transitions, either $0 \rightarrow 1$ or $1 \rightarrow 0$. If we configure the plaintext in a way that for certain clock cycles the power consumption of functions f_a and f_b contributes most, then the power information extracted from the supply current can be maximally related to the key information. More specifically, we can selectively configure the plaintext to be consecutive zeros or ones. Therefore, the power consumption of KATAN32 highly depends on functions f_a and f_b, because the power cost of the left-shift operation is negligible in each clock cycle.

To further implement the power analysis methods, four selected plaintexts are initially loaded into the two registers as given in (13.2) and the 80-b keys are set to all zeros. Note that in real cases, the key is the attackers' target and is unknown to attackers.

$$\begin{aligned} P1 &= \text{x}00000000 \Rightarrow p[18] = 0, p[31] = 0 \\ P2 &= \text{x}80000000 \Rightarrow p[18] = 0, p[31] = 1 \\ P3 &= \text{x}00040000 \Rightarrow p[18] = 1, p[31] = 0 \\ P4 &= \text{x}80040000 \Rightarrow p[18] = 1, p[31] = 1 \end{aligned} \tag{13.2}$$

However, the chosen input values are not constrained to (13.2), as long as the plaintext interacts mostly with the subkeys. When the start signal is received, KATAN32 begins encryption. Figure 13.5 shows the proposed CPA attack flow on KATAN32. Each selected plaintext and the hypothetical subkeys K_a and K_b are calculated to achieve the intermediate values "v" matrix. Then, intermediate results are further calculated by the power model, which is defined as the Hamming weight

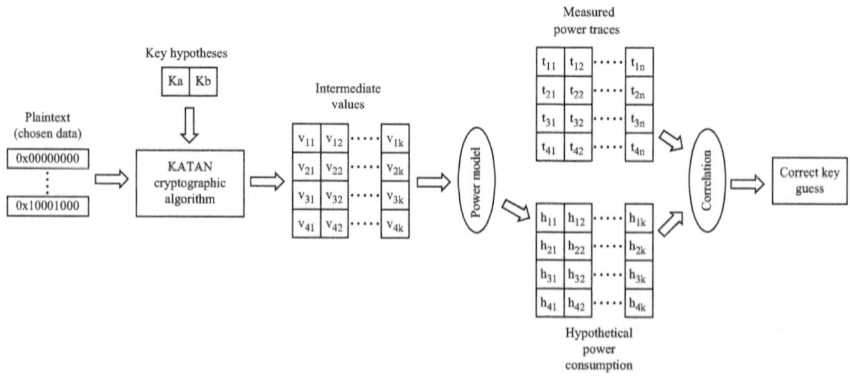

Figure 13.5 The correlation power analysis flow on KATAN cipher

model. The results from the Hamming weight model are defined as the hypothetical power consumption. Based on our chosen plaintexts, the matrix of hypothetical power consumption is given in the following equation:

$$\text{Hypothetical power consumption} = \begin{bmatrix} 0 & 1 & 1 & 2 \\ 1 & 0 & 2 & 1 \\ 1 & 2 & 0 & 1 \\ 2 & 1 & 1 & 0 \end{bmatrix} \tag{13.3}$$

$$\text{Corr. coef.} = \frac{\sum_{i=1}^{4} (t_i - \bar{t}) \cdot (h_i - \bar{h})}{\sqrt{\sum_{i=1}^{4} (t_i - \bar{t})^2 \cdot \sum_{i=1}^{4} (h_i - \bar{h})^2}} \tag{13.4}$$

The predicted power consumption is then compared with the measured real power consumption by the correlation coefficient formula as given in (13.4). The highest correlation coefficient result stands for the correctly guessed keys. In this case, the keys "00" reflect the largest correlation coefficient value. The next round follows the same mechanism, but with slightly different ciphertext, which is generated by the last round. Figure 13.6 shows the detailed correlation power analysis for the respective TFET static KATAN32 and TFET CML KATAN32 on one clock cycle. The black line describes the correct key value for subkeys K_a and K_b (="00"), which are the two most significant bits of the key. It is apparent that the correlation coefficient is largest for a static, TFET-based KATAN32 implementation when the correct keys are applied as shown in Figure 13.6(a). By comparison, the correlation coefficient of TFET CML KATAN32 is more significant, and all four hypothetical keys are similarly distributed as shown in Figure 13.6(b). Consequently, the TFET CML KATAN32 implementation is capable of successfully counteracting the correlation power analysis. Because the power consumption is mainly determined by AND/XOR logic gates of two nonlinear functions—and the effect of CPA is maximized—the correlation coefficients for KATAN32 are higher on average than other block ciphers, e.g., CPA on S-box [17].

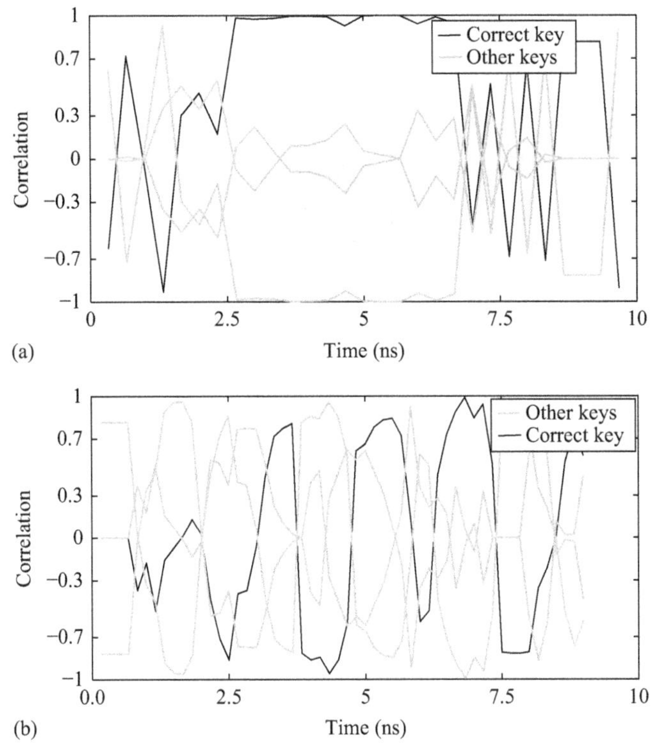

Figure 13.6 CPA attack on one clock cycle (a) TFET static KATAN32 vs. (b) TFET CML KATAN32

13.5 Discussion

Besides the introduced tunnel FET, other emerging device can also be employed for the implementation of CML. One example of that is polarity-controllable FETs, such as silicon nanowire (SiNW) FETs.

SiNW FETs: Transistors with a controllable polarity have already been experimentally fabricated in several novel technologies, such as carbon nanotubes [36], graphene [37] and silicon nanowires (SiNWs) [38,39]. An illustration of a silicon nanowire structure first fabricated in [40] is depicted in Figure 13.7. Given an additional gate, the operation of these FETs is enabled by the regulation of Schottky barriers at the source/drain junctions. The polarity gate (PG) regions closer to the drain and source contacts adjust the Schottky barrier heights deciding the channel's carrier type and thereby configuring the device's polarity. The control gate provides conventional gate control over the device. Schottky D/S devices have recently been exploited for achieving steep subthreshold in FinFET-like structures [41].

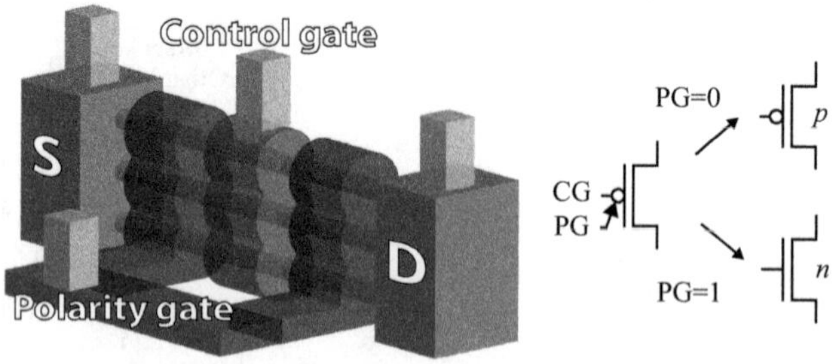

Figure 13.7 3D sketch of the SiNW FETs featuring two independent gates and its associated symbol [42]

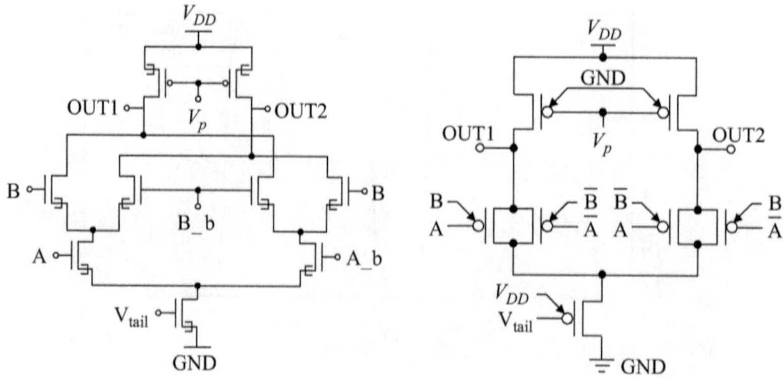

Figure 13.8 Two CML XOR schematics using (a) TFET (b) SiNW FET

Figure 13.8[1] depicts two CML examples of exclusive-OR (XOR) gate using TFET and SiNW FET, respectively. The two-input TFET CML XOR gate includes two level and three differential pairs, similar to the conventional CMOS counterpart. On the other hand, the SiNW FET CML XOR gate is enabled by only one level and two differential pairs. The polarity-controllable feature of SiNW FET can further lower the area consumption of circuit implementation, especially for the cryptographic system, where the XOR gates are heavily adopted.

Also, the authors in [43] applied sleep technique for the dynamic CML implementation using SiNW FET. Putting CML circuits into sleep mode can lead to further power saving. Figure 13.9[2] depicts new power gating technique employing onto TFET and SiNW FET-based CML designs. For the TEFT-based CML, a sleep transistor is incorporated at the bottom of the schematic. However, no extra sleep transistor is needed for the SiNW FET-based CML. Based on the applied power gating technique, the cryptographic system can work into two modes, standby mode and operation

[1]With permission from EDAA.
[2]With permission from EDAA.

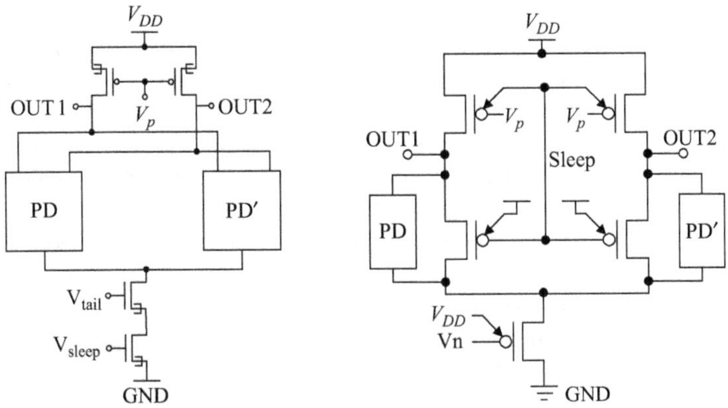

Figure 13.9 Power gating techniques on (a) TFET (b) SiNW FET

mode, respectively. The cryptographic system is only turned on when the encryption enable signal is on. In our future work, we would like to include both the other emerging technologies and power gating methods into our study to achieve a secure cryptography with even better performance.

13.6 Conclusion

In this chapter, we have demonstrated that the usage of emerging transistors, i.e. TFETs, can help improve circuit design resilience against CPA attacks while still preserving low-power consumption compared to their CMOS counterparts. Additionally, besides the traditional criteria for emerging devices such as area, power, delay and non-volatility, security may serve as a new criterion to thoroughly judge the advantages and disadvantages of emerging devices. Using this new standard, we plan to revisit existing emerging transistors to have a full comparison between emerging technologies and CMOS technology. Meanwhile, we believe that more research outcomes are expected in this area where unique properties of emerging transistors can help enhance the security of circuit designs.

Acknowledgment

Dr. Sharon Hu and Dr. Michael Niemier were supported in part by the Center for Low Energy Systems Technology (LEAST), one of the six SRC STARnet centers sponsored by MARCO and DARPA.

References

[1] *Advanced Encryption Standard (AES) FIPS Pub 197 (2001, Nov).* [Online]. http://csrc.nist.gov/publications/fips/fips197/fips-197.pdf.

[2] A. Bogdanov, L. R. Knudsen, G. Leander, *et al.* Present: an ultra-lightweight block cipher. In *Proceedings of the 9th International Workshop on Cryptographic Hardware and Embedded Systems*, CHES '07, pages 450–466, 2007.

[3] G. Leander, C. Paar, A. Poschmann, and K. Schramm. New lightweight DES variants. In Alex Biryukov, editor, *Fast Software Encryption*, volume 4593, pages 196–210. Springer, Berlin, Heidelberg, 2007.

[4] R. Beaulieu, D. Shors, J. Smith, S. Treatman-Clark, B. Weeks, and L. Wingers. The SIMON and SPECK families of lightweight block ciphers. Cryptology ePrint Archive, Report 2013/404, 2013.

[5] C. Cannière, O. Dunkelman, and M. Knežević. KATAN and KTANTAN – a family of small and efficient hardware-oriented block ciphers. In *Proceedings of the 11th International Workshop on Cryptographic Hardware and Embedded Systems*, CHES '09, pages 272–288. Springer-Verlag, 2009.

[6] P.C. Kocher, J. Jaffe, and B. Jun. Differential power analysis. In *Proceedings of the 19th Annual International Cryptology Conference on Advances in Cryptology*, CRYPTO '99, pages 388–397, 1999.

[7] P. Kocher. Design and validation strategies for obtaining assurance in countermeasures to power analysis and related. In *Attacks, in the Proceedings of the NIST Physical Security Workshop*, 2005.

[8] M.-L. Akkar and C. Giraud. An implementation of DES and AES, secure against some attacks. In *Cryptographic Hardware and Embedded Systems – CHES 2001*, volume 2162, pages 309–318. 2001.

[9] S. Yang, W. Wolf, N. Vijaykrishnan, D.N. Serpanos, and Y. Xie. Power attack resistant cryptosystem design: a dynamic voltage and frequency switching approach. In *Design, Automation and Test in Europe, 2005. Proceedings*, pages 64–69 Vol. 3, Mar 2005.

[10] K. Tiri, M. Akmal, and I. Verbauwhede. A dynamic and differential CMOS logic with signal independent power consumption to withstand differential power analysis on smart cards. In *Solid-State Circuits Conference, 2002. ESSCIRC 2002. Proceedings of the 28th European*, pages 403–406, Sep 2002.

[11] Yu Bi, P.-E. Gaillardon, X.S. Hu, M. Niemier, J.-S. Yuan, and Y. Jin. Leveraging emerging technology for hardware security – case study on silicon nanowire FETs and graphene SymFETs. In *Test Symposium (ATS), 2014 IEEE 23rd Asian*, pages 342–347, Nov 2014.

[12] Y. Bi, K. Shamsi, J.-S. Yuan, *et al.* Emerging technology-based design of primitives for hardware security. *Journal on Emerging Technologies in Computing Systems*, 13(1):3:1–3:19, Apr 2016.

[13] Y. Bi, K. Shamsi, J. S. Yuan, F. X. Standaert, and Y. Jin. Leverage emerging technologies for DPA-resilient block cipher design. In *2016 Design, Automation Test in Europe Conference Exhibition (DATE)*, pages 1538–1543, Mar 2016.

[14] M. Yamashina and H. Yamada. MOS current mode logic MCML circuit for low-power GHz processors. *NEC Research & Development*, 36:54–63, Jan 1995.

[15] K. Baddam and M. Zwolinski. Evaluation of dynamic voltage and frequency scaling as a differential power analysis countermeasure. In *VLSI Design, 2007. Held Jointly with 6th International Conference on Embedded Systems., 20th International Conference on*, pages 854–862, Jan 2007.

[16] S. Badel, E. Guleyupoglu, O. Inac, *et al.* A generic standard cell design methodology for differential circuit styles. In *Design, Automation and Test in Europe, 2008. DATE '08*, pages 843–848, Mar 2008.

[17] A. Cevrero, F. Regazzoni, M. Schwander, S. Badel, P. Ienne, and Y. Leblebici. Power-gated MOS current mode logic (PG-MCML): a power aware DPA-resistant standard cell library. In *Design Automation Conference (DAC), 2011 48th ACM/EDAC/IEEE*, pages 1014–1019, Jun 2011.

[18] E. Trichina, T. Korkishko, and K.H. Lee. Small size, low power, side channel-immune AES coprocessor: design and synthesis results. In *Advanced Encryption Standard – AES*, volume 3373, pages 113–127. 2005.

[19] F. Macé, F.-X. Standaert, J.-J. Quisquater, and J.-D. Legat. A design methodology for secured ICs using dynamic current mode logic. In *Integrated Circuit and System Design. Power and Timing Modeling, Optimization and Simulation*, volume 3728, pages 550–560. 2005.

[20] A. Tajalli, E. Vittoz, Y. Leblebici, and E.J. Brauer. Ultra low power subthreshold MOS current mode logic circuits using a novel load device concept. In *Solid State Circuits Conference, 2007. ESSCIRC 2007. 33rd European*, pages 304–307, Sep 2007.

[21] W.-Y. Tsai, H. Liu, X. Li, and V. Narayanan. Low-power high-speed current mode logic using tunnel-FETs. In *Very Large Scale Integration (VLSI-SoC), 2014 22nd International Conference on*, pages 1–6, Oct 2014.

[22] B. Sedighi, X.S. Hu, H. Liu, J.J. Nahas, and M. Niemier. Analog circuit design using tunnel-FETs. *IEEE Transactions on Circuits and Systems I: Regular Papers*, 62(1):39–48, Jan 2015.

[23] D. Hisamoto, W.-C. Lee, J. Kedzierski, *et al.* FinFET-a self-aligned double-gate MOSFET scalable to 20 nm. *IEEE Transactions on Electron Devices*, 47(12):2320–2325, Dec 2000.

[24] A.C. Seabaugh and Q. Zhang. Low-voltage tunnel transistors for beyond CMOS logic. *Proceedings of the IEEE*, 98(12):2095–2110, Dec 2010.

[25] H. Lu and A. Seabaugh. Tunnel field-effect transistors: state-of-the-art. *IEEE Journal of the Electron Devices Society*, 2(4):44–49, Jul 2014.

[26] U.E. Avci, R. Rios, K. Kuhn, and I.A. Young. Comparison of performance, switching energy and process variations for the TFET and MOSFET in logic. In *VLSI Technology (VLSIT), 2011 Symposium on*, pages 124–125, Jun 2011.

[27] G. Zhou, R. Li, T. Vasen, *et al.* Novel gate-recessed vertical InAs/GaSb TFETs with record high ion of 180 μa/μm at VDS=0.5 V. In *Electron Devices Meeting (IEDM), 2012 IEEE International*, pages 32.6.1–32.6.4, Dec 2012.

[28] H. Lu, J.W. Kim, D. Esseni, and A. Seabaugh. Continuous semiempirical model for the current-voltage characteristics of tunnel FETs. In *Ultimate Integration on Silicon (ULIS), 2014 15th International Conference on*, pages 25–28, 2014.

[29] H. Lu, D. Esseni, and A. Seabaugh. Universal analytic model for tunnel {FET} circuit simulation. *Solid-State Electronics*, 108:110–117, 2015.

[30] V. Saripalli, G. Sun, A. Mishra, Y. Xie, S. Datta, and V. Narayanan. Exploiting heterogeneity for energy efficiency in chip multiprocessors. *IEEE Journal on Emerging and Selected Topics in Circuits and Systems*, 1(2):109–119, Jun 2011.

[31] S. Badel, I. Hatirnaz, and Y. Leblebici. Semi-automated design of a MOS current mode logic standard cell library from generic components. In *Research in Microelectronics and Electronics, 2005 PhD*, volume 2, pages 155–158, Jul 2005.

[32] N. Debande, Y. Souissi, M. Abdelaziz El Aabid, S. Guilley, and J.-L. Danger. Wavelet transform based pre-processing for side channel analysis. In *Proceedings of the 2012 45th Annual IEEE/ACM International Symposium on Microarchitecture Workshops*, pages 32–38, 2012.

[33] J.G.J. van Woudenberg, M.F. Witteman, and B. Bakker. Improving differential power analysis by elastic alignment. In *Proceedings of the 11th International Conference on Topics in Cryptology: CT-RSA 2011*, CT-RSA'11, pages 104–119, 2011.

[34] H. Liu, V. Saripalli, V. Narayanan, and S. Datta. III–V Tunnel FET Model. Apr 2015 [online]. Available from: https://nanohub.org/publications/12/2.

[35] Arizona State University. PTM Model, 2014, [online]. Available from: http://ptm.asu.edu/

[36] Y.-M. Lin, J. Appenzeller, J. Knoch, and P. Avouris. High-performance carbon nanotube field-effect transistor with tunable polarities. *IEEE Transactions on Nanotechnology*, 4(5):481–489, 2005.

[37] N. Harada, K. Yagi, S. Sato, and N. Yokoyama. A polarity-controllable graphene inverter. *Applied Physics Letters*, 96(1): 012102-1-012102-3, 2010.

[38] J. Appenzeller, J. Knoch, E. Tutuc, M. Reuter, and S. Guha. Dual-gate silicon nanowire transistors with nickel silicide contacts. In *Electron Devices Meeting, 2006. IEDM '06. International*, pages 1–4, 2006.

[39] A. Heinzig, S. Slesazeck, F. Kreupl, T. Mikolajick, and W.M. Weber. Reconfigurable silicon nanowire transistors. *Nano Letters*, 12(1):119–124, 2012.

[40] M. De Marchi, D. Sacchetto, S. Frache, *et al.* Polarity control in double-gate, gate-all-around vertically stacked silicon nanowire FETs. In *2012 International Electron Devices Meeting*, pages 8.4.1–8.4.4, Dec 2012.

[41] J. Zhang, M. De Marchi, P.-E. Gaillardon, and G. De Micheli. A Schottky-barrier silicon FinFET with 6.0 mV/dec subthreshold slope over 5 decades of current. In *Proceedings of the International Electron Devices Meeting*, 2014.

[42] M. De Marchi, D. Sacchetto, S. Frache, *et al.* Polarity control in double-gate, gate-all-around vertically stacked silicon nanowire FETs. In *Electron Devices Meeting (IEDM), 2012 IEEE International*, pages 8.4.1–8.4.4, Dec 2012.

[43] L. Amaru, P.-E. Gaillardon, J. Zhang, and G. De Micheli. Power-gated differential logic style based on double-gate controllable-polarity transistors. *IEEE Transactions on Circuits and Systems II: Express Briefs*, 60(10):672–676, Oct 2013.

Index